# The Biology of Alpine Habitats

# THE BIOLOGY OF HABITATS SERIES

This attractive series of concise, affordable texts provides an integrated overview of the design, physiology, and ecology of the biota in a given habitat, set in the context of the physical environment. Each book describes practical aspects of working within the habitat, detailing the sorts of studies which are possible. Management and conservation issues are also included. The series is intended for naturalists, students studying biological or environmental science, those beginning independent research, and professional biologists embarking on research in a new habitat.

# The Biology of Alpine Habitats

Laszlo Nagy and Georg Grabherr

OXFORD

UNIVERSITY PRESS

# OXFORD

UNIVERSITY PRESS

Great Clarendon Street, Oxford OX2 6DP

Oxford University Press is a department of the University of Oxford.
It furthers the University's objective of excellence in research, scholarship,
and education by publishing worldwide in

Oxford  New York

Auckland  Cape Town  Dar es Salaam  Hong Kong  Karachi
Kuala Lumpur  Madrid  Melbourne  Mexico City  Nairobi
New Delhi  Shanghai  Taipei  Toronto

With offices in

Argentina  Austria  Brazil  Chile  Czech Republic  France  Greece
Guatemala  Hungary  Italy  Japan  Poland  Portugal  Singapore
South Korea  Switzerland  Thailand  Turkey  Ukraine  Vietnam

Oxford is a registered trade mark of Oxford University Press
in the UK and in certain other countries

Published in the United States
by Oxford University Press Inc., New York

British Library Cataloguing in Publication Data

Data available

Library of Congress Cataloging in Publication Data

Data available

Typeset by Newgen Imaging Systems (P) Ltd., Chennai, India
Printed on acid-free paper by
the MPG Books Group in the UK

ISBN 978–0–19–856703–5 (Hbk.)    978–0–19–856704–2 (Pbk.)

10 9 8 7 6 5 4 3 2 1

# Preface

Habitat is one of the oldest concepts of ecology—and an important one, as R. H. Yapp devoted his presidential address to the British Ecological Society in 1921 to discussing this concept in detail (Yapp 1922). Reading his paper nearly 90 years on, the meaning associated with the term habitat has not changed; it amalgamates the place and environmental factors that characterize the growing and reproductive conditions of the location where organisms—individuals or assemblages—live. The Biology of Habitats series has been making a compilation of how distinct habitats are structured and their biota function. Some of the previous titles imply collections of habitats, such as those in polar regions, while others are more homogeneous, e.g. aquatic habitats. The biology of alpine habitats, in broad terms, deals with a narrow spectrum of life conditions that are best described as cold. However, even a cursory look will lead to the realization that under the umbrella term alpine, lie a number of geographies with regional (and local) peculiarities with regard to growing season length, or altitude and associated atmospheric phenomena, just to mention a few. We therefore, in giving an account of the biology of alpine habitats, have concentrated on giving general outlines, without generalizations, and reminding the reader time and time again of the multitude of factors that we thought were important for alpine habitats world-wide and in individual regions at the same time.

In the Northern Hemisphere the majority of high mountain (alpine) environments has been better preserved in a state that resembles naturalness than surrounding lowlands, which are the main centres of settlement and production, and where natural ecosystems have largely been converted since prehistoric times, especially after the advent of sedentary agriculture. Conversely, in tropical and subtropical climates, mountains have provided a healthier environment than the disease-prone lowlands and this has led to more intensive use of the high reaches, too. The traditional use of mountains has much changed recently with the rapid development of communication and transport networks that are increasingly connecting mountains to the global economy.

Our main objective in writing this book was to enable the reader to become acquainted with the alpine environment globally and inspire further reading that leads to understanding particular aspects of specialized interest. Our task is accomplished if such understanding results in the appreciation of, and respect for these systems, whether as students, hill walkers, or resource managers.

# Acknowledgements

We would like to thank a number of colleagues who directly, or indirectly helped the writing along the way. The list of names is incomplete, as always. D. Gómez and M. Ohsawa kindly supplied unpublished material; A.N. Kuypers ensured that the photographs conveyed their message in black and white. Field visits by L.N. and accompanying field discussions were inspirational to the content: Andes (S. Beck, A. Grau, S. Halloy, A. Seimon, A. Tupayachi), Alaska (S. Talbot), Alps (M. Gottfried, H. Pauli), Pyrenees (D. Gómez, L. Villar), and the Himalayas (D. Karki, S. Schmidt). Over the years we had many useful discussions with colleagues within the scope of ALPNET, GLORIA (http://www.gloria.ac.at/), and GMBA (http://gmba.unibas.ch/index/index.htm), and in particular with C. Körner, S. Dullinger, A. Mark, and K. Green.

L.N. is grateful for the encouragement provided by R.M.M. Crawford and J. Grace. G.G. dedicates this book to his teachers at the University of Innsbruck—W. Larcher, H. Reisigl, W. Moser and H. Pitschmann, whose teaching and encouragement were essential for becoming interested in the fascinating world of alpine environments. G. G. thanks his wife Traudl for all her support. Finally, we warmly thank Ian Sherman, commissioning editor, for his trust, patience, and help, H. Eaton and all the assistant biology editors who have come and gone over the years, and the production team at the Press.

Laszlo Nagy and Georg Grabherr
Glasgow and Vienna
May 2008

# Contents

# 1 Introduction: What is alpine?

This book introduces the reader to a special mountain environment that can be relatively well defined by its characteristic physiognomy and its relative position along a bioclimatic gradient, which is reflected by the vertical zonation of the vegetation. Alpine environments are found where, along an altitude gradient, natural woody vegetation changes from various lowland and montane forest formations and gives way to dwarf-shrubs, various grass–heath, sedge–heath, and moss–heath formations, and finally, to open frozen ground. This book is about the upper, treeless end of this vertical gradient. The transition from closed forest to treeless vegetation—the upper montane zone with the treeline ecotone—marks the upper limits of tree growth. More precisely, an imaginary line connecting the highest outposts, forming clumps of small-stature trees, called the treeline is the lower limit of the alpine zone; areas lying above it are called alpine (Grabherr *et al.* 2003; Körner 2007a). According to such criteria, overall, about 3% of the terrestrial surface of the Earth is covered by alpine ecosystems, where about 4% of the Earth's flora is found (Heywood 1995). This definition of alpine is relatively easy to apply to the Holarctic mountains (found between approximately 30°N and 70°N), where about three-quarters of the world's alpine areas are found (Ozenda 2002). They are all characterized by seasonal changes in climate and a growing season limited by winter snow cover; they also fulfil the criteria of *Hochgebirge* in the German geo-ecological tradition (mountains that have an alpine zone with landforms, soil processes, and plant cover characteristics similar to those in the high alpine zone of the European Alps; e.g. Gerrard 1990; Richter 2001).

Humid tropical alpine areas are quite specific in that they have large daily temperature amplitudes, ranging from early morning frost to full 'summer' heat on a clear early afternoon; however, they do not experience growing season shortenings through seasonality. The resulting vegetation is characterized by giant rosettes, a life form not encountered

**Table 1.1** Terminology, as applied by various sources to alpine environments across the world

| Geographic area | Term | Characteristics | Usage | Reference |
|---|---|---|---|---|
| Arctic | Arctic tundra | Fell-field and arctic heath vegetation on mountains | General descriptive, phytogeographical | Wielgolaski (1997a,b) |
| Northern Eurasia: Iceland, Scotland, Scandinavia, Urals, and Siberian ranges | Mountain tundra | Vegetation above the treeline, characteristically formed by extensive belts of dwarf-shrub heaths | Phytogeographical, and physiognomic; applied to distinguish the northern alpine vegetation from the alpic (=pertaining to the Alps) alpine, or temperate type on the one hand, and to distinguish arctic tundra (lowland or mountain) vegetation from tundra-like vegetation on boreal zone mountains, on the other hand | Ozenda (1994); Wielgolaski (1997a,b) |
| British Isles | Montane vegetation | Heath (dwarf-shrub, grass, and moss) vegetation above the treeline, which is largely absent, in the Scottish Highlands, Pennines, and Snowdonia | An imprecise term widely used to describe mountain vegetation (often applied to all mountains, except the Alps in Europe) | Horsfield and Thompson (1996) |
| North American temperate | Alpine tundra | Vegetation complex above the tree limit on high mountains | Connotation with the vast tundra region in the north of the continent (cold-limited vegetation) | See overview in Bowman (2001) |
| Middle Europe | Alpine | Vegetation above, but not including the treeline ecotone | Applied to mountains between approximately 40°N and 50°N | Ozenda (1994); Ellenberg (1996); Richter (2001) |
| Southern Europe | Cryo-oro-mediterranean | Non-arborescent vegetation, composed of dwarf-shrubs, and sedge and grass heaths | Phytogeographical | Rivas-Martínez (1981); Richter (2001) |
| North Africa | Cryo-oro-mediterranean/altimediterranean | Non-arborescent vegetation, composed of dwarf-shrubs, and sedge and grass heaths | | Quézel (1957) |

| Region | Name | Description | Approach | Reference |
| --- | --- | --- | --- | --- |
| East and north-east tropical Africa along the Great Rift Valley | Afro-alpine | Species-poor flora characterized by giant rosette life forms occurring on East African tropical high mountains (>3000 m) | Phytogeographic, chorological, palynological | Haumann (1933), cited in Hedberg (1961) |
| South Africa | Austro-afro-alpine | Southern African afro-alpine | Phytogeographical | Killick (1978) |
| | Alpine | As in European alpine usage | Altitude belts/zones | Killick (1978) |
| New Guinea (Malesia) | Topical alpine | Areas above and outside the mountain forests; other terms:tropical alpine tundra, upper mountain grassland | Physiognomic distinction of vegetation types | Smith (1974) |
| New Guinea | Alpine | A treeless zone bordering a transitional subalpine zone and with the last vascular plants at its upper limit; in New Guinea, grasslands alternating with tundras and heaths, composed of herbs; ground coverage (5–100%); (3000 m) 3300–4100 (–4600 m) | | van Royen (1980) |
| Northern Andes | Páramo | Diverse ecosystem complex above the tree limit in the humid Andes, characterized by the presence of giant Espeletias | | Monasterio (1980) |
| Central Andes | Puna | Arid and semi-arid dwarf-shrub vegetation | Phytogeographical, physiognomic | Cabrera (1976) |
| Central Andes | Alti-Andino | High Andean grasslands, largely dominated by bunch grasses | Altitude zone, physiognomic | Navarro and Maldonado (2002) |
| All humid tropical (meso America, northern Andes, East Africa, Malesia) | Páramo | Tropical alpine ecosystems of the world | Geographical | Hofstede et al. (2003) |
| All tropical | *Altotropisch/* subnival, nival | Tropical ecosystems above the treeline | Geographical | Richter (2001) |

**Table 1.2** Definitions of terms associated with bioclimatic and physiognomic elevation zonation in high mountains (modified after Grabherr *et al.* 2003)

| Feature | Definition |
| --- | --- |
| *Linear* | |
| Line of closed arborescent vegetation | The line where the closed forest (cf. timber line or forest line) or abutting scrub ends, as seen from a distance |
| Treeline | The line that connects the highest growing groups of trees of approximately 3 m height (incl. prostrate trees or scrub). This may not be readily apparent on many high mountains with a long history of land use. |
| Tree species line | The line beyond which no adult trees of a species (including prostrate ones or scrub) occur |
| *Altitude zone or belt* | |
| Treeline ecotone | The zone between the forest line and tree species line (sometimes termed as upper subalpine) |
| Alpine | The zone above the treeline and the upper limit of closed vegetation (cover >20–40%); vegetation is a significant part of the landscape and its physiognomy. Sometimes divided into lower and upper alpine. Lower alpine—where dwarf-shrub communities are a major part of the vegetation mosaic (incl. the thorny cushion formations of the Mediterranean mountains). Upper alpine—where grassland, steppe-like and meadow communities are a significant part of the vegetation mosaic |
| Alpine-nival ecotone (subnival) | The transition between the upper alpine zone and the nival zone |
| Nival | The zone of scant patchy vegetation above the upper alpine zone; no predominating life form with frequent cushion and small rosette plants; vegetation is not a significant part of the landscape and its physiognomy |
| Aeolian | The highest reaches of mountains where no vascular plants grow; dominated by ice and rocks. Wind (hence aeolian) is important in providing nutrient input and maintaining food chains. |

outside tropical alpine environments. An overarching definition of alpine therefore has some ambiguities, arising from the existence of life forms that have no equivalents in both the tropics and the Holarctic. A further complicating factor is the case of arid mountains where treelines—if they occur at all—may well be limited by the lack of available water alone rather than by temperature, or where tree belts may appear at suitably moist high elevations, above treeless zones. These have led to a varied terminology (see Table 1.1); however, for all the differences that are addressed later on, we term all this variety uniformly alpine. An additional glossary is provided on recurring terminology relating altitude zones (Table 1.2).

Following a latitude pattern, the lower limits of the alpine zones are found at lower elevations at high latitudes and higher elevations closer to the Equator, such as can been seen along the extensive cordilleras of the Andes from approximately 55°S to approximately 12°N. The highest elevation of tree growth is not reached in the humid tropics, where cloud formation suppresses treelines, but in the seasonal subtropical mountains. In many areas of the world, mountains with an alpine zone do not form a continuous landscape and alpine habitats occur sporadically and in isolation— particularly striking when the East African volcanic cones and the Ruwenzori Mountains are compared with the Himalayas or the Andes.

Large areas lie at high altitudes that lack the typical alpine landforms. For example, high plateaux extensively occur in the Ethiopian Highlands, Pamir, Tibet, and in the Andes. Most of these areas offer better conditions to human occupancy than steep high mountainsides and have a long history of land use. From a topographical point of view these areas are not mountains; however, from an ecological point of view they form part of alpine ecosystems and are included in this account.

# 2 High mountains in latitude life zones: a worldwide perspective

## 2.1 Introduction

High mountains occur across all continents, from the Arctic to as far as 55°S, disregarding Antarctica (Fig. 2.1), in all broadly defined latitude zones (arctic, boreal, temperate, subtropical, and tropical) (Table 2.1). The definition of these broad latitude zones follows a simple model of a presumed climate effect, temperature, in particular, on living organisms from the Equator to the poles. However, there are difficulties in generalizing even simple temperature–vegetation latitude zone models. For example, a well-fitting Europe-wide model that uses annual mean temperatures to delineate vegetation zones (Ozenda 1994) has little applicability when extended to the scale of Eurasia, or beyond (Fig. 2.2). Temperature depends not only on the distance from the Equator, but indirectly, on the distance from the oceans; continentality is an important factor that results in such cold winters in Siberia that the annual temperatures average below zero there. According to the simple model in Fig. 2.2, this would cause a temperature climate akin to that in the Arctic. A much-refined pattern in variability is observed when a measure of growing season length is introduced (Fig. 2.3). In addition to temperature, the amount of precipitation, both in absolute terms, and relative to temperature and vegetation cover is important for living things. The main areas that have an arid or desert climate are shown in Fig. 2.4.

Climate is reflected in the natural vegetation, which has been used to classify life zones (e.g. Holdrige 1967; Walter 1973). Such climate life zones refer to lowlands. Mountains, emerging from their surrounding lowlands, transpose the latitude climate zones vertically and thereby make the Earth's surface a three-dimensional mosaic of climates and life zones, and corresponding vegetation. The life zone classification system proposed by Holdridge (1967), widely used internationally, is a complex one; the categories are defined by indices derived from temperature and humidity, and in the resulting cartographic representation latitude climate zones are applied to mountains. For example, forest tundra, the zonal ecotone between polar tundra and boreal forest, shows up in most latitude zones, representing the altitude treeline ecotone on high mountains. Similarly, arctic tundra is equated with the alpine zone on high mountains in Holdridge's system. Walter's long-standing standard reference system in the European literature defines nine zones and recognizes separate mountain life zones within each of the climatically defined latitude life zones (Table 2.2; see also Breckle 2002). For the purposes of this book, we discuss mountains in a framework that combines elements from the various life zone classification systems and geographical location. The altitude zonation of mountains is discussed in detail in Chapter 3.

## 2.2 Arctic mountains

As polar ecosystems have been described elsewhere (e.g. Thomas et al. 2008; Fogg 1998), arctic mountain areas are not discussed in detail. The Arctic forms a contiguous circumpolar zone across large expanses of Eurasia and North America. The peculiarity of arctic mountain ecosystems is that they lie beyond the northern latitude limit to tree growth, and hence are inseparable from the zonal tundra, the diversity of which is enhanced by mountain topography.

The circumpolar mountain areas, beyond the extent of the northern tree limit (see fig. 2 in Payette *et al.* 2002; Fig. 2.5), encompass northern Iceland, Svalbard, Franz Josef's Land, the northern Scandes in Europe, Novaya Zemlya, Severnaya Zemlya, Taymir Peninsula (Byrranga Mountains), and the Chukhotsk, Koryakovsk, and Sredinnyy ranges in the Russian northern Far East in Eurasia; and north and north-eastern Alaska, the Canadian Arctic islands (Ellesmere, Axel Heiberg, and Baffin) in North America, and Greenland. The common feature of these areas is that the mean annual temperature is below 0°C, mean annual precipitation is low (largely <300 mm year$^{-1}$, and in large areas <100 mm year$^{-1}$), and the length of growing season is short (typically 2, and up to 3 months). Snow accumulation and the inversely related depth of permafrost are important ecological factors. The typical vegetation is polar desert, polar tundra, with its dwarf-shrub, fell-field, and wet tundra formations.

## 2.3 Boreal mountain regions

The boreal zone lies south of the northern treeline, and its southern limit is at about 56° in Scotland, 48° in the Altai, 50° in the Baikal Mountains,

**Fig. 2.1** Arctic and alpine high mountain areas in different latitude life zones, overlain on the world vegetation map, where mountain areas in different climate zones, according to the classification by Fedorova & Volkova (1992) are shaded. Black dotted, subarctic and boreal mountains (see Figs 2.5–2.6 for detail); black, temperate; solid grey, temperate—subtropical; vertical grey hatching, subtropical; white dot over grey, tropical; horizontal grey hatching, equatorial.—

(a) arctic mountain areas (in Roman numerals): i, northern Iceland; ii, Svalbard; iii, Novaya Zemlya; iv, Severnaya Zemlya; v, Gory Byrranga (Taimyr); vi, Chukhotskoye Nagorye; vii, Koryakskoye Nagorye; viii, Sredinnyy Khrebet (N Kamchatka); ix, Brooks piedmont; x—xiii, the mountains of Ellesmere, Axel Heiberg; Baffin; and Greenland;

(b) boreal zone mountains (numerals in white-filled circles): 1, mountains of Iceland; 2, Scottish Highlands; 3, Scandes; 4, Chibiny (Kola); 5, Polar and Northern Urals; 6, Putorana; 7, Verkhoyansk; 8, Chersk; 9, Kolmsky; 10, Sredinnyy Khrebet (S Kamchatka); 11, Dzhungdzur; 12, mountains of the Baikal region; 13, Sayan; 14, northern Altai; 15, Brooks Range; 16, Alaska Range; 17, Wrangell – St Elias Range; 18, Mackenzie Mts; 19, Canadian Rockies; 20, Labrador Mts; 21, Newfoundland; 22, Laurentian Mts; 23, southern Cordillera Patagonica;

(c) temperate forest zone mountains (triangle): 24, Cantabrian Mts—Pyrenees; 25, Alps; 26, Carpathians; 27, Dinarids; 28, northern and central Apennines; 29, Balkan Ranges; 30, Caucasus; 31, north-eastern Anatolian Mts; 32, Changbaishan; 33, Sikhote-Alin; 34, Japanese Alps; 35, Cascades; 36, Rocky Mts; 37, North Appalachians; 38, south Chilean Andes; 39, New Zealand Alps (South Island);

(d) arid mountains of temperate latitudes (white-filled diamonds): 40 southern Altai; 41, Tianshan; 42, Pamirs; 43, Karakoram; 44, Kunlunshan and the Plateau of Tibet;

(e) Himalayas (subtropical—temperate inter-zonal)

(f) subtropical sclerophyllous forest zone (Mediterranean type) mountains (empty circles): 46, Hindukush; 47, Sierra Nevada of Spain; 48, High Atlas; 49, Hellenids; 50, Maloti—Drakensberg; 51, southern Taurus; 52, Lebanon; 53, Australian Alps; 54, Sierra Nevada, California; 55, central Chilean Andes; 56, Zagros; 57, Elburz;

(g) seasonal tropical, and subtropical arid and desert zone mountains (empty diamonds): 58, Sierra Madre Oriental; 59, Sierra Madre del Sur; 60–61, central Andes, 62, Cordillera de la Atacama;

(h) aseasonal tropical mountains (empty squares): 63, Mt Cameroon; 64, Ruwenzori; 65, Ethiopian Highlands; 66, Kilimanjaro and other volcanoes; 67, Mt Kinabalu; 68, New Guinean Highlands; 69, Sierra Nevada de Santa Marta; 70, Cordilleras Occidental, Central and Oriental of Colombia; 71, Andes of Ecuador and northern Peru (sometimes treated as humid puna); 72, Cordillera de Merida, Venezuela.

In addition, the Saharan Tibesti and Hoggar (unnumbered dashed line squares) are sometimes listed as high mountains.

**Table 2.1** The percentage of the Earth's surface classified as mountain ecosystems (comprising the bioclimatic zones of montane, alpine and nival) in latitude zones FAO (2000)

| Continent | Tropical (>1000 m) | Subtropical (>800–1000 m) | Temperate (>800 m) | Boreal (>600 m) | Arctic (>0 m) | Total |
|---|---|---|---|---|---|---|
| Eurasia | 0.7 | 2.7 | 3.8 | 3.8 | 1.5 | 12.6 |
| North America | – | 0.3 | 1.5 | 0.9 | 2.6 | 5.3 |
| Northern Hemisphere | 0.7 | 3.0 | 5.3 | 4.7 | 4.1 | 17.9 |
| Africa | 1.1 | 0.3 | – | – | – | 1.4 |
| Central and South America | 1.6 | 0.3 | 0.1 | – | – | 1.7 |
| Australasia | 0.1 | – | 0.1 | – | – | 0.2 |
| Southern Hemisphere | 2.8 | 0.6 | 0.2 | – | – | 3.3 |
| Total | 3.5 | 3.6 | 5.5 | 4.7 | 4.1 | 21.2 |

Note that 1000 m is usually considered as the upper limit of lowland tropical evergreen forest and the lower limit of montane forest at tropical latitudes is at approximately 1500 m, with 500 m in between being considered as a transition zone–sometimes called submontane. For the arctic region there are no separate estimates for mountains. In the FAO statistics of life zones many mountain areas that vegetation scientists classify as boreal (e.g. Fig. 2.1) are accounted for as temperate.

and about 49° in Canada. Boreal high mountains are special in that they grade into zonal arctic tundra at high latitudes; at the other extreme they give way to dry high mountain formations at their southern limit in central Asia. Boreal mountains follow a circumboreal distribution and comprise the mountains of northern Europe, Siberia, and extend from the Alaskan ranges, through the Canadian Rockies to the Laurentian Mountains in eastern North America (Fig. 2.6). The European ranges include the Caledonids, which were all formed during the Caledonian folding (Scottish Highlands, Scandes and the Polar and North Urals), the Chibiny Mountains of the Kola Peninsula and the volcanic mountains of Iceland). In Asia, the north Siberian mountains (Putorana, Verkhoyansk, Tchersky, Kolyma, Kamchatka), Anadyr, and the north-east–south-westerly diagonal chains of the Djugdjur, Baikal, Sayan, and the northern Altai are the main ranges with an alpine zone. In North America, the Alaskan high ranges (Brooks, Alaskan, St Elias, and Wrangell), some smaller ranges in central Alaska, the Canadian Rockies (including the Mackenzie Mountains), and the mountains of Labrador, Newfoundland, and the Laurentian Plateau in eastern North America complete the full suite in the Northern Hemisphere. The southernmost Andes in Patagonia (between approximately 49 and 56°S) lie in a comparable climate with that of the boreal forest zone of the Northern Hemisphere.

Boreal alpine environments, similarly to arctic ones, receive moderate snow in winter and are characterized by severe frosts leading to cryoturbation,

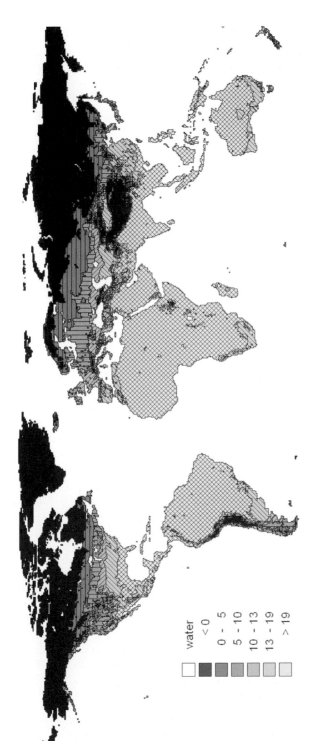

**Fig. 2.2** Annual mean temperature and the latitude zonation of the vegetation (after Ozenda 1994). The classes shown are annual mean temperatures: <0°C, arctic and subarctic (solid black); 0–5°C, boreal (horizontal stripe); 5–9.5°C northern nemoral (vertical stripe); 10–13°C, sub-Mediterranean (backward diagonal); 13–19°C, Mediterranean (forward diagonal); >19°C, other subtropical and tropical (diagonal cross hatched). The corresponding lengths of growing season are: <60 days, arctic and subarctic; 60–120 days, boreal; 120–180 days, northern nemoral; 180–240 days, sub-Mediterranean; >240 days, Mediterranean, other subtropical and tropical. Data source for map: Leemans & Cramer (1992)

This model was proposed by Ozenda (1994, pp. 21–22) and it relates annual mean temperatures to observed vegetation latitude zones in Europe. This scheme is applicable to Europe, but cannot be extended globally, as most of Siberia and the North American boreal zone would become aggregated within the arctic.

water
< 0
0 - 5
5 - 10
10 - 13
13 - 19
> 19

**Fig. 2.3** Global map of the number of months with a mean monthly temperature of >3°C. The number of months with a mean monthly temperature >3°C is an indicator of the period during which plant growth occurs in the alpine zone in temperate mountains. The number of months >3°C is indicated on a scale from black (1) to light grey (12).

As opposed to a clear growing season temperature latitude zonation in the Northern Hemisphere, little such zonation is seen south of the temperate forest zone. Growing season limitations are observable in the High Andes south of c. 10°S, and in New Zealand. Some high mountain areas in the Southern Hemisphere e.g. New Guinea Highlands are notably missing because of the coarse resolution (50 km) of the base data. Data source: Leemans & Cramer (1992).

**Fig. 2.4** Global distribution of desert (precipitation <100 mm year$^{-1}$, black) and arid (precipitation <300 mm year$^{-1}$, dark grey) areas. Data source: Leemans & Cramer (1992).

**Table 2.2** Life zone and climate classifications

| Zonobiome | Walter's Life Zones (Breckle 2002) | Walter's climates | Holdridge–Olson Life zone–ecosystem classes (Leemans 1992) |
|---|---|---|---|
| I | Evergreen tropical rainforest | Equatorial humid diurnal | Tropical rainforest |
| I–II | Semi-evergreen forest | ecotonal | Tropical seasonal forest |
| II | Savannas, deciduous forests and grasslands | Humido-arid tropical summer rain | Tropical dry forest |
| II–III | | ecotonal | Tropical semi-arid |
| III | Hot deserts | Subtropical arid | Hot desert |
| IV | Sclerophyllic woodlands (Mediterranean) | Arido-humid winter rain | Chapparal |
| V | Laurel forests | Warm-temperate humid | Warm temperate forest |
| VI | Deciduous forests | Temperate nemoral | Temperate forest |
| VI–VII | | ecotonal | Cold parklands (including upper montane groves on arid mountains) |
| VII | Steppes and cold deserts | Arid-temperate | Steppe, cool desert |
| VIII | Taiga | Cold-temperate boreal | Boreal forest (incl. upper montane) |
| VIII–IX | | ecotonal | Forest tundra (incl. alpine treeline ecotone) |
| IX | Tundra | Arctic | Tundra (including alpine and nival) |

solifluction, and gelifluction, resulting in patterned ground over large areas. In the summer, long days cause an extended light period that may selectively favour certain adapted species. Mixed dwarf-shrub heath and dwarf birch–willow scrub cover dominate in the treeline ecotone. Above this, ericaceous dwarf-shrub heath grows, replaced at higher elevations by fell-fields with small cushion plants and prostrate dwarf-shrubs, sedges, and rushes. In continental areas a cold desert-like formation with extensive boulder fields (goltsy) is a typical feature of these mountains. Glaciers cover most of the nival zone (Gorchakovsky 1989; Billings 2000).

## 2.4 Temperate regions

The most extensive mountain areas are found in the Holarctic (Eurasia and North America) temperate zone *s. l.* (Table 2.1). The temperate zone is different from the arctic and boreal zones—both of which form more or less contiguous bands centred around the North Pole—in that the distance

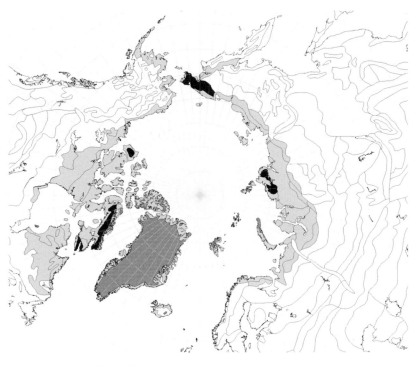

**Fig. 2.5**     Mountain areas in the arctic zone.

Black, mountain areas and mountain vegetation; dark grey, glaciers; light grey, arctic, including forest-tundra ecotone vegetation types and polar desert. Data sources: Fedorova & Volkova (1990) and Fedorova *et al.* (1993).

from the oceans becomes a major climate and vegetation determining factor at temperate latitudes. This oceanity versus continentality, in effect, separates largely humid to mesic temperate areas, such as found in the nemoral zone in Europe from the dry to arid central Asian and southwest North American steppes and deserts, where a large proportion of the Asian temperate mountain ranges in Table 2.1 are found.

## 2.4.1  Mesic temperate zone mountains (temperate deciduous forest zone)

Large expanses and the highest mountains of the temperate zone are in the Himalayas. The Himalayas are often considered separately from the temperate zone because of their transitional position, separating the tropical and subtropical Indian subcontinent from the cold and desertic mountain areas to the north. A large and somewhat heterogeneous group of mesic temperate mountains comprises mountains that lie between the boreal and Mediterranean zones in Europe, the Far Eastern Ranges

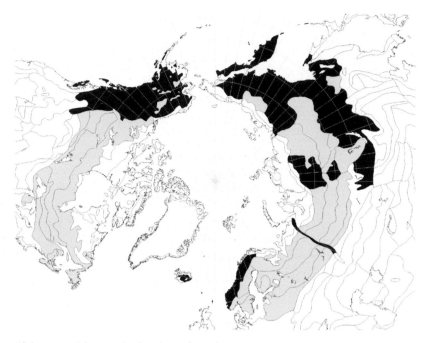

**Fig. 2.6**     Alpine mountain areas in the circum-boreal zone.

Black, mountain areas and mountain; grey, boreal zone lowland vegetation. For names of main ranges see Fig 2.1. Sources: Fedorova & Volkova (1990) and Fedorova *et al.* (1993).

of Changbaishan, Sikhote-Alin, and the Japanese Alps, and the North Appalachian Mountains in eastern North America. The European mid-latitude high mountains include the Cantabrian Mountains, Pyrenees, Alps, Carpathian Mountains, Dinarids, Rila, and Pirin of the Rhodopean system and the north Balkan Mountains, the north and central Apennines, and the Corsican massifs. Some smaller massifs, such the Massif Central, Jura, Vosges, and the central European Middle Mountains (e.g. Giant Mountains) occasionally bear alpine-like vegetation that is sometimes referred to as a pseudo alpine area. Further to the east, the Pontic system (*sensu* Ozenda 2002) comprises the Caucasus and the coastal mountains in northern Turkey of the Anatolian Range, of which only those in the north-east reach alpine altitudes. In the Far East, the Korean mountains, peaking at the Changbaishan in south-eastern China, the Sikhote-Alin in south-western Siberia, and the Japanese island of Honshu, and, to a lesser extent, Hokkaido fall into a temperate forest zone. Another hetero-geneous group comprises the western North American ranges, principally the Cascades; however, various ranges of the Rocky Mountains are often considered here too (despite being surrounded at low elevation by steppe and desert), and the northern Appalachians in eastern North America.

At southern latitudes, the temperate ranges of the Andes in Chile and Argentina (between approximately 35 and 49°S), and the New Zealand Alps of South Island (between 40 and 47°S) can be considered as temperate forest zone mountains. The Atlantic face of the Cantabrian Mountains of northern Spain, the Cascades, the southern Andes and the New Zealand Alps, as well as the Japanese Alps of Honshu and the Taisetsushan on Hokkaido have strong oceanic influences. For this reason they are sometimes considered as forming part of the warm temperate forest zone.

Temperate mountains are snow-rich, providing protection from deep frost; however, late-lying snow may shorten the growing season. Snow accumulation and melt pattern result in a distinctive mosaic of plant communities. Temperate alpine vegetation consists of dwarf-shrub communities at the treeline, while the alpine zone is characteristically covered by a variety of graminoid-dominated sedge heath or grasslands. In Europe, nival summits are concentrated in the Alps and the Caucasus, along with a few peaks in the Pyrenees. There are extensive glacial areas in the Himalayas, and in New Zealand, with lesser ones in the Far East and in western North America. In the nival zone, scattered cushion and rosette plants and low-stature graminoids grow in sheltered and favourable sites.

### 2.4.2 Dry mountains of temperate latitudes (steppe and cool desert zone)

The majority of these mountains are found in central Asia at the interface between the southern limit of temperate and subtropical climates: most of the Altai (with the exception of the northernmost parts), the Tianshan, Pamir, Karakoram, Tibetan Plateau (Kunlunshan), and some of the Great Basin ranges in North America.

The central Asian mountains are characterized by an arid continental climate; however, because of their size and extent they are a major regional climate-modifying factor and orographic precipitation may locally reach levels that permit lush vegetation. The vegetation of the alpine zone is characterized by *Kobresia* sedge heath, which forms near closed stands in the more humid ranges but is confined to small favourable areas in the drier ranges where xeromorphic dwarf-shrubs dominate. Scattered herbaceous species form very sparse vegetation near the upper limit to plant life. There are large glaciers in the Tianshan and the Pamir.

## 2.5 Subtropical mountains

The largest montane and alpine areas in the subtropics are found in Eurasia, with far smaller areas in North and South America, and Africa

(Table 2.1). Many (seasonally) arid high mountains belong to subtropical high mountain environments. Subtropical mountains form at least two different groups, one that may be called Mediterranean, or sclerophyllous forest zone type, and another, more arid continental desert type.

### 2.5.1 The Mediterranean type scelrophyllous forest zone mountains

In Europe and North Africa, the most notable mountain areas that extend beyond the potential treeline include the Sierra Nevada, High Atlas, Mount Etna, some peaks of the Hellenids, including the mountains of Crete, the Taurus, and the Lebanese Mountains. In North America, the Sierra Nevada of California can be considered as a Mediterranean zone mountain range, while in Asia the Zagros, Elburz, and Hindukush.

In the Southern Hemisphere, the Maloti-Drakensberg of southern Africa, the Australian Snowy Mountains (approximately 33–38°S), and the central Chilean Andes (approximately 33–35°S) belong to this category, while the highest reaches of the mountain peaks in Tasmania (40–43°S) are more in the temperate forest zone than in the Mediterranean.

The high seasonality experienced in the temperate zone is accentuated in the sclerophyllous forest zone and even the high mountains may experience drought in the summer months, a trend perceptibly increasing from north to south and from west to east in the Mediterranean proper. As most precipitation falls in the winter, a considerable amount of snow may accumulate and persist into the summer in sheltered areas. Such areas are important sources of summer irrigation from melt water and are home to snowbed communities, commonly associated with more northerly latitudes. The most notable feature of the Mediterranean-type alpine (or cryo-oromediterranean) zone is the presence of thorny cushion communities (with many geophyte genera, such as *Eremurus*, *Iris*, and *Tulipa*). Nival plant assemblages of endemic cushion plants and short grasses are found in the Sierra Nevada of Spain above 2800 m; and on the Zagros, Elburz, and the Hindukush.

### 2.5.2 Arid subtropical mountains (hot desert zone mountains)

Arid zone mountains of subtropical latitudes include some of the ranges the south-central Andes. The eastern ranges of the south-central Andes (approximately 22–25°S) are of true desert type, being the driest in the world. Some authors include in this category the high mountains of the Sahara, such as the Tibesti and Hoggar (Quézel 1965) with their characteristic mountain steppes at middle elevations, becoming desertic again on high grounds (scattered vegetation of xeromorphic shrubs and grasses).

# 2.6 Tropical regions

The mountains of the tropics encompass a wide variety of climate, geology, geomorphology, soils, and natural vegetation. They range from comparatively flat highland plateaux (Altiplano, Ethiopia) through the eroded glaciated volcanic peaks of East Africa and acid high ranges (Andes) to limestone, ultramafic, and granitic high mountains with patchy scant vegetation (Mount Kinabalu). There are striking differences in climate from wet tropical to arid within the Andes chain alone. According to estimates published by FAO (2001), in the tropics, 3.5% of the Earth's area falls in the montane and alpine (including nival) zones (Table 2.1). The largest areas are found in South America, followed by Africa and Asia.

Tropical latitudes are characterized by very little seasonality. Such a general statement applies to the humid tropics, where the zonal lowland vegetation is rainforest. Seasonality increases with distance from the Equator. A subdivision of tropical latitude mountains may be made on the basis of seasonality in mountain climate and not on the fact whether the zonal lowland vegetation is rainforest or not. Thus, although most of the East African high mountains lie in the tropical dry and semi-arid zone towards the east with seasonal tropical forest to the west, they are discussed together with the other aseasonal tropical mountain areas of the world.

Seasonally dry tropical–subtropical mountain areas are found towards the Tropics of Cancer and Capricorn and form a link with the subtropical sclerophyllous zone (or hot deserts in arid areas). Such seasonally dry tropical–subtropical mountain areas are found in Meso-America, in the Sierra Madre Oriental and Sierra Madre del Sur of Mexico (approximately 18–24°N), and in South America in the central Andes approximately 14–30°S), the latter with its the so-called *puna* zone. The puna itself is a semi-arid type in the west, while humid puna is found in the eastern ranges.

The aseasonal tropical zone in South America encompasses: the Cordillera Talamanca of Costa Rica; in the tropical Andes the Cordillera de Merida of Venezuela, Sierra Nevada de Santa Marta and the Cordilleras Occidental, Central, and Oriental of Colombia, the Ecuadorian Andes and the northern third of the Peruvian Andes. The characteristic alpine vegetation of these areas is termed *páramo*, with its altitude variants ranging from the largely man-made stands abutting the treeline and typically displaying giant rosettes of *Espeletia* spp. to the *superpáramo*, with its scant plant cover in the subnival zone. The single West African mountain that reaches an alpine elevation is Mount Cameroon; however, its continuing volcanic activity leaves little room for uninterrupted development of vegetation.

The East African mountains (Mount Kenya, Mount Kilimanjaro, Mount Meru, Mount Elgon, and Aberdares), the mountains of the Ethiopian Highlands (Simen, Abuna Yosef, Abuya Meda, Guna, Choke, Mangestu, Bada, Kaka, Bale, and Guge), but not the wet Ruwenzori–Virunga ranges, have been much affected by fire, which has caused the expansion of the high-altitude vegetation. The East African mountains, similarly to the aseasonal tropical Andes, feature giant rosettes (e.g. *Lobelia*, *Senecio*). The typical vegetation sequence from the treeline upwards is from scrublands (*Alchemilla*, *Erica*, *Helichrysum*, *Hypericum*), giant rosette formations to a subnival open vegetation. Even at the highest peaks vascular plants can grow (e.g. the grass *Poa ruwenzorensis* at about 5000 m on the Ruwenzori), some on volcanic hot spots, such as in the Kibo crater on Kilimanjaro.

In the South-east Asian tropics, Mount Kinabalu is often cited to have an alpine vegetation however sparse. The highest mountains in New Guinea (e.g. Mount Jaya, Mount Trikora, Mount Wisnumurti, Mount Madala, and Mount Wilhelm) harbour an alpine zone. In contrast to East Africa and the Andes, there are no giant rosettes in New Guinea; however, similarly to the Andes, there are hard cushion wetlands.

# 3 Elevation gradients

## 3.1 Introduction

There are striking changes in vegetation from lowlands to high mountain-tops, similarly to that from the Equator towards the poles. These changes entail a reduction in the structural and floristic diversity of ecosystems and a reduction in biomass from the tall lowland forests to low stature alpine heaths, or arctic tundra. The simplification in structure and the trends in species richness and biomass have been related primarily to a reduction in temperature alone, or in combination with other climatic and physiographic factors in alpine environments (e.g. Theurillat *et al.* 2003). Harsh mountain climate is locally ameliorated by glacial geomorphic landscape features, which contribute much to habitat diversity, and vegetation and species diversity patterns.

To appreciate fully the scale of changes in vegetation (and climate) along an altitude gradient there is no better place to start than in the rainforest, at the foot of an equatorial high mountain (Table 3.1). From evergreen rainforest at near sea level, one can scale the range of montane forest and treeless alpine formations and reach high mountain tops, capped by glacial ice. All this is possible within a horizontal distance of less than 30 km, such as in the Caribbean Andes in northern Colombia (Nevada de Santa Marta, 5775 m; 73.7°W, 10.8°N), or in Papua, Indonesia (Mount Jaya, 4884 m; 137.1°E, 4.1°S). While the above are equivalent to no more than a 0.2°C temperature change in a north–south direction, they involve between about a 25 and 29°C temperature difference vertically. The vertical arrangement of vegetation formations along tropical high mountains (which mirrors, to some extent, temperature regimes found from the tropics to the Arctic, noted already by Humboldt (1817)) has led Holdridge (1967) to equate life zone categories between arctic tundra and any non-woody temperature-limited high mountain vegetation types. However, behind the apparent similarities there are large differences, which are principally in terms of seasonality of temperature, precipitation, and radiation.

**Table 3.1** Altitude zonation in Malesia (including New Guinea), modified after van Steenis (1972)

| Zone | Altitude (m) | Physiognomic vegetation |
|------|--------------|-------------------------|
| Tropical rainforest | 0–1000 | Closed tall forest with emergents |
| lowland | 0–500 | |
| hill | 500–1000 | |
| Lower montane (submontane) | 1000–1500 | Closed tall forest |
| Montane | 1500–2400 | Closed tall forest—decreasing stem diameter and increasing bryophyte cover with increasing elevation |
| Upper montane | 2400–3600 | Dense low forest, often mossy, conifers usually present |
| Treeline ecotone | 3600–4000 | Low scrub; trees isolated, or in groups |
| Alpine | 4000–4600 | Low stature, open to closed herbaceous vegetation |
| Nival | >4600 | Snow, ice, rock |

This chapter discusses the relationship between elevation and climate as a premise to interpret patterns of change in soils, vegetation, and plant and animal species richness in terrestrial and aquatic ecosystems in the mountains of the world. The multiple causes of the shift from wooded to treeless landscapes are elaborated, the treeline being considered as the dividing line between alpine and non-alpine environments. The use and reliability of simple bioclimatic models (based on the altitude zonation of vegetation) for predicting future vegetation changes under forecast climate scenarios is discussed (see also Chapter 9). Finally, the impact of human land use on modifying natural elevation zonation and its effects on vegetation dynamics are highlighted (see also Chapter 10).

# 3.2 Climate and elevation

## 3.2.1 Temperature

Temperature decreases with elevation; however, there is a large variation with latitude, weather circulation systems, and distance from the oceans, which are the main determinants of regional climate. For example, diurnal variation in temperature is smaller in the forested tropics than over deserts because of cloudiness and high air humidity in the tropics. The annual amplitude of temperature is also lowest in equatorial humid climates (see Chapter 4), while it can reach 60°C in continental Siberian climate. The annual temperature amplitude narrows with altitude, i.e. high mountains are 'cold'.

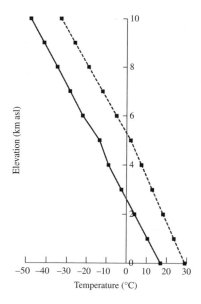

**Fig. 3.1**     Temperature decrease with elevation in the temperate (solid line) and tropical (broken line) zones. Standard atmosphere values in free air were increased by 2 °C to reflect Earth surface temperatures (International Organisation of Civil Aviation and Pisharoty 1959, both cited in Rudloff (1981).

The generality of the statement, while it is true in global comparisons, must not disguise the fact that diurnal amplitudes can be high; therefore, it can expose living organisms to alternating stresses of cold and heat, both in tropical and temperate alpine environments.

Temperature decrease (or lapse rate) in free air is on average about −0.6°C per 100 m elevation (Fig. 3.1), with extreme ranges of about −0.2 and −0.8°C per 100 m (Dillon *et al.* 2006); for comparison, the corresponding decrease is about 0.75°C per degree latitude, or by approximately 111 km. The vertical lapse rate in temperature is modified by other factors such as exposure to wind; for example, a wet windward mountain side can have a lapse rate of 0.55°C versus 0.8°C on the leeward side (Rudloff 1981). Slope exposure can modify surface temperature, and, as a result, the lower limit of the alpine zone is usually lower on north-facing slopes in the Northern Hemisphere and on south-facing ones in the Southern Hemisphere than on isolated slopes. For example, in Corsica, the upper distribution limit of *Alnus viridis* ssp. *suaveolens* (considered as the lower limit of the alpine zone) is approximately 2100 m in northerly exposures; on southerly slopes the species is replaced by *Juniperus communis* ssp. *alpina,* marking the lower limit of the alpine zone at approximately 2300 m (Gamisans 1999). The 200m difference in elevation is equivalent to about a difference in temperature of just over 1°C. A recent study of alpine summits in European mountains has shown that, indeed, slopes in different exposures vary in their soil temperatures 10 cm below ground surface by a maximum of about 1°C. This implies that the vertical zonation of vegetation patterns may be rather fuzzy; different plant community types and associated animal assemblages may inhabit contrasting slopes (east versus west). Such differences may influence plant species richness on alpine summits (Pauli *et al.* 2009).

The lapse rate in air temperature and changes in other climate components refer to those measured in free air above ground level and does not necessarily describe adequately the environment that plant tissues are exposed to along elevation gradients; for example, whereas free air temperature can decrease by 0.9°C per 100 m elevation from montane *Pinus sylvestris* forest at 450 m asl to alpine *Loiseleuria procumbens* prostrate dwarf-shrub heath at 850 m in the oceanic, boreal-temperate Cairngorm Mountains, Scotland, the corresponding lapse rate in apical meristem and shoot temperature is about zero (Wilson *et al.* 1987). In other words, the tissue temperature of plants over an air temperature gradient of –3.6°C can remain the same. This is achieved through decreasing stature (from approximately 16 m to <0.1 m) and by corresponding changes in aerodynamic resistance. The patchy forest has a low aerodynamic resistance and the plant surface to air differential is high, while in the low stature alpine heath vegetation there is a high aerodynamic resistance that results in high temperature differentials between plant tissue and free air, especially on sunny days. This is so despite much higher wind speeds and reduced irradiation with increasing altitude. A particularly striking phenomenon is the reversal of cooling immediately above the timberline and treeline ecotone that may be recorded by heat sensors over alpine vegetation (mean 14.2°C between 2300 and 2600 m) in comparison with upper montane forest (mean 7.6°C between 2000 and 2250 m) on a sunny summer day, as has been reported by Körner (2007a).

## 3.2.2 Atmospheric pressure

Atmospheric pressure decreases with altitude. The decrease is not linear and the higher a location is the more altitude difference is required to achieve a unity drop in pressure (e.g. about 8 m elevation at sea level with approximately 1000 hPa pressure versus 16 m at 500 hPa pressure; Fig. 3.2). The drop in atmospheric pressure causes the partial pressure of component gases to decrease with elevation, in constant proportion to each other. However, this is unlikely to limit species distributions or the upper limit of plant life. The decrease in the partial pressure of carbon dioxide with elevation has been suggested as a potential limit to plant growth

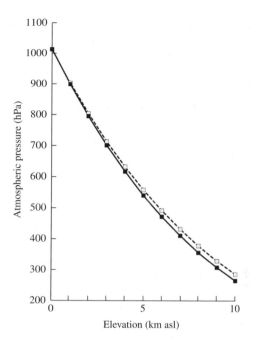

**Fig. 3.2**     Decrease in atmospheric pressure with elevation. Solid line, temperate latitudes; broken line, tropical latitudes. Data from Rudloff (1981).

(see Körner 2003, pp. 173–175); however, for example, the ability of *Polylepis tarapacana* trees in the central Andes to grow to an elevation over 5000 m points to other controlling factors (see also Section 4.3.2). On the other hand, there are indications that some animals may indeed be limited by factors related to atmospheric pressure, for example, the maximum altitude for nesting birds. The porosity of eggshells to increase gas exchange increases with decreasing partial pressure of oxygen, which, in turn, is directly related to atmospheric pressure. It is less clear, however, how insects in general respond to the combination of decrease with altitude in atmospheric pressure (less convective heat loss; Fig. 3.3), temperature (reduced thermoregulation efficiency), and partial pressure of oxygen (may affect development and physiology) (Dillon *et al.* 2006). Decreased atmospheric pressure, meanwhile, have been hypothesized to have physiological consequences on plant growth, such as it may increase frost tolerance (Halloy and González 1993) and thereby influence the survival and fitness of species. How, and if, atmospheric pressure affects species richness, distribution, and elevation zonation is not clear. One may hypothesize that low oxygen partial pressure is likely to affect animals and their diversity more at high elevations, relative to plant diversity. Such perceptions in temperate alpine environments are probably supported by the reported large herpeto

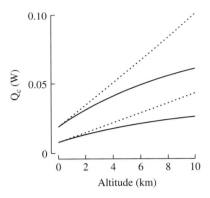

**Fig. 3.3**    Modelled instantaneous convective heat loss ($Q_c$) for an idealised homeothermic insect as calculated in constant atmosphere (broken line) and estimate true heat loss, taking into account the decrease in air density (solid line) (redrawn from Dillon *et al.* 2006). Note the non-linear relationship between heath loss with altitude, as a result of decreasing air pressure/density.

faunal similarity over altitude (a single frog species in the Alps); however, the pattern in tropical mountains is different for this group of animals (Ghalambor *et al.* 2006; Navas 2006). See more on atomspheric pressure in Section 4.3.1.

### 3.2.3 Cloudiness and precipitation patterns

Cloudiness patterns usually, but not always, parallel those of precipitation. One notable exception is that occurs in coastal areas of Chile at 25°S (the latitude of the Cordillera Atacama high altitude desert) where precipitation is close to zero; however, cloudiness ranges between 35 and 85%. High cloudiness is characteristic of tropical (10°N–10°S) and middle latitudes (50–60°N; 45–60°S). Especially low cloudiness is found between 20° and 30°N and 20° and 30°S. There is a marked absence of cloudiness at subtropical latitudes in the summer and high cloudiness may occur in the winter. Some subtropical areas lie in the monsoon zone, where cloudiness is similar to that in the tropics. As a result, the altitude distribution of precipitation shows a great variation with latitude. Where cloudiness increases with elevation it causes a decrease in net radiation (Table 3.2), which in turn makes the lapse rate steeper in atmospheric temperature. In oceanic mountains, such as in Scotland or western Norway this may result in lapse rates of –0.9°C per 100 m elevation.

Precipitation usually increases with elevation, but it may, after reaching a maximum in the cloud zone, decrease above. In boreal and temperate mountains, precipitation increases with altitude up to about 3500–4000 m. At other latitudes, the pattern is rather complex and depends on the

**Table 3.2** Climate variables along an altitude gradient from tall montane forest to wind-clipped prostrate dwarf-shrub heath in the Cairngorm mountains, Scotland. Source Wilson *et al.* (1987). Measurements were taken at 15-minute intervals; values are daily means over the month of June 1985

| Vegetation | *Pinus sylvestris* forest | *Pinus sylvestris* treeline | *Arctostaphylos uva-ursi* heath | *Loiseleuria procumbens* prostrate alpine heath |
|---|---|---|---|---|
| Altitude (m) | 450 | 600 | 650 | 850 |
| Air temperature (°C) | 12.8 | 10.3 | 8.7 | 9.1 |
| Net radiation (W m$^{-2}$) | 195 | 181 | 176 | 152 |
| Water vapour pressure (kPa) | 0.75 | 0.98 | 1.12 | 1.90 |
| Wind speed (m s$^{-1}$) | 4.3 | 5.8 | 8.4 | 9.4 |

location of the mountains within the general circulation system and on climatic differentiation caused by local exposure. For example, on the glacier-capped Sierra Nevada de Santa Marta, Colombia hydrological zonation follows a sequence from a dry Caribbean coastal zone at sea level through to about 1700 m asl, where rainfall peaks, to approximately 2300 m in the cloud forest zone, where mist drip maximizes the hydrological input (Reimer 1970). In contrast, on Mount Cameroon the highest precipitation on the coastal side is at sea level at approximately 5000 mm year$^{-1}$, which is estimated to decrease to <2000 mm year$^{-1}$ on the summit at c. 4000 m (Embrechts and Tavernier 1987; Proctor *et al.* 2007). For hydrological and ecological conditions water loss via evaporation and transpiration are just as important as precipitation. The change with elevation in temperature, precipitation, air humidity, radiation, and atmospheric pressure influences evapotranspiration and as such has direct physiological consequences for the gas exchange balance, especially transpiration of plants (e.g. Leuschner 2000; Körner 2003). Evapotranspiration has also been found to account for most variation in the distribution of species-richness patterns at coarse resolution (5 × 3 arc minutes) across the Austrian Alps (Moser *et al.* 2005). A general evaporation–elevation relationship for many areas of the tropics and subtropics, using temperature and humidity profiles may be characterized as follows (Nullet and Juvik 1994): (1) a general decline in evaporation with elevation (for most mountainous areas); (2) an initial decrease in evaporation with elevation to a minimum and then a continual increase above that level (for trade wind islands, such as Hawaii, and possibly the subtropical Andes); and (3) an increase in evaporation with elevation in a shallow layer near sea level and a continual decrease above that (for dry, coastal, subtropical west coasts).

## 3.3 Elevation pattern in weathering and soils

Studies on weathering rates and element losses in the Alps have shown that they are not linear in relation to elevation (950–2440 m) and are highest at and near to the treeline or timberline, where a pronounced podsolization occurs (e.g. Egli *et al.* 2003). Soil types change with elevation, which is caused partly by climate and partly by the fact that different vegetation types have had different lengths of time to influence soil formation and development processes (Fig. 3.4). One measure of this is the change in soil organic matter with elevation. Temperature, the main ecologically important climatic factor that changes with elevation, and precipitation have been shown to directly relate to soil organic matter content. Surface soil organic matter concentration is, in general, negatively correlated with annual mean temperature and positively correlated with annual mean precipitation (and in many mountain ranges with elevation) (e.g. Dai and Huang 2006). This is likely to be correlated with decreasing mineralization rates with elevation of organic matter. Low soil organic carbon mineralization is paralleled by low decay rates of dead plant material on the soil surface, which results in an increase with elevation, similar to that in soil organic matter, of undecomposed plant material; e.g. dead wood biomass increased by 75 kg ha$^{-1}$ m$^{-1}$ elevation, or 13 Mg ha$^{-1}$ per 1°C decrease in mean air temperature in an upper montane/treeline forest in the Colorado Rockies (Kueppers *et al.* 2004). However, elevation is not a reliable predictor, as habitat diversity (e.g. waterlogging) can much influence patterns on the ground.

Soil nutrient limitation is a cause often evoked to explain alpine plant growth. Soil nutrients do change along elevation: generally, there are decreasing values of nutrient concentration, nutrient stock, nutrient flux, and nutrient turnover up to the treeline, where these trends reverse as, for example, Frangi *et al.* (2005) have reported for a *Nothofagus pumilio* forest in Tierra de Fuego, Argentina. In the alpine zone, total nitrogen and organic carbon can be significantly higher, reflecting greater belowground production and lower rates of organic matter turnover than in soils at lower elevations (Bockheim *et al.* 2000). However, it is important to bear in mind that total soil nutrient concentrations do not reflect nutrient availability and, therefore, alpine zone soils are not nutrient rich *per se* (see Chapter 5).

## 3.4 Elevation and vegetation zonation

While climate changes gradually with elevation, vegetation is generally described in distinct belts, in terms of life form (functional type) composition and the physiognomy of dominant species that characterize the

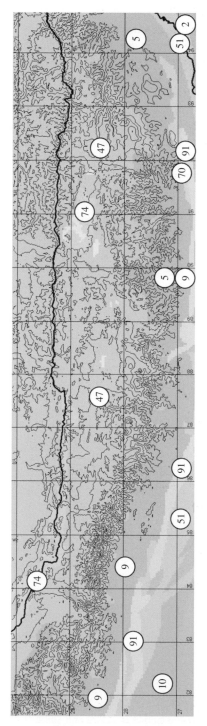

**Fig. 3.4** Elevation sequence of soil types (grey-scale polygons with numbers) in the Himalayas from the lowlands to the alpine zone. The lowest contour is 2000 m; contour around the upper valley of the River Ganges (bold black line) is 4000 m; the lower Ganges plain (lower right corner) is below 100 m. Numbers refer to legend in the FAO soil map. Average altitude values (m asl) were extracted from Global TOPO-30 DEM for FAO soil map (FAO/UNESCO 1992) categories present:

| Code | Soil type (FAO) | Soil Type description/equivalent | Mean elevation (m) | Area (%) |
|------|-----------------|----------------------------------|--------------------|----------|
| 10 | Eutric Cambisols | Brown earths | 110 | 5.7 |
| 51 | Eutric Fluvisols | Alluvial soils | 120 | 4.5 |
| 2 | Ferric Acrisols | Ultisols (leached tropical lowland soil) | 134 | 0.4 |
| 91 | Dystric Regosols | Regosol | 342 | 3.6 |
| 70 | Dystric Nitosols | Similar to acrisols, red clay, less leached | 835 | 0.9 |
| 9 | Dystric Cambisols | Brown earths (normal or podsolised) | 1736 | 12.7 |
| 5 | Orthic Acrisols | Ultisols | 2069 | 10.9 |
| 47 | Lithosols | Shallow skeletal soil | 4644 | 55.4 |
| 74 | Dystric Histosols | Organic soils (peat) | 4809 | 3.3 |

landscape (e.g. Tables 1.2 and 3.1). However, species richness and species composition do not necessarily reflect such clear-cut perceptions of zonation, and in most cases they change gradually. In general, lowland belts (e.g. termed planar, colline, and eumediterran in Europe) at the base of a mountain are replaced by a montane zone at middle altitudes up to the treeline (Fig. 3.5). Beyond the treeline, the treeless alpine zone extends up to the subnival zone where vegetation becomes patchy, and in the so-called nival (and aeolian) zones, life is restricted to a few favourable locations in an environment dominated by rock and ice. Such a simple mountain elevation model, however, in reality is confronted by a number of modifying factors, such as shortage of precipitation, excessive cloudiness, or windiness that may limit certain types or living organisms.

Elevation zonation above the forest zone has been the subject of numerous studies (e.g. Grabherr *et al.* 1995; Theurillat *et al.* 2003). The alpine zone

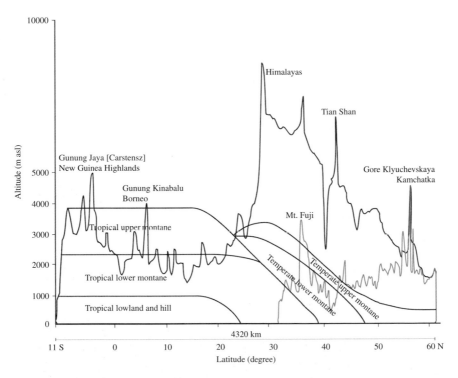

**Fig. 3.5**     Altitude zonation of forest ecosystems in Asia from tropical south-east Asia to north eastern Siberia. Approximate elevation belts are based on Ohsawa (1990,1995) and are redrawn from Hassan *et al.* (2005). The main mountain profile in black outline encompasses areas between latitudes 11°S and 60°N and longitudes 62°E and 180°E. In the foreground the east Asian mountain profile (in grey outline) was generated from a 1-km resolution DEM between latitudes 31°N and 60°N and longitudes 130°E and 165°E.

has traditionally been subdivided into successive altitude belts: treeline ecotone, and low, middle, and high alpine (subnival). The definitions of altitude belts, despite varied traditions worldwide, can be reconciled and show a good overall correspondence (Table 1.2).

The treeline ecotone is the area between the timberline (closed upper forest line) and the tree species line (the elevation where the highest established tree seedlings and sapling are observed). It is characterized by a mosaic of trees of varying stature, growing in clumps or scattered, scrub, dwarf-shrubs, and grass heath (Figs 3.6, 3.7). The majority of the extant treeline ecotones is a product of natural processes, but, in some regions, have been heavily altered by long-term human land use. This fact warrants caution in interpreting observed changes with regard to treelines and climate as their presumed driving force, as it is sometimes difficult to separate cause and effect occurring contemporaneously, such as climate warming and a decrease in livestock grazing, and even more so when the historical causes are undocumented.

The low-alpine belt, where scrub and/or dwarf-shrubs dominate, is sand-wiched between the treeline ecotone, from which an indeterminate and fuzzy boundary separates it, and the middle-alpine belt of closed swards of non-woody plants, which are often sedge heath, or extensive grass-lands. The humid tropical mountains of equatorial South America and East Africa, however, are a different world, being characterized by their unique giant rosette formations changing gradually into a subnival belt with scattered individual plants.

**Fig. 3.6**     Upper montane forest with treeline in the Colorado Rocky Mts. Observe the patchy outposts of trees, interspersed with alpine vegetation in the lower right-hand of the photograph. (Photo: G. Grabherr)

**Fig. 3.7**     Treeline with *Picea schrenkiana* in the Tianshan, east of Lake Issyk Kul, Kyrgyz Republic. The pattern is influenced by permafrost (striking north-south effects). Observe the high density of animal tracks in the lower left of the photograph. (Photo: G. Grabherr)

The middle-alpine belt of the non-tropical mountains is what is commonly known as alpine grasslands. Grasses (Poaceae) form an important structural and functional component of these formations in the Southern Hemisphere; in Northern Hemisphere alpine communities, it is the sedges (Cyperaceae) and rushes (Juncaceae) that dominate the vegetation (sedge and rush heaths). It is commonly held that these heaths are among the least impacted ecosystems by land use. However, many alpine areas have been subject to summer grazing for as long as sedentary agriculture has been established in the lowlands and in the montane forest zone, and were part of wide-ranging transhumance systems.

The high-alpine (subnival) belt is above the closed swards where vegetation becomes patchy, confined to favourable sheltered locations, near the nival/aeolian zone. The vegetation patches either appear as fragments of the middle-alpine sedge heath, or show a considerable shift to dominance by bryophytes and lichens, at the expense of vascular plants (Ozenda 1985; Reisigl and Keller 1994). None the less, fragmented vegetation patches form well-defined 'communities', which reflect environmental conditions (Pauli *et al.* 1999). This is a dynamic belt where species respond to the availability of a colonizable area (snow and ice free) in a largely predictable manner (Gottfried *et al.* 1998).

### 3.4.1 Treelines or upper limits to crop cultivation?

Studies on the temperature characteristics in relation to the distribution of high-altitude treelines (Körner and Paulsen 2004) and alpine areas

(Körner *et al.* 2003) have served both to draw up general patterns and highlight latitudinal differences. The equivalent air temperatures of the growing season mean of measured soil temperatures at the treeline is 6.7 ± 0.7°C globally, with the warmest month means being mostly below 10°C. The distribution of such areas shows a good correspondence with that of areas with a monthly mean air temperature below 10°C during the warmest month, as suggested by earlier work, and is a good approximation of the world-wide distribution of alpine regions (Fig. 3.8).

The causes of treelines include constraints to physiological processes, growth, reproduction, dispersal, and survival. The physiological constraints appear to be less linked to gas exchange related assimilative capacity rather than to the inability to fully use photosynthates for growth (limited sink capacity; e.g. Hoch and Körner 2003). Nevertheless, there is some evidence that some species may respond by increased growth to ameliorated resource availability, just as 30-year-old deciduous conifer *Larix decidua* trees did (but not the evergreen *Pinus uncinata*) in the Swiss Alps when exposed to increased experimental atmospheric carbon dioxide enrichment *in situ* (Handa *et al.* 2005). Although much is known

**Fig. 3.8**     The potential global distribution of the limit to tree growth, based on the mean temperature of the warmest month of the growing season (July-August in the Northern Hemisphere and January in the southern) temperature being T ≤10°C. Black, both July and January mean is below 10°C (= alpine); dark grey, either July (Southern Hemisphere) or January (Northern Hemisphere) mean monthly temperature is below 10°C; light grey, both July and January mean monthly temperature are above 10°C. South of the arctic, the black areas (embedded in a grey matrix) highlight most of the major mountain systems that have an alpine zone (compare Fig. 2.1). Most of the subtropical (including Mediterranean type) mountains fall in the grey shaded area (and do not show up as areas above the treeline), where the mean temperature of the coldest month (January in the Northern Hemisphere and July in the southern) is ≤10°C. Equatorial alpine areas show up as black dots in a light grey matrix (Andes, east Africa, New Guinea). As the spatial resolution of the temperature dataset is 0.5 degree, or c. 50 km², some mountain areas are not visible at this scale. For detailed information on temperature limits to tree growth see Körner & Paulsen (2004). Data source: Leemans & Cramer (1992).

about the limitations imposed by climate on the above-ground parts to tree growth, it is soil temperatures at rooting depth that can also limit overall tree growth, as has been observed in agricultural crops, where low temperatures can induce relative nutrient deficiencies. The soil temperature effect has been shown for many treeline-forming species. For example, the altitude distribution of *Betula pubescens* ssp. *tortousa*, the treeline-forming species in northern Europe may be limited by low soil temperature, especially by its effect on nitrogen uptake (Karlsson and Nordell 1996). Treeline trees, for the most part, rely on episodic clement weather that can result in sufficient seed set. This alone is no guarantee that new trees will spring up unless there is adequate seed dispersal. This, until recently, has received little attention by modellers in predicting treeline changes in the face of climate change (but see Dullinger *et al.* 2004). Equally important are ecological conditions for seeds for successful establishment after dispersal: a complex matrix of abiotic (e.g. substratum, microclimate) and biotic factors (herbivory, soil microbiology), resistance by existing vegetation (Dullinger *et al.* 2004) and innate seedling viability are at work.

The occurrence or absence of a treeline in a particular location, otherwise comparable in elevation and latitude to another one, depends on a number of climatic, topographic and associated factors, such as air humidity, precipitation and its seasonal distribution, availability of soil and availability of a topographic shelter, or the presence of a barrier (Körner 2007a). Boreal and temperate humid climate altitude treelines, where precipitation does not decrease below a certain threshold with elevation, in a sense represent the maximum potential elevation for tree growth. Accordingly, the highest elevation treelines would be expected to occur in those parts of the Himalayas where neither precipitation nor the availability of high mountain ground limit. However, this is not the case; trees reach their highest elevations on Nevado Sajama, Bolivia in the semi-arid puna region of the subtropical central Andes where *Polylepis tarapacana* peaks at about 5300 m. Stands of old *P. tarapacana* grow from approximately 4300 m on mountains up to a maximum altitude of 5300 m on Nevado Sajama. It has been suggested that these stands are surviving remnants of previously abundant forests (e.g. Kessler 1998), while others maintain that tree growth is confined to those high-altitude stands on volcanic slopes where orographic precipitation provides sufficient water for tree growth (Fig. 3.9). Treeline elevations, to some extent, depend on the species that form them. For example, a relatively low treeline in New Zealand, formed by *Nothofagus menziesii* (Fig. 3.10) (Mark and Dickinson 1997) might be a result (other than climate) of the lack in New Zealand of deciduous *Nothofagus* species, which are widespread and treeline forming in Patagonia. It is well documented that exotic species can surpass local treeline species by several hundred metres in altitude (Körner 2003).

**Fig. 3.9**     *Polylepis tarapacana* (along slopes in middle plane, above the black guide line) forms rather regular rows on gentle slopes at c. 4600 m near Nevado Sajama (snow and ice covered peak in the background), Bolivia. Since the seminal paper by Ellenberg (1979) the view has been that such stands are surviving remnants of previously abundant forests and would have covered most of the Altiplano of the Andes which are now covered by non-forest vegetation, such as the shrubby tussock grass area (*Parastrephia* cf. *lucida—Festuca orthophylla*) in the foreground of the photograph. Some workers, on the other hand, attributed the local patchy occurrence of *Polylepis* stands to climatic factors and suggested that tree growth was confined to those high altitude stands on volcanic slopes where orographic precipitation provided sufficient water for tree growth. Are the regular appearing rows perhaps the result of tree planting undertaken in pre-Hispanic times? Or are they simply arranged along fault lines in the bedrock? (Photo: L. Nagy)

In dry climates, the lack of available water can cause treelines to occur below an altitude where low temperatures would limit tree growth. In arid climates, it is not uncommon to find an upper, as well as a lower treeline. Such girdles of forest form in the moist zone on mountain slopes, with climate being too dry both above and below, such as in the Atlas, parts of the Rocky Mountains, or the mountains of Central Asia (Altai, Tianshan). Under extreme drought conditions, forests do not develop at all (e.g. mountains of the Sahara).

The role of human land use and herbivores (wild or domesticated) can be over-riding for treelines. In many areas, such as the Altiplano in the Andes, or much of the Tibetan Plateau, which have been populated since prehistoric times, no contemporary potential treeline exists. However, in the Andes it is perhaps possible to estimate potential limits to tree growth at the limits of potato cultivation. Such estimates would considerably increase the perceived lower limit of the alpine zone to about 4600 m. This may well be more accurate than current consensus. Recently, Körner (2008) in exploring potential common causes for

**Fig. 3.10**     *Nothofagus menziesii* treeline at Arthur's Pass, South Island, New Zealand. The sharp treeline has not been affected by human influence in recent history. In the Southern Hemisphere temperate zone in New Zealand, evergreen *Nothofagus* or *Podocarpus* forests occur in the lowlands, followed by evergreen montane *Nothofagus* forests, sharply transiting into a narrow low alpine shrub belt. There are extensive tussock grasslands with the endemic genus *Chionochloa*, above which fell-fields, with a highly diverse cushion vegetation give way to the nival zone. (Photo: G. Grabherr.)

growth limitation in cold-adapted plants, has compared similarities and differences between winter crops and treeline tree responses to low temperatures.

## 3.5  Species richness in relation to elevation

Despite the large geographical variation in high mountain vegetation physiognomy (Fig. 3.11) and species (Figs 3.12, 3.13) across the globe, there are general patterns for species richness in relation to elevation.

### 3.5.1 Plants

Vascular plant species richness decreases with elevation (Figs 3.14, 3.15). The trend is general; however, the absolute number of species can vary many-fold. A good example to illustrate this is Ecuador (Fig. 3.14; one of the most species-rich areas of the world) in contrast to Norway (Fig. 3.15; where the present-day flora is rather low as a result of repeated glaciations). The trend in decrease from sea level to the limit of plant life is not linear, but sometimes with peaks of species richness at ecotonal positions, such as the treeline, where montane forest and alpine species meet (e.g. Grabherr *et al.* 1995). In the alpine zone itself, the decrease in plant species is perceived as close to linear (e.g. Theurillat *et al.* 2003). A recent

**Fig. 3.11**    Hemispherical thorny shrub vegetation is characteristic on dry subtropical mountains in Eurasia, such as *Astracantha cretica* in the Lefka Ori, Crete. Subtropical mountains have a highly seasonal climate, with periodic drought. Depending on the severity of the drought, the vegetation is largely sclerophyllous evergreen forest or scrub in the lowlands, followed by deciduous and coniferous montane forest, upper montane scrub and prostrate thorny scrub. The alpine zone (cryo-oromediterranean) is often dominated by hemispherical thorny cushion forming dwarf-shrubs, particularly in Eurasian mountains; herbaceous vegetation dominates in snow accumulation patches and where snowmelt provides summer irrigation; non-irrigated areas may have grasslands that become scorched in the summer months. (Photo: G. Grabherr)

Europe-wide assessment of alpine summit habitats has shown that the net rate of decrease is about three species per 1°C temperature decrease (or approximately 200 m increase in elevation; Gottfried *et al.* 2008). This is a global figure for areas in the treeline ecotone and above and is likely to be different from place to place from the Mediterranean to the subarctic. The net change of three species per 200 m elevation increase is the result of a mean loss of 13 species that reach their uppermost occurrence at each 200-m elevation window against the accession of 10 species with their lowermost occurrence in that elevation window (M. Gottfried *et al.* unpublished).

While maximum species richness is often observed in the lower montane zone, such as at 1500–2500 m in the Nepalese Himalayas (Vetaas and Grytnes 2002), the maximum endemic species richness is usually found at higher altitudes (near to the alpine zone, at approximately 4000 m in the Himalayas), which is the zone where topographic fragmentation is highest, such as has been reported from the Ecuadorian Andes (Kessler 2002). The proportion of endemics is also highest in the alpine zone of the European mountains (see chapters 3.1–3.10 in Nagy *et al.* 2003; Pauli *et al.* 2003).

Trends in cryptogam species richness in a montane forest show a decrease, partly because of loss of habitat (tree stems, branches, and leaves of the

**Fig. 3.12**     Silversword (*Argyroxiphium sandwicense*) desert on Mt Haleakala, Hawaii at c. 2800 m asl. (Photo: G. Grabherr.)

montane forest) above the treeline. This is particularly well documented in the humid tropics of the Andes. For example, Gradstein (1995) has reported the richness of non-moss bryophytes (Hepaticae and Anthocerotae) at Santa Rosa de Cabal, Colombia at 192 species (2400–3500 m), decreasing to 88 at the upper reaches of the montane forest (3600–3700 m), with over half the species being epiphytes. For lichens Sipman (1995) has listed 415 species between 2400 and 3200 m, 232 between 3200 and 4000 m, and 101 between 4000 and 4800 m in Colombia. Neotropical bryophytes that inhabit an open soil and rock surface have low cover in lowland forest, increase in the montane zone and reach their maximum in the alpine zone (e.g. van Reenen and Gradstein 1983). In the alpine zone itself, bryophytes show a similar decrease in species richness to that observed for vascular plants, such is the Alps (Fig. 3.16).

**Fig. 3.13** *Echium wildpretii* growing up to 2 m tall in the Caldera of Mt Teide, Tenerife (Cañadas) at about 2100 m. Island mountains display a great diversity in climate and vegetation. For example, the altitude zonation on Taiwan represents a very humid subtropical type with a dense endemic dwarf bamboo-vegetation in the alpine zone; that on the Canary Islands follows a pronounced dry-wet-dry pattern with its treeline far below the potential low temperature determined one. The alpine vegetation is a unique endemic semi-dry shrubland. (Photo: G. Grabherr.)

The decrease in vascular species in lakes along an elevation gradient follows a decrease similar to that in terrestrial species. For example, aquatic plant species richness and diversity show an approximately linear decrease with increasing altitude examined along a tropical (77 m) to alpine (4750 m) altitude gradient in the Nepalese Himalayas, the main driving factors being water temperature, substratum quality, altitude, pH, transparency, and conductivity (Lacoul and Freedman 2006).

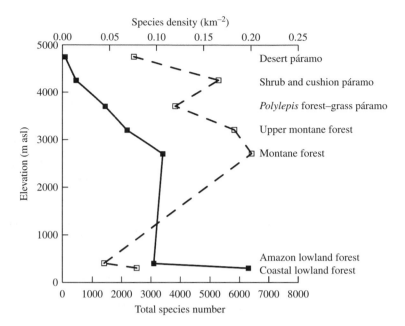

**Fig. 3.14**    Vascular plant species richness (solid line) and density (species km⁻², broken line) patterns in elevation vegetation zones in Ecuador. Data from Jørgensen *et al.* (1995). Species density is calculated as the total number of species in a vegetation zone/belt divided by the total area of that vegetation zone/belt.

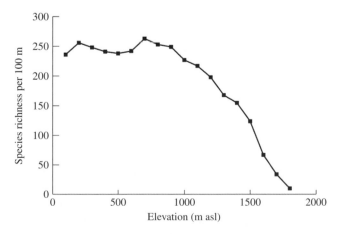

**Fig. 3.15**    Vascular plant species richness in 100 m elevation belts in Aurland, Scandes, Norway (Odland & Birks 1999).

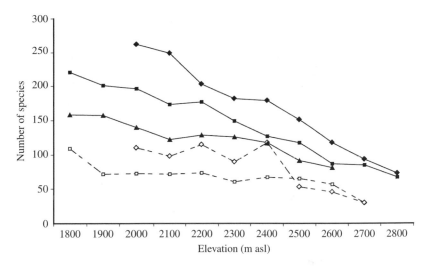

**Fig. 3.16**     Vascular plant (solid line) and bryophyte (broken line) species richness in 100 m elevation intervals along transects from the lower edge of the treeline ecotone to alpine summits, Valais Alps, Switzerland (from Theurillat *et al.* 2003)

## 3.5.2 Changes with elevation in morphology, phenology, seed bank formation, and seed survival

Not only does species richness decrease with increasing elevation, but plants become smaller: leaves grow smaller and rounder, less rigid, more often pubescent, and folded into crypts or similar structures, as has been shown, for example, for New Zealand alpine plants along a low- to high-alpine gradient that entailed reductions of 2–3°C in mean annual air temperature and approximately 10% in mean minimum relative humidity (Halloy and Mark 1996). In a similar fashion to morphological changes, the life form spectrum of vascular plants changes with elevation and becomes dominated by hemicryptophytes, many of them cushions, as has been shown in the highest reaches of the alpine zone in the Caucasus, Georgia (Nakhutsrishvili 2003). Typical alpine plants thus take advantage of the favourable microclimate close to the surface, enhanced by a high density of leaves and branches that act as heat collectors. For example, the leaf area index in prostrate dwarf-shrubs, such as *Loiseleuria procumbens* carpets may exceed a value of 2, which is equivalent to that found in tall hay meadows (Cernusca 1976). In cushions, where leaves are closely packed, but with sufficient space to trap warm air, temperature may reach well above that measured in standard meteorological Stevenson screens 2 m above ground level. Such cushions provide a clement climate and habitats for invertebrates, ectothermic insects in particular, in the harsh subnival zone.

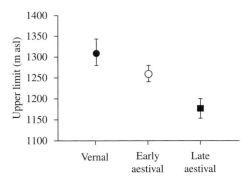

**Fig. 3.17**　　The altitude distribution of alpine species at Latnjanvagge (18°29′E; 68°21′N), according to their flowering phenology. Early flowering species have a low heat sum requirement and are the highest growing, as opposed to late summer flowering species that reach lower elevations. Source: Molau *et al.* (2005).

Morphological changes are paralleled by reductions in growing season length and changes in phenology along altitude gradients. Among others, species differ in the heat sum they require for prompting flowering. This has consequences for the altitude distribution of species with regard to their ability to flower and reproduce. Species with a low heat sum requirement flower early in the season and reach higher altitudes than late flowering species (Fig. 3.17).

The overall size of soil seed bank (transient and persistent), along an altitude sequence is likely to decrease from montane forest to the alpine zone, such as in the Cairngorms, Scotland, where *Pinus sylvestris* woodland had 83,000 seeds m$^{-2}$ versus 200 seeds m$^{-2}$ in *Racomitrium lanuginosum*-dominated alpine moss heath above 1000 m (Miller and Cummins 2003). Some alpine plants have, however, relatively large persistent seed banks (Onipchenko *et al.* 1998), perhaps related to, or reflecting plant strategies in different alpine habitats. Seed bank formation and persistence within species may increase with elevation, probably as a result of a combination of selection/strategy and soil conditions amenable to seed survival. Half-lives of seed buried at the source elevation were found to increase with elevation within the distribution range of *Phacelia secunda* (Hydrophyllaceae) in the Andes, central Chile, from 188 days at 1600 m (upper montane forest zone) to 354 days at 3400 m (high Andean or alpine zone) (Cavieres and Arroyo 2001). Natural densities of persistent seed banks (>1 year) of the same species have also been found to be larger in the alpine zone than in the montane forest with 20 seeds m$^{-2}$ at 1600 m versus 190 m$^{-2}$ at 3400 m (Cavieres and Arroyo 1999). However, these may well illustrate the response of the species to growing in forest versus open alpine vegetation, rather than an 'altitude effect' *per se*.

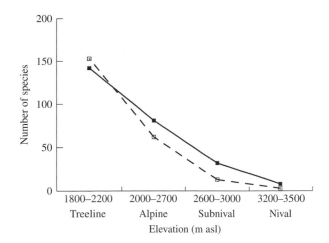

**Fig. 3.18**    The total number of Araneida (solid line) and Coleoptera (broken line) in altitude belts at Obergugl, Ötztaler Alps, Austria (data from Meyer & Thaler 1995)

## 3.6  Invertebrates

Animals, just like any other organism groups show distinct altitude distribution patterns (Meyer and Thaler 1995; Fiedler *et al.* 2008); both the numbers of taxa and their total biomass reduce with increasing altitude (Fig. 3.18, Table 3.3); especially marked is the change from the upper montane forest to the open alpine habitats. For example, in eastern Tibet, the soil arthropod community composition is dominated by mites (Acari), which changes from predominantly Prostigmata at the montane forest site (3800 m) to a mixture of Prostigmata and Oribatida at the treeline (4000 m), and then to Oribatida with 19 genera in the alpine zone (sampled at 5000 m). Collembola are most abundant at the treeline and insects in the forest. Diptera larvae, Protura, and Homoptera are the most abundant insect taxa all along the elevation gradient, while Hemiptera, Thysanoptera, and Protura occur only in the forest (Jing *et al.* 2005).

The determinants of glacier-fed streambed macro-invertebrate communities are essentially the same as those of alpine–subnival terrestrial ecosystems-temperature is a major driver together with energy-related perturbation (Fig. 3.19). Water temperature and ice-free season (and importantly, streambed environment stability) increase with decreasing altitude. Accordingly, the abundance and diversity of zoobenthic taxa increases with distance from the glacier snout downstream (Fig. 3.20), in a similar fashion to the elevation distribution of terrestrial plants and animals.

Terrestrial vegetation and productivity in lakes are more closely linked than they are in streams, and as such the productivity of alpine lakes along

**Table 3.3** The main animal groups and their altitude distribution limits from the treeline to the nival zone in the Austrian Alps (after Meyer and Thaler 1995)

| Altitude belt | Groups reaching their upper limit | Dominant groups |
| --- | --- | --- |
| Nival (2900–3100 m) | | Protozoa, Rotifera, Nematoda, Tardigrada, Collembola, Acari, Enchytraeidea, Araneida, Diptera, Lepidoptera |
| Subnival (2600–3000 m) | Gastropods (Gastropoda), centipedes (Chilopoda), millipedes (Diplopoda), symphylans (Symphyla), most beetles (Coleoptera), sawflies, wasps and bees (Hymenoptera) | Mammals: small mammals, snow finch (*Montifringilla nivalis*) |
| Alpine (2000–2700 m) | Earthworms (Annelida), pauropods (Pauropoda), proturans (Protura), diplurans (Diplura), grasshoppers (Caelifera), true bugs (Heteroptera), cicadas (Cicadina), ants (Formicidae) | Vertebrates: *Salamandra atra, Triturus alpestris, Rana temporaria, Lacerta vivipara, Vipera berus, Lagopus mutus, Pyrrhocorax graculus, Anthus spinoletta, Prunella collaris, Oenanthe oenanthe, Montifringilla nivalis, Capra ibex, Rupicapra rupicapra, Marmota marmota, Lepus timidus*, Insectivora, Rodentia; parasites of small mammals: Sporozoa, trypanosomes, helminths, lice, fleas |
| Treeline (2000 m) | Scorpions (Scorpiones), woodlice (Isopoda), cockroaches (Blattidea) earwigs (Dermaptera), psyllids (Psyllina), lacewings (Planipennia), scorpionflies (Mecoptera) | |

altitude gradients well reflects vegetation zonation. There is an increasingly lower input of nitrogen and organic carbon from vegetation into lakes with increasing elevation. A temperature gradient of about 6°C (which corresponds to about 600–900 m in altitude) may cause a 10-fold change in productivity, such as has been shown in a subarctic alpine lake system in northern Sweden (Karlsson *et al.* 2005).

# 3.7 Vertebrates

It is far more difficult to establish elevation patterns for vertebrate species than for invertebrates or plants. Large vertebrates can have extensive ranges associated with shelter seeking and feeding behaviour, and these may change with season. They may use a wide range of forested montane and open alpine habitats. The vertical distribution amplitude of birds can be especially difficult to define for species other than specialist feeders, such as humming birds or galliforms. For example, a study in the Pyrenees

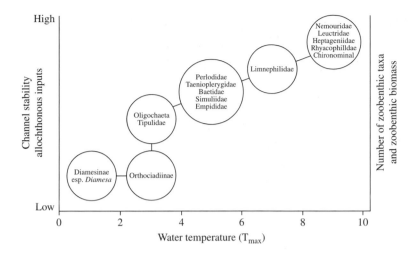

**Fig. 3.19** A suggested conceptual model of changes in macroinvertebrate taxa and their biomass along a water temperature gradient from a glacier snout (high altitude) to lower altitude riverbeds in the Alps (Milner *et al.* 2001). This example well illustrates the way which environmental factors may be correlated with altitude (temperature and level of disturbance in this example) and where altitude alone would be a poor proxy to characterise the environment (see e.g. Körner (2007b).

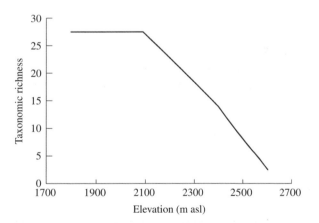

**Fig. 3.20** Trend in taxonomic richness of benthic macroinvertebrates along glacier-fed streams. As in terrestrial ecosystems, taxonomic richness decreases with elevation. The 800 m altitude difference is equivalent to 3600 m distance between glacier snout at 2600 m and sampling station in upper montane forest at 1800 m in the Swiss Alps (Lods-Crozet *et al.* 2001). The rate of decrease in water temperature is about 1°C per 100 m elevation between 2100 m and 2600 m, which is paralleled by a steep decrease in taxa. The treeline is at about 2000 m.

between 1400 m and 2921 m comparing two galliforms, the ptarmigan (*Lagopus mutus pyrenaicus*) and the grey partridge (*Perdix perdix hispanensis*) showed that most individuals of the ptarmigan occupied land above 2300 m (lower limit of the alpine zone) all year round; the grey partridge stayed well below the alpine zone in winter and spring, but to some extent overlapped in altitude with the ptarmigan in summer and autumn. Despite some overlap in altitude, the feeding ranges of the two species were largely separate, the ptarmigans predominantly using north to easterly slopes and exposed high ground above the mean elevation of the southerly slopes, preferentially used by partridges (Gonzalez and Novoa 1989). Small vertebrates, whose range is small in comparison with that of large herbivores or carnivores may also have a considerable altitude distribution amplitude such as the alpine meadow lizard (*Algyroides alleni*) that occurs on Mount Kenya from 3400 to 4600 m. For most species observed there are no detailed ecological studies available and the observed altitude ranges are sometimes based on single sightings. As a consequence, alpine in most cases is applied to species, whose mean elevation distribution is known to be above the treeline. Most alpine species also occur in the upper montane zone. The vertebrate species observed above the treeline (approximately 3000 m) on Mount Kenya include 50 residents, 20 visitors, 13 migrants, three introduced salmonid fish species, five accidentals, and 21 species with little known status (Young and Evans 1993). However, the alpine status of most of these species is difficult to establish, as the lower limit of their distribution range is largely unknown. A notable exception on Mount Kenya, in addition to the alpine meadow lizard, is the rock hyrax (*Procavia johnstonii*), a mammal species that occurs on scree between 3450 and 4700 m. In the Andes, there is a high herpetological diversity that decreases with altitude similarly to other taxa. However, for all the decrease, a large number of anuran and lizard genera make it into the páramo and the puna (Navas 2002), with about 12 species restricted to the treeline ecotone and above in the Colombian páramo (Fig. 3.21).

In general, vertebrate species richness and abundance decrease with elevation, especially sharply so in the montane zone (Fig. 3.22). For example, in the Pyrenees, 15 species have a mean altitude distribution above 2000 m, with only six of them having their mean distribution above the treeline (approximately 2300 m). In the Sierra Nevada, only the alpine accentor (*Prunella collaris*) and the European snow vole (*Chionomys nivalis*) are restricted to the alpine zone. In the Alps only the snow vole could be considered as true alpine (Allainé and Yoccoz 2003); the ibex (*Capra ibex*), and few birds, such as the snow finch (*Montifringilla nivalis*) and the ptarmigan (*Lagopus mutus*) also largely occur in the alpine zone. Similarly, few all year round residents are in the alpine zone in the Colorado Rockies, e.g. the white-tailed ptarmigan (*Lagopus leucurus*) (Armstrong *et al.* 2001), and the pika (*Ochtonoa princeps*).

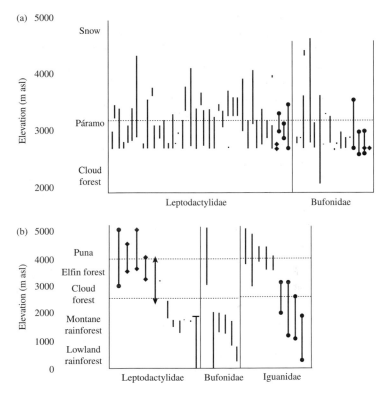

**Fig. 3.21**    Altitude ranges of two families of anurans (a) in the páramo, Colombia and (b) the puna, southern Peru (after Navas 2002). The puna figure also includes a lizard family Iguanidae. Each line represents a genus within a family.

## 3.8  Geography and vegetation altitude boundary shifts

The geographic range and associated exposure to global and regional weather systems along a mountain range is well exemplified in the Andes, which run from approximately 12°N to about 55°S. Not only they cover approximately 67 degrees of latitude, with a perhumid topical climate at one extreme through arid tropical and subtropical to subantarctic (boreal type) climate at the other, their western Pacific and eastern Amazonian slopes are exposed to highly contrasting influences by circulation systems. The western Pacific macro-slope is dry to arid. A phytogeographical cross-section profile of Chile from the Pacific coast at Arica (18°28′S; 70°22′W) to Volcán Parinacota (6342 m asl) c. 130 km inland illustrates the altitude sequence of vegetation types (Fig. 3.23). In contrast to the arid Pacific west the eastern macro-slope receives sufficient precipitation from the Amazon basin to support a mesic vegetation (Fig. 3.24).

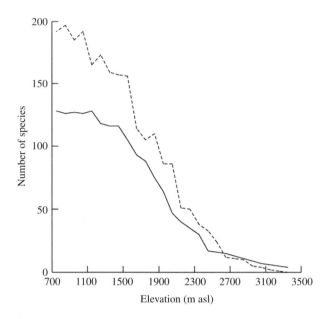

**Fig 3.22**     The decrease in vertebrate species richness with increasing altitude in the Pyrenees (dashed line) and the Sierra Nevada (solid line), Spain (Martinez-Rica 2003). Note the low species richness in the alpine zone (>2300 m in the Pyrenees and >2600 m in the Sierra Nevada).

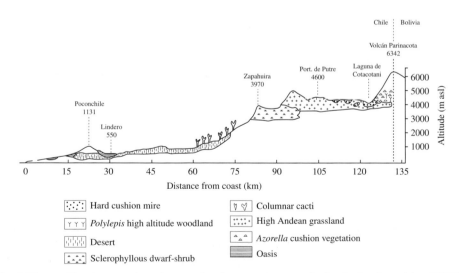

**Fig. 3.23**     The cross-section of vegetation from west to east in the Andes from Arica 18°28'S: 70°22'W on the arid Pacific coast to Volcán Parinacota (6342 m). Horizontal scale about 1:750 000; vertical scale c. 1:150 000 (redrawn from Pérez 1983)

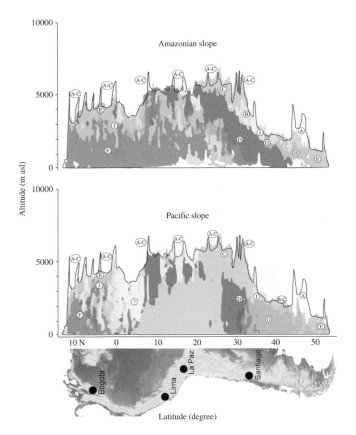

**Fig. 3.24** The land use—land cover of the Andes viewed from the eastern humid Amazonian side (top figure) and from the western arid Pacific side (middle figure), with the elevation map of the Andes (bottom figure). A, snow and ice; B, alpine heath; C, barren or sparse, desert-like vegetation; D, shrubland; E, grassland; F, evergreen broadleaf forest; G, mixed forest; H, mixed shrubland/grassland; I, savannah; J, grassland/woodland mosaic. Global Land Cover data (http://edcdaac.usgs.gov/glcc/globe_int.html) at an approximate resolution of 1 km.

## 3.9 Conclusions

This chapter has illustrated the large variety in regional and global climate patterns and ecosystem properties in relation to altitude. Although the examples were all drawn together along the common theme of altitude patterns, it is important to remember that the only consistent altitude patterns are those of temperature and atmospheric pressure and related concentration of gases in the air (for a review see Körner 2007b). All other climate variables and ecosystem properties show a

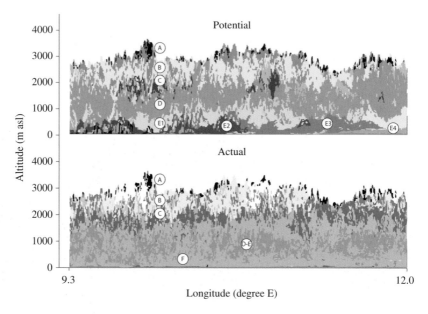

**Fig. 3.25**     Vertical zonation of vegetation in the Alps, (9.3°E, 48.5°N; 12°E, 43.3°N) viewed from the south (Po Valley, Italy). The potential vegetation is reconstructed after Bohn *et al.* (2004); actual vegetation is derived from Global Land Cover data (http://edcdaac.usgs. gov/glcc/globe_int.html) at an approximate resolution of 1 km. Note that the pattern appears rather similar between the potential and actual vegetation from the higher reaches of the upper montane zone upwards (A–C). In contrast, most of the lowland and hill (E), and large parts of the montane forest zone (D) form today the basis of agricultural production (F).

A         snow, ice and sub-nival open vegetation (*Androsace* spp., *Saxifraga* spp., *Draba* spp., lichens);

B         alpine sedge heaths (*Carex curvula*, *Oreochloa disticha*, *Juncus trifidus*, *Festuca halleri*, *Nardus stricta* on acid substratum; *Kobresia myosuroides*, *Sesleria albicans*, *Carex ferruginea*, *C. firma* on calcareous substratum in the peripheral ranges—in front);

C         treeline ecotone, open forest and scrub (*Pinus cembra*, *P. mugo*, *Larix decidua*, *Rhododendron hirsutum*) and upper montane forest (*Abies alba*, *Picea abies*);

D         montane forest (*Fagus sylvatica*);

E1, E3   lower montane forest (*Quercus petraea*, *Q. cerris*);

E2        lowland and hill forest

E4        lowland and hill forest (*Fraxinus ornus*, *Quercus petraea*, *Q. pubescens*)

F         cropland and pasture

considerable variation, according to geographical position in relation to circulation systems. Observed elevation zonation of the vegetation represents a simple model of climate (temperature) driven vegetation distribution that has been widely used for forecasting changes caused by climate change, or more broadly by global change (see Chapter 9).

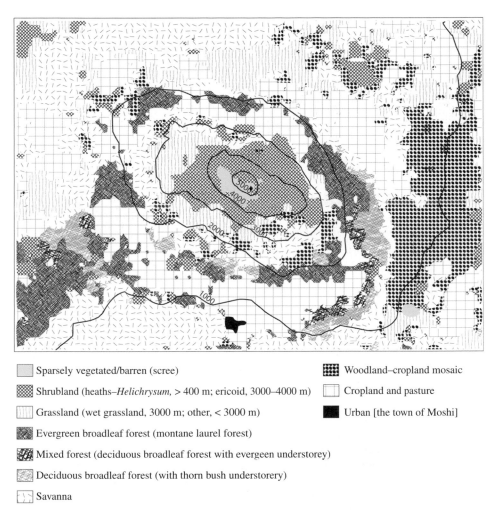

Sparsely vegetated/barren (scree)

Shrubland (heaths–*Helichrysum,* > 400 m; ericoid, 3000–4000 m)

Grassland (wet grassland, 3000 m; other, < 3000 m)

Evergreen broadleaf forest (montane laurel forest)

Mixed forest (deciduous broadleaf forest with evergeen understorey)

Deciduous broadleaf forest (with thorn bush understorery)

Savanna

Woodland–cropland mosaic

Cropland and pasture

Urban [the town of Moshi]

**Fig. 3.26**   Altitude zonation of natural vegetation and human land use on Mt Kilimanjaro. Concentric lines are contours at 1000 m intervals. The figure illustrates the reduction in area with elevation. As in the majority of mountains, especially in the tropics, the montane forest zone between c. 1000 and 3000 m is largely cultivated and used for grazing by domestic livestock. For a conventional altitude diagram see Klötzli (2004). The contours were generated from a 1-km DEM and the land cover from Global Land Cover data (http://edcdaac.usgs.gov/glcc/globe_int.html) at an approximate resolution of 1 km.

In addition to geographic position, the interpretation of the effect of complex climate and weather systems on the elevation zonation of natural vegetation is further complicated by millennia of human land use (e.g. Figs 3.25 and 3.26). In relation to the alpine zone, such impacts are discussed in Chapter 10.

# 4 The alpine environment: energy and climate

## 4.1 Introduction

Radiation and thermal energy in alpine environments is as important for life conditions as in any other life zone, and in addition, mountain environments are characterized by a high kinetic potential, or relief energy. The complex alpine landscape results in a distribution pattern in direct incident solar irradiance that creates a broad range of thermal conditions within the alpine zone. Relief energy manifests in high-energy alpine mass movement events. These either develop over a long time period (non-catastrophic events) and form permanent landscape features, such as scree, rock glacier, or block fields, or are caused by a sudden descent of material that radically alters landscapes and ecological conditions in the short-term (catastrophic events, such as rock slide, debris slide, mudflow). The historic or contemporaneous movements of high-energy material, including water and ice have traversed the altitude gradients of mountains and contributed to the creation of azonal habitats (see Chapter 6).

Topography related phenomena and climate determine largely the alpine life zone and its habitats. The discussion of habitats and their underlying ecological conditions (energy and climate, hydrology, soils, and vegetation) follows a model framework that builds up from mechanistic energy budgets to vegetation feedbacks and considers habitat-forming factors alone, and in combination for habitat characteristics. The logical framework follows that of layers and overlays of layers applied in geographical information systems.

In the first part of this chapter, solar irradiance in relation to slope angle and direction, along a latitude gradient is explained, following Kumar *et al.* (1997). This serves as a basis for understanding the relationship between irradiance and energy budgets in complex high mountain topography. The second part of the chapter summarizes climatic factors and observed climate in high mountain areas in latitude life zones.

# 4.2 The atmosphere–surface system

## 4.2.1 Solar irradiance

The amount of solar irradiance reaching the surface of the Earth varies with latitude and with season. Solar angle is largest and least variable near the Equator and, on an annual basis, the Equator receives 43% more direct irradiance than the North Pole (e.g. fig. 1.2 in Thomas *et al.* 2008). This is based on clear day radiation, such as in Fig. 4.1; however, frequent cloudiness at equatorial latitudes, or in oceanic climates greatly restricts direct irradiation. The differences in solar irradiance influence the length of the growing season, which is nearly all year round at equatorial latitudes and reaches near zero at the Poles. What happens to incident solar irradiation depends on the quality of the surface it reaches: ice and snow reflect most of it back into the atmosphere, while rocks, bare ground, vegetation, and water largely absorb it, which leads to their heating up. Some of the absorbed heat is re-radiated and is absorbed again by the surface, as well as by clouds, water vapour, and various gases in the atmosphere. Of the re-radiated heat loss, about three-quarters is returned to the Earth's surface.

Several aspects of the radiation budget in alpine environments are similar to those in non-alpine ones, such as the controls exerted on net radiation by atmospheric and surface conditions outlined above, and the strong

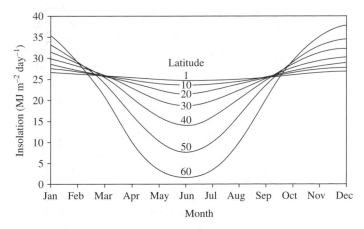

**Fig. 4.1**     Clear day mean solar radiation at various southern latitudes on the 21st day of each month (redrawn from Kumar *et al.* 1997).

relationship between global solar radiation and net radiation (Saunders and Bailey 1994). In complex alpine landscapes, incident direct irradiation, as well as all surface radiative fluxes greatly vary with slope aspect and angle, elevation, albedo, shading, sky view factor (the angular exposure to the radiating sky), and vegetation cover or leaf area index (e.g. Matzinger *et al.* 2003; Oliphant *et al.* 2003). A strong site-to-site variation in net radiation occurs at clear skies in the daytime, largely because of differences in the incoming direct radiation. The topographic complexity of mountain environments creates gradients in insolation, which, as will be shown in

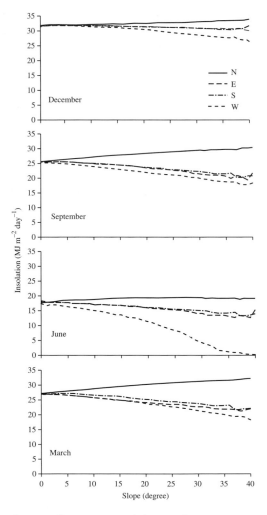

**Fig. 4.2**     The influence of seasonality, steepness of slope and aspect on potential insolation as modelled for a site in the Southern Hemisphere at 36.5°S latitude (from Kumar *et al.* 1997).

later chapters, greatly determines meso- and microclimatic differences, energy budgets, and ultimately, habitat conditions. The insolation gradients are determined by the direction (aspect) and the steepness (angle) of the slope, and by the time of the year. A thorough simulation study on insolation in relation to topography has shown that solar irradiance has least variation over the course of the year on insolated slopes (south-east– south-west in the Northern Hemisphere and north-east–north-west in southern latitudes) and that differences in solar irradiation among aspects are smallest in summer and largest in winter (Fig. 4.2). Similarly, slope angle has little effect on incipient solar radiation on insolated slopes in the summer; however, there are increasingly large differences in winter; east- and west-facing slopes are intermediate. The position of a mountain in relation to the Equator greatly influences the annual course of differences in insolation with regard to slope aspect (Fig. 4.3).

While high altitudes experience larger inputs of solar radiation than lower elevations, orographic clouds, generated by the uplift of moist air along mountain masses, restrict the time available for such high irradiation (Saunders and Bailey 1994, 1996). At night, net outgoing radiation increases with elevation because the sky view fraction increases with elevation too.

In addition to cloudiness (seasonal, regional, or orographic), there is a considerable amount of shading from neighbouring topographic features in a complex high mountain landscape (Fig. 4.4). This can be appreciated by taking into account topography (digital terrain models) in modelling actual insolation received under clear skies (e.g. Kumar *et al.* 1997; I. Mészároš, SOLEI 32 model – http://147.213.145.2/solei/index.asp).

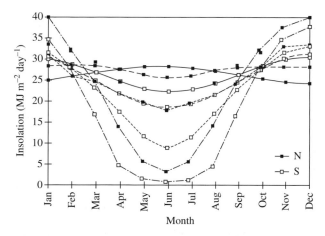

**Figure 4.3**   Solar radiation received by the north and south aspects at various latitudes —1°S; ----20°S; - - - -40°S; —·—60°S in the Southern Hemisphere (after Kumar *et al.* 1997).

**Figure 4.4**     In complex alpine country shading from neighbouring landscape features may much change the amount of potential solar irradiance received, irrespective of slope aspect and angle.  The actual amount received under clear skies can be modelled by using high resolution digital elevation models. (Photo: Khumbu Himal, Nepal; L. Nagy)

## 4.2.2 Energy balance and energy budgets

The overall surface heat or energy balance is related to radiation, sensible heat loss, heat transfer from the Earth's mass, latent heat of vaporization, the mass of water evaporated, and biological processes, and can be expressed as:

$$Rn = H + G + L \times Et + p,$$

where:

$Rn$ (net radiation) is the sum of all incoming short- and long-wave radiation from the sun and sky, less the reflected short-wave radiation and the emitted long-wave radiation;

$H$ is the convective, sensible or thermal heat exchange of the soil with the atmosphere;

$G$ is the conducted heat through the ground (soil);

$L$ is the latent heat of vaporization (586 cal $g^{-1}$ at 20°C);

$Et$ is the mass of water evaporated; and

$p$ is the energy from biological processes (photosynthesis, respiration).

Energy from net radiation drives fluxes of sensible heat exchange, evaporation and condensation, and soil heat fluxes during the snow-free period, and snow melts from sensible heat and from energy from net radiation. In view of the importance of slope aspect and angle, as shown above, it is easy to see how important net radiation is for determining habitat conditions, especially temperature and evapotranspiration.

Heat balance is connected to climate at various spatial scales (Greenland and Losleben 2001). At the macro-scale, a general heat balance range, which occurs as a function of latitude and altitude, controls the existence of an alpine zone in any one location from the Equator to the Arctic (alpine climate range, see Section 4.5). Within the alpine zone, meso-scale factors, such as topography and prevailing wind determine snow distribution and soil moisture availability. At the micro-scale, the vegetation types themselves are important components of heat (and water) budgets as, for example, Greenland (1991, 1993) has demonstrated it by examining differences in the summer heat budgets for six vegetation types along a moisture gradient at Niwot Ridge, Colorado Rocky Mountains (Table 4.1). Meso- and microclimate, therefore, determine habitat and vegetation conditions within any alpine climate zone.

Latent heat (of vaporization) is of importance for both the heat and water balances, two factors that are critical for defining plant growth conditions. Evapotranspiration summarizes combined water loss by evaporation from the soil and through transpiration by plants, the latter being influenced by soil water status, among others. As discussed earlier, while all high mountain environments are cold, in many areas water deficit is as important a factor for limiting plant growth or species distributions as are low temperatures, or high diurnal variation in temperature. Water deficit arises in some tropical and subtropical high mountain environments, e.g. the oceanic tropical mountains of Hawaii, where a low cloud base,

**Table 4.1** Field estimated surface energy balance values (MJ m$^{-2}$ day$^{-1}$) in six vegetation types in the alpine zone of the Niwot Ridge, Colorado Rocky Mountains, USA (after Greenland 1991, 1993)

|  | Cushion and prostrate dwarf-shrub fell-field | *Kobresia myosuroides* dry meadow | *Acomastylis rossii– Deschampsia caespitosa* moist meadow | *Carex scopulorum– Caltha leptosepala* wet meadow | *Salix planifolia– S. villosa* moist low scrub | Snow (bed) |
|---|---|---|---|---|---|---|
| Net radiation[1] | 11.2 | 11.7 | 11.8 | 12.3 | 11.9 | 10.3 |
| Soil heat | −0.2 | −0.2 | −0.7 | −0.7 | −0.7 | (0.1) |
| Sensible heat | −4.5 | −4.6 | −3.8 | 0.1 | −5.1 | (−1.4) |
| Latent heat | −6.5 | −6.9 | −7.3 | −11.7 | −6.1 | (−6.9) |

[1]Net radiation=[incoming short-wave radiation (19.3 MJ m$^{-2}$)] + [incoming long-wave radiation (26.2 MJ m$^{-2}$)] − [outgoing short-wave radiation (surface dependent)] − [outgoing long-wave radiation (surface and soil dependent)]. Values in parentheses are modelled.

forming from advective moisture, leaves the mountain tops with little precipitation (Leuschner 2000), or in mountains where there is little rising moist air to form clouds, such as the subtropical Andes or the Pamirs. In addition, at high elevation stations (approximately >3000 m) there is a high evaporative demand for plants, despite low temperatures, because of low air pressure. For these reasons, evapotranspiration may decrease (e.g. on wet tropical mountains, humid middle latitude mountains such as the Alps), not change, or increase with elevation, in the latter case following the availability of moisture in the soil instead of the decreasing temperature gradient (e.g. on dry mountains in East Africa, mountains of Hawaii). Despite regional trends, daily evapotranspiration rates on sunny days in the alpine zone of the Alps (3.1–3.9 mm day$^{-1}$; Körner 2003), in the arid central Andes (3.5 mm day$^{-1}$; Halloy 1991), in the Ethiopian Highlands (approximately 4.5 mm day$^{-1}$; Ayenew 2003) appear similar. (For comparison, daily evapotranspiration rates in the lowlands of the Alps region are 4.5 mm day$^{-1}$ and in the Ethiopian Rift Valley it ranges from 4.9 to 5.9 mm day$^{-1}$ over water bodies, but it is only 0.2 mm day$^{-1}$ in rift lacustrine soils and salty lakeshores. Evapotranspiration in the foothills of the Brooks Range, Alaska is 1.0–2.1 mm day$^{-1}$; Hinzmann *et al.* 1996.)

The sensible heat transfer between air and soil and the conductive heat transfer from soil particle to particle in the soil results in a much higher daily temperature amplitude in insolated stony bare ground than in irrigated soil under closed snowbed vegetation (Table 4.2).

The overall atmosphere–plant–soil temperature differentials may be quite small on an overcast day. However, on sunny days plant tissue may reach much above that of the air (Buchner and Neuner 2003), or that of the top

**Table 4.2** Soil and vegetation/substratum surface properties in six vegetation types in the alpine zone of the Niwot Ridge, Colorado Rocky Mountains, USA (after Greenland 1991)

| | Cushion and prostrate dwarf-shrub fellfield | *Kobresia myosuroides* dry meadow | *Acomastylis rossii– Deschampsia caespitosa* moist meadow | *Carex scopulorum– Caltha leptosepala* wet meadow | *Salix planifolia– S. villosa* moist low scrub | Snow (bed) |
|---|---|---|---|---|---|---|
| Soil heat capacity (MJ m$^{-3}$ K$^{-1}$) | 1.28 | 1.28 | 2.96 | 2.96 | 2.96 | 0.84 |
| Soil thermal conductivity (W m$^{-1}$ K$^{-1}$) | 0.30 | 0.30 | 2.20 | 2.20 | 2.20 | 0.42 |
| Surface resistance (s m$^{-1}$) | 8.30 | 8.30 | 4.50 | 10.00 | 28.00 | 0.10 |
| Albedo | 0.169 | 0.182 | 0.182 | 0.182 | 0.182 | 0.416 |

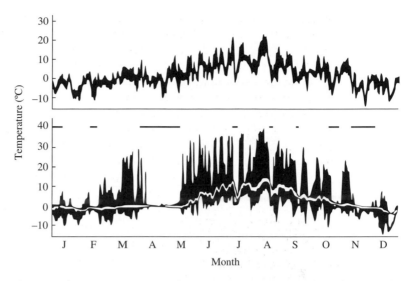

**Fig. 4.5**     Daily range of temperature measured in a Stevenson screen 2 m above the vegetation (top); in the canopy of *Loiseleuria procumbens* prostrate dwarf-shrub heath (bottom, black); and 10 cm below the soil surface (bottom, white) in 1972. Periods of snow cover are indicated by horizontal bars in the bottom panel. Data from Cernusca (1976).

soil (0–10 cm) at a given point (Fig. 4.5). Topographic landscape variability and heat balance variability are highly connected and are largely responsible for determining where and when such differences in ambient air, plant tissue, and soil (rooting zone) exist and how steep they are. Alpine landscape-scale variation may be visualized by using heat sensors (see e.g. fig. 7 in Körner 2007a).

# 4.3  Other atmospheric physical and chemical factors related to climate

## 4.3.1  Atmospheric pressure

Atmospheric pressure and the partial pressure of gases in the air decrease with altitude, the only environmental factor specific to mountains. The rate of decrease is related to altitude (Fig. 3.2). While the partial pressure of component gases in the air decreases, their ratio to each other remains unchanged. The impacts of a decrease in atmospheric pressure on living organisms have been investigated in detail, especially on animal and human physiology and adaptation (Table 4.3). A variety of physiological

**Table 4.3** Examples of the impacts of decreased atmospheric pressure on animals

| Organism group | Impact/adaptation | Reference |
| --- | --- | --- |
| *Invertebrates* | | |
| Insects | Reduced convective heat loss because of low atmospheric pressure (predicted from models). Low oxygen partial pressure may affect development and physiology. Decreased air density compensated for by flying insects with increased stroke amplitude (not frequency); evolutionary adaptation by increased wing size relative to body. | Dillon *et al.* (2006) |
| *Vertebrates* | | |
| Anurans | Little impact of low oxygen partial pressure on adult amphibians and lizards because of their low energetic expenditure for thermoregulation. Phenotypic plasticity in physiological adaptation to hypoxia. | Hutchinson (1982); Ruiz (1987); Ruiz *et al.* (1989); Navas (2002) |
| Birds | In response to hypoxia above 3000 m adaptations in the pulmonary, cardiovascular, and muscular systems. Low air pressure (density) may help forward flight, but requires more energy for hovering. Compensation mechanism in hovering birds are increased wing size and larger stroke amplitudes. | Altschuler and Dudley (2006) |
| Small mammals | Low partial pressure reduces pulmonary oxygen loading, while cold increases the demand for heat production. Selection for high maximal rate of oxygen consumption may be detected to maintain thermo-regulation. | Hayes and O'Connor (1999) |
| Humans | Hypobaric hypoxia increases haemoglobin concentration in the high Andes but not on the Tibetan Plateau or the Ethiopian Highlands. Increased percent oxygen saturation of haemoglobin in Ethiopian population but not in the Andes, or Tibet. | Beall (2006); West (2006); see also issues of the journal *High Altitude Medicine and Biology* |

and evolutionary adaptations have helped the colonization of high-altitude habitats by flying insects (Dillon *et al.* 2006), anurans and reptiles (Navas 2002, 2006), birds (Altschuler and Dudley 2006), and mammals (Hayes and O'Connor 1999). For plants it has been observed that some biophysical–physiological phenomena are connected to air pressure, such as transpiration, which increases with a decrease in atmospheric pressure Körner (2003). Low atmospheric pressure has also been related to putative morphological adaptations in the form of mucilage production and low stomata numbers in high Andean plants (González 1985). There has been some recent experimental work on plant physiology in extreme hypobaric environments (e.g. Paul *et al.* 2004). The interesting finding from these studies was that low pressure caused the expression in *Arabidopsis* of about 200 genes, of which less than half were prompted by hypoxia alone, too. This indicates that hypobaria has its own impact, largely related to plant water economy, irrespective of the partial pressure of oxygen. However, as the above experiments used one-tenth of the pressure at sea level—much lower than any plant growing at extreme alpine altitudes can experience—results are difficult to relate to real-life high mountain conditions. There are some complex and less well explained phenomena related to low atmospheric pressure and related hypoxic conditions, such as the high survival rate of common garden lettuce (*Lactuca sativa*) at 6000 m, as observed by Halloy and González (1993). Frost damage, appeared to decrease under low atmospheric pressure (approximately 472 hPa versus 1013 hPa pressure at sea level) at temperatures (night–day) that would have prevented the plants from surviving at normal atmospheric pressure. Halloy and González (1993) explained the increased frost/cold tolerance by the reduced oxidative damage to cells after thaw at a low ambient oxygen partial pressure, less than half of that experienced at sea level. The question remains, however, as to what extent cold tolerance is shaped by selection of taxa, or physiological adaptation in hypobaric (and hypoxic) high mountain environments.

## 4.3.2 Atmospheric chemistry

The radiative balance of the atmosphere is modified by changes in the component gases. Some, the so-called greenhouse gases (e.g. $CO_2$, $CH_4$, $N_2O$, CFCs), reduce radiative heat loss via re-radiating long-wave radiation into the atmosphere, while aerosol-forming ones, such as sulphates reflect a relatively large fraction of incoming radiation, thereby cooling the atmosphere. Besides impacting on the radiative heat balance, changes in the chemical composition of the atmosphere affect biogeochemical cycles and can become driving forces of change (see Chapter 5 on soil nutrients and Chapter 9 on global change). For example, atmospheric nitrogen deposition provides fertilizer input and is thought to be a regionally important factor for biodiversity change in the alpine zone (Walker *et al.* 2001a).

Elevation-related changes in atmospheric (biosphere, near land surface) chemistry include a reduction in the partial pressure of component gases (in constant proportion to each other) in the air. Plant physiologists have argued that as a gamut of factors is involved in carbon gas exchange a decrease in the partial pressure of $CO_2$ may or may not result in resource limitation for photosynthesis. Reductions with altitude in photosynthetic efficiency observed in some species (e.g. Körner 2003, pp. 173–177) are not only caused by decreased availability of $CO_2$, but are the net outcome of changes in the whole plant–atmosphere–soil system with altitude. In fact, Crawford *et al.* (1995) have argued, based on comparing carbon isotope values from soil respiration (approximately –19‰) with that of *Saxifraga oppositifolia* tissue (–19.51 ± 0.32‰; n = 8) on Svalbard, that $CO_2$ limitation is unlikely for low stature, ground-hugging arctic (and alpine) species because their main source of $CO_2$ is from soil respiration, which should more than compensate for reduced atmospheric concentrations of that gas. However, such a pattern was not confirmed in the Alps, where a sample of eight cushion-forming species, including *S. oppositifolia*, gave a mean value of –26.38 ± 1.10‰ (Körner 2003, p. 177).

### 4.3.3 Ultraviolet radiation

One often cited adaptation of high-altitude biota is in connection with ultraviolet (UV) radiation and ozone (Fig. 4.6). Ozone production is a naturally occurring phenomenon, prompted by solar UV radiation, which results in a photochemical reaction that causes oxygen molecules to dissociate. The produced ozone, in turn, reduces solar UV radiation into the biosphere. There exists a gradient in UVB radiation with its highest values on tropical high mountains and lowest (by about 10-fold in comparison) in the Arctic (Fig. 4.7). Gaseous nitrogen compounds in the atmosphere not only cause a fertilization effect once they reach vegetation/soil, but they catalyse the destruction of the ozone molecules in the atmosphere and thereby increase incident UV radiation.

The impacts of increased UV radiation were limited in experiments in subarctic/arctic environments (Callaghan *et al.* 2004). On the other hand,

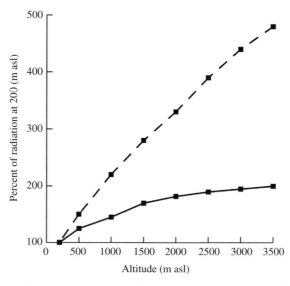

**Fig. 4.6**     Seasonal changes in UVB radiation with altitude in the Alps: summer (solid line); winter (broken line) (Source: table 4 in Viet 2002).

Sullivan *et al.* (1992) and Ziska *et al.* (1992) have shown experimentally that natural plant populations along a 3000 m altitude gradient in Hawaii exhibited a wide variation in sensitivity to UVB radiation and that this variation was related to the ambient UVB radiation environment in which the plants naturally occurred. In particular, plants growing in a naturally high UVB environment may have evolved mechanisms in relation to reproductive phenology and carbon uptake, which may maintain their productivity in a high UVB environment. In addition, ecotypic differentiation may have occurred in response to increasing UVB radiation over an altitude gradient. Such differentiation could influence the responses of plant communities and animal assemblages to other global change drivers. The main adaptation to UV radiation is the build up of flavonoids in the epidermis of plants (Caldwell 1968, 1979) and pigments in animals. Pigmentation (melanism) in animals reduces mortality caused by UVB, as has been shown for *Daphnia* by Zellmer (1995), or for the zooplankton rotifer *Hesperodiaptomus arcticus* (Vinebrooke and Leavitt 1999); see also Sømme (1997) for a review. The impact of UVB radiation on the abundance of individuals for several species has been demonstrated: in the Australian Snowy Mountains the exclusion of UVB significantly enhanced the survival of *Litoria verreauxii alpina*, a frog species that has

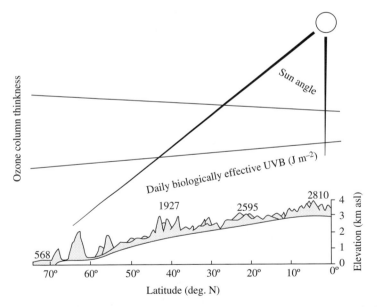

**Fig. 4.7**   The reduction in UVB radiation from the equatorial Andes to the arctic in Alaska. The decrease in effective UVB irradiance is caused by (a) latitudinal gradient in the thickness of the ozone layer, (b) changes in solar angle—high at the Equator, low in the arctic, and (c) by elevation differences. Redrawn and with UVB values taken from Caldwell *et al.* (1980).

been declining (Broomhall *et al.* 2000). Navas (2002, 2006) has suggested that frog eggs are indeed susceptible to UVB radiation; however, it is unclear if the jelly coating of the eggs provides protection from it. Algal growth and aquatic invertebrate abundance may be heavily inhibited by UVB, especially at summer low water conditions (Kiffney *et al.* 1997). None the less, some organisms are unaffected, or may even respond positively, such as some cyanobacteria (Vinebrooke and Leavitt 1999). In general, ambient UVB has significant effects on heterotrophic organisms, including marked inhibitory effects on insect herbivory (e.g. Ballare *et al.* 2001; Rousseaux *et al.* 2004).

## 4.4 Relief energy in alpine landscapes

As opposed to radiation energy, which is present in all parts of the alpine landscape, phenomena related to relief energy occur where slopes are steepest and where resistance to it is lowest (e.g. formation of water courses). Relief energy is highest in young, rugged mountain chains, such as the Alps, Himalayas, or the Andes, whereas in old worn-down ranges, such as the Scandes or the Appalachians, it is largely confined to certain glacial landforms only. Relief energy is manifested by the movement of high-energy material: rock falls, avalanches, debris flow, landslides, and watercourses, and less and less frequently these days, slowly advancing glaciers often interrupt the climatic altitude zonation of the vegetation in mountains.

Rock falls occur most often when temperatures increase or the thaw of permafrost in rock walls upsets their stability (e.g. Gruber *et al.* 2004). Rock falls may be incidental events in regions with a long history of relative stability, following the thawing of permafrost. They may cause local disturbance to plant populations or localized habitats. Where rock fractioning and rock falls are a regular event talus or scree forms. Particles are sorted along the scree with fine material accumulating near the top and a block or boulder field forming at the bottom. The fine material in the upper scree is more prone to cryo-turbation caused by freeze–thaw cycles than the coarser material at the bottom of the slope. Screes may span several hundreds of metres of elevation, crossing climate-determined bioclimatic elevation zones. With their particular structure, radiative, thermic, and chemical properties they offer a specialized habitat for plants and animals (see Chapter 6).

In extreme cases avalanches leave behind tracks that are devoid of vegetation and with the soil or subsoil exposed. There is physical sorting of particles along avalanche tracks similar to that found on screes (e.g. Jomelli and Francou 2000); however, avalanche tracks contain more fine soil-forming material than screes which makes avalanche tracks readily available for colonization by plants. Mass movement affects many avalanche tracks in a less destructive, but annually recurring manner. They are most visible in the montane forest zone where they show up as treeless scars running down mountain-sides. Perturbation is most intensive at the highest elevation points and in the centre of the avalanche paths. As a result the vegetation of avalanche tracks has progressively taller vegetation away from the centre. While screes are considered to be unstable habitats because of their structure, they are entirely different from avalanche tracks, which are prone to periodic scouring by new avalanche events. Established screes offer relatively stable ecological conditions in comparison with most avalanche tracks.

All the above are related to coarse debris, which together with the fine sediment and geochemical wasting allow the calculation of sediment budgets and the corresponding mechanical energy released (Table 4.4). It appears that (1) coarse debris movement in the Colorado Rockies is about fivefold when compared with fine sediment, and (2) that the highest energy component (rock fall) releases half the total energy per annum but shifts only about $1.5 \times 10^{-5}$ of the total volume. In contrast, mass movement involving large amounts of material, such as scree creep or rock glacier flow releases/requires little specific energy per volume. The relative amounts of mass wasting are small when compared with those that can be estimated to have occurred in glacial and inter-glacial periods. The linear rate of wasting is about 0.02 cm year$^{-1}$ of rock face, 3 cm year$^{-1}$ scree creep

**Table 4.4** Budgets of material transport and energy in a high alpine watershed above a glacial lake in the Colorado Rockies. Data from Caine (2001)

| | Volume (m$^3$ year$^{-1}$) | Energy (MJ year$^{-1}$) | Energy/unit volume (MJ year$^{-1}$ m$^3$) |
|---|---|---|---|
| *Coarse debris* | | | |
| Rock fall | 10.00 | 44.60 | 4.46 |
| Scree accumulation | 1.50 | 2.87 | 1.91 |
| Scree creep | 206,000.00 | 14.43 | <0.01 |
| Debris flow | 9.40 | 15.32 | 1.63 |
| Rock glacier flow | 373,000.00 | 3.59 | <0.01 |
| *Fine sediment* | | | |
| Soil creep and solifluction | 95,440.00 | 0.90 | <0.01 |
| Surface wasting | 240.00 | 0.03 | <0.01 |
| Lake sedimentation | 13.50 | 1.00 | 0.07 |
| Fluvial sediment export | 0.38 | 0.05 | 0.13 |
| Aeolian transport | 20.00 | ? | ? |
| *Geochemical* | | | |
| Solute transport | 1.90 | 13.36 | 7.03 |

downhill, 2–20 cm year$^{-1}$ rock glacier movement, and 0.1–1.0 cm of solifluction, the latter occurring on wet sites (Caine 2001). Sediment transport out of the alpine zone is low, not exceeding approximately 15% of
wasting that occurs in the alpine zone. In historic terms, this is a relative
geomorphic stability that has contributed to the development of habitats
and their distinctive assemblages of plants and animals.

Debris flows consist of a mixture of water and sediments of various sizes.
Avalanche tracks are especially prone to debris flows, usually prompted by
high intensity rains, and the downslope transport of material (Bardou and
Delaloye 2004) may contribute to the redistribution of plant and animal
propagules. The small-scale surface soil and turf slides that occur on
steep slopes are related to slope angle, aspect, soil depth, and the type of
vegetation cover, primarily through root length and density in the soil
(Fig. 4.8; Tasser *et al.* 2003).

### 4.4.1 Vegetation processes and distribution on land disturbed by mass movement

Mass movements can greatly modify vegetation and its distribution, and
override factors such as temperature, slope angle, or aspect. For example,
*Populus tremuloides* and *P. balsamifera* on Kathul Mountains (Alaska)

**Fig. 4.8**     Small-scale erosion scars from turf slides (*Blaiken*) on steep slopes, Montafon, Northern
Alps, Austria. (Photo: G. Grabherr)

have been shown to grow mostly in areas prone to frequent, catastrophic landslide disturbance, and associated substratum (Lewis 1998).

Habitats related to complexes of landslide relief elements reflect the varied conditions that mass movements cause by exposing bedrock, transporting, and mixing soils, and by changing local hydrology. For example, in the Polish Carpathians 16 types of natural habitats were identified in landslide areas (Alexandrowicz *et al.* 2003). The importance of avalanche tracks and other areas affected by regular disturbance from mass movement has also been highlighted for the survival of arctic–alpine species in the central European Middle mountains (Jeník and Stursa 2003). Avalanche tracks themselves show a clear zonation (see e.g. Erschbamer 1989) from non-woody vegetation in the centre (where snow persists longest) to prostrate woody shrubs (e.g. *Pinus mugo* and *Alnus viridis* in the Alps, but also deciduous tree species elsewhere, such as *Fagus* or *Nothofagus* forming prostrate multi-stemmed shrubby individuals).

Overall, the different forms of mass movement cause a range of disturbance regimes. This, in turn, results in a range of vegetation dynamic processes from re-colonization of bared areas (primary succession) to recovery of vegetation after lesser perturbations (secondary succession). These perturbation gradients may overlay, or be embedded inside the mosaic of alpine habitats along topographic and related abiotic gradients and catalyse vegetation dynamics, alone, or in combination with changes in the use of habitats by animals.

## 4.5 Alpine zone climates

Temperature and precipitation are the most important descriptors of climate for biological systems. Precipitation makes all the difference for vegetation distribution and associated fauna within the alpine temperature regime. There are some general trends in precipitation, such as its decrease away from the coasts, as can be seen, for example, in the Scandes, Norway (Holten 2003) or in New Zealand, where the coastal ranges receive in excess of 4000 mm precipitation annually as opposed to inland mountains, where the rain shadow restricts it to approximately 1000–1500 mm (Mark and Dickinson 1997). Trends in precipitation are also found within inland mountain systems (e.g. sub-oceanic northern Alps versus continental central Alps, or the southern versus northern slope of the Himalayas). These general trends in precipitation, however, are more subject to exceptions than those in temperature. The reason for this is that mountains may not only have a range of latitudinal climate zones arranged along their altitude axis, and have a diverse array of microclimates arising from their varied topography, but they can locally modify climate. Some extensive highland plateaux such as the Tibetan and the Andean are large enough to have their own climate. High mountains can intercept moist air masses and create a rain shadow in areas lying opposite from the prevailing wind direction. For example, in the East African mountains south- and south-east-facing slopes receive about twice as much precipitation than leeward slopes. Most notably, the monsoon rain is intercepted by the Himalayas, which create a rain shadow for the Trashimalayan ranges to the north, reducing precipitation there to 100–400 mm year$^{-1}$. In addition, mountains can generate local wind systems (Barry 1992), which can affect diurnal and seasonal wind patterns and precipitation. The basic pattern of diurnal wind patterns is that of valley or anabatic uphill flow of air during the day and mountain or katabatic air flows at night as a result of differential warming and cooling of mountain slopes in relation to the free atmosphere over lowlands or tablelands. Anabatic winds can cause or contribute to orographic rainfall on the windward side of the mountains, often with an early afternoon maximum.

Objective bioclimatological systems, based on measured meteorological data have been proposed at various spatial resolutions to classify climatic conditions for life worldwide (e.g. Köppen 1931; Thornthwaite 1933; Holdridge 1967; Walter 1973; Rivas-Martinez 1995; Kottek *et al.* 2006). The life zone system of Holdridge (1967) is for example, based on mean annual biotemperature (T) for values between 0 and 30°C, mean annual total precipitation (P), and on the potential moisture deficit or excess, or evapotranspiration ratio (ER, ratio of mean annual potential evapotranspiration to mean total annual precipitation). The Holdridge system defines alpine areas (3 < T > 1.5°C), along a precipitation gradient from 1500 < P > 750 mm and 0.25 < ER > 0.125 at the wet end, to P < 125 mm and 2 < ER > 1 at the arid end (Holdrige 1967, fig. 1, p. 17). Rivas-Martinez (1999) has refined previous systems, which distinguished a single high mountain bioclimate, and expressed mountain climates as bioclimatic altitude zones, using various calculated climatic indices. The application of these indices to define bioclimatic ranges in the Bolivian Andes is illustrated in Table 4.5. The recent bioclimatic characterization of the treelines worldwide by Körner and Paulsen (2004) suggested that the treeline lies at the +5°C to +8°C growing season mean soil temperature isocline worldwide, with 6–7°C in the subarctic zone and 7–8°C in the temperate (Fig. 4.9) and non-European Mediterranean zone, the tropical treeline being at approximately 5°C.

There are constraints for a widespread acceptance of any of the bioclimatological classifications. However, the use of the meteorological data, which these classifications are built on, permits a cursory exploration of the temporal and spatial dynamics of bioclimatic conditions. For example, Yue *et al.* (2005) used data from 735 meteorological stations to model decadal changes in the extent and position of Holdridge life zones in China. They found that the potential extent of the nival zone, which was calculated as the single

**Table 4.5** Bioclimatic indices calculated for the Andes, in the Peruvian puna and Altiplano biogeographic regions of Bolivia. Source: table I.12 in Navarro and Maldonado (2002)

| Altitude zone | Elevation (m asl) | Vegetation | Bioclimatic index | Frost incidence | Bioclimatic zone |
|---|---|---|---|---|---|
| Nival | >5200 | – | Tp = 0 | All year round; part permanently frozen, part night frost only | Athermic |
| Subnival | 4700–5200 | Open | Tp = 1–450 | Night frost all year round | Cryo-orotropical |
| Altoandino (upper montane to alpine) | 4000–4600 | Grassland | Tp = 450–1050 | Night frost, except 1–3 months in the summer | Orotropical |
| Puna (upper montane to alpine) | 3200–4000 | Dwarf shrub | It = 160–320 | Night frost from few days to months of the year | Supratropical |

$It = (T + M + m)*10$, where $T$, annual mean temperature; $M$ mean of the maxima of coldest month; $m$, mean of the minima of the coldest month. $Tp$ = sum of mean monthly temperatures of those months where the mean is >0°C. Where the value of $It$ is <120 $Tp$ is calculated instead.

largest zone in the 1960s (covering 12.75% of China) was 25% smaller using the data from the 1990s and the mean centre of distribution of the nival zone has shifted to the north-east. During the same period, the alpine bioclimatic zone expanded by 21% relative to its initial cover; however, the displacement of the centre of the alpine types was to the south in seven out of 12 cases, i.e. opposite to the direction of shift of the nival zone. The general implications of the above are that peripheral alpine and nival areas (at the edge of large core areas, or in locations where they have a marginal existence) experience larger climatic fluctuations than core areas. The dynamics of alpine vegetation in relation to climatic factors is discussed in Chapters 8 and 9.

One of the major motives for delineating bioclimatic boundaries have arisen from the striking differences in macroclimatic conditions from equatorial latitudes to polar ones. For example, in the alpine zone of the wet tropics, the daily temperature amplitude is three to 10 times higher than the annual one, while at subtropical, temperate, and higher latitudes seasonality becomes very marked (compare Figs 4.10 and 4.11). The following section gives a brief overview of some of the climatic variety that alpine habitats are exposed to globally (see also Chapter 2).

Arctic, boreal, temperate and subtropical seasonal alpine climates characteristically allow a very short (typically <3 months) to short (3–4 (6) months) growing season and vary a lot in the way of precipitation,

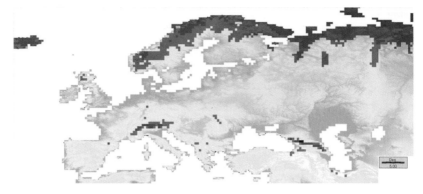

**Fig. 4.9**    Mean growing season air temperature at the treeline across the sites in Europe from Northern Scandinavia to the Alps appears to vary between c. 6.0–9.3 °C (calculated from Körner & Paulsen, 2004), with a mean of 6.1°C near Abisko (two measurements, 68°N, 19°E) and with a mean of c. 7.6°C in the Alps (12 measurements, 46°–47°N, 7°30′–11°30′ E). Map: 0.5 degree cell mean June-September air temperature (<6°C, dark grey raster; <9°C, medium grey raster) from data by Leemans & Cramer (1992).

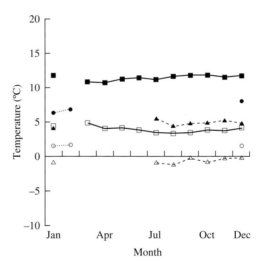

**Fig 4.10**    The annual course of mean monthly minimum (empty symbols) and mean maximum (filled symbols) temperatures in the mountains of New Guinea. The alpine zone is approximately above 4000 m. Data from: Allison & Bennett (1976) for Mt Jaya, 4250 m (circles); McVean (1974) for Mt Wilhelm, 3480 m (squares), 4380 m (triangles).

both in quantity and seasonal distribution (e.g. Bliss 1985; Chapin III and Shaver 1985; Billings 2000; Shahgedanova *et al.* 2002a,b; Veit 2002). The arid zone continental mountains of temperate latitudes, such as the Pamir, the southern Altai, or the Tianshan are far from being uniform in their

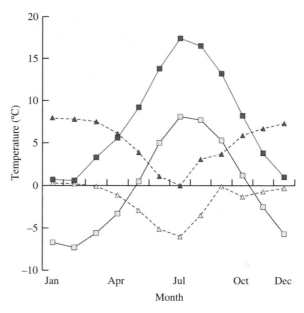

**Fig. 4.11**  The annual course of mean monthly minimum (grey symbols) and maximum (black symbols) temperature in the Pyrenees (42°30′N, 1°00′E), Spain 2174 m (solid line) and on Volcán Parinacota (18°12′S, 69°08′W), Chile 4390 m (broken line). Both show a high degree of seasonality. Data for the Estany Gento, Pyrenees is courtesy of L. Villar.

climate. The large topographic contrasts result in local and regional temperature and precipitation patterns, and the large extent and height of the massifs influence large-scale circulation (Merzlyakova 2002). Temperature is highly seasonal, with locations at high elevation experiencing annual temperature amplitudes of up to 30°C. Maximum precipitation occurs between 2000 and 4000 m asl. The Himalayas form a physical and climatic boundary between temperate continental in the north and subtropical monsoon in the south, and accordingly, they are affected by contrasting influences. The northern parts grade into the arid continental alpine climate of the Tibetan Plateau (Wang 1988). In contrast, in the Himalayan ranges from west to east an increasingly humid alpine climate, determined by the monsoons, is found. There is a sharp decline in precipitation with elevation and the Himalayas experience extremely cold winters in the alpine zone (see climate diagrams in Miehe 2004, p. 328).

The subtropical sclerophyllous 'Mediterranean' forest zone mountains (Fig. 2.1) in the Southern Hemisphere have their precipitation peak in the southern summer months e.g. about 80% in the Drakensberg, eastern Lesotho falls between October and March. Both north-westerly and south-easterly winds produce orographic rainfall; however, the interior is

**Fig. 4.12**    Snowfall is not exceptional in the (sub)tropical Andes. Snow usually lies a short time and the morning sun melts it early. In exceptional years, snow cover may last for weeks which can decimate local ungulates. In seasonal climates, snowlie determines growing season length and local long-lying patches of snow contribute to vegetation type diversity. (Photo: L. Nagy.)

in a rain shadow, which causes a considerable geographical variation in precipitation (Sani Pass summit (2865 m) 995.8 mm, Organ Pipes Pass (2927 m) 1609 mm, and Letsen-la-Draai (3050 m) 713.6 mm). Also, there can be large year to year differences: Sani Pass 1441.6 mm in 1938–1939 versus 439.3 mm in 1944–1945 (Killick 1978). Snowfall (mainly in July) can lie up to 2 months on southern slopes.

Snow-lie is an important factor for determining the length of the growing season, vegetation types and their distribution in all seasonal climates. In contrast, aseasonal tropical climates are little influenced by snow-lie on the ground, which is largely ephemeral, and overnight snow usually melts early in the day (Fig. 4.12).

At equatorial latitudes, on an annual basis, the minimum and maximum daily irradiance differs by about 13%, increasing to about 60% at the Tropics of Cancer and Capricorn (List 1971, cited in Rundel 1994). This results in little variation in temperature near the Equator over the whole year (e.g. Fig. 4.10), but with relatively large diurnal ranges; the occurrence of night frost is frequent (e.g. Monasterio and Reyes 1980).

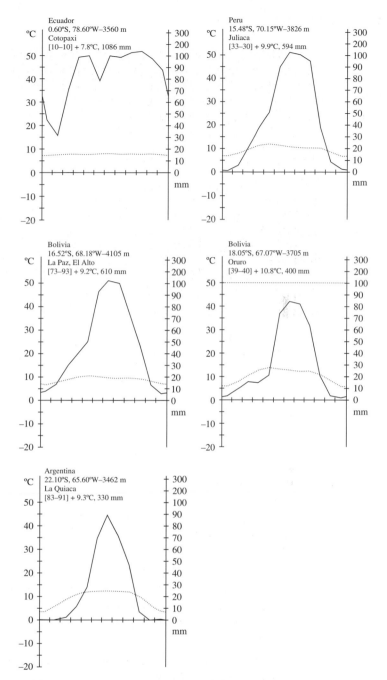

**Fig. 4.13**     Climate diagrams arranged in a decreasing north–south precipitation gradient in the Andes. Note the increasing seasonality in temperature with distance from the Equator and the change in precipitation pattern from a brief dry season on Cotopaxi to a marked unimodal peak in the other localities. Solid line, mean annual precipitation (right axis); broken line, mean annual temperature (left axis). Along the x-axis are the months, with January in the centre; the values in square brackets show the number of years over which temperature and precipitation are averaged. Source: Lieth *et al.* (1999).

The equatorial monsoon climate dominates in the south-east Asian alpine areas in New Guinea and Borneo. There is remarkably little variation in monthly mean temperature over the course of a year (1.0–1.5°C; Fig. 4.10) and the differences between monthly minima and maxima of about 7°C at 3500 m decrease to approximately 3°C at 4300 m. There is a high rainfall, with a decrease in total precipitation that is apparent from west to east; for example, Mount Kinabalu in Borneo is estimated to receive an annual total of 5000 mm and Mount Wilhelm, Papua New Guinea 2900 mm (Hnautiuk *et al.* 1976), with the high summits in the western part of New Guinea receiving intermediate values (Allison and Bennett 1976). The over-riding monsoon effect is absent from the climate of equatorial Africa, and the climate is more of a continental highland type. On the continent scale, West Africa is perhumid and humidity decreases from west to east. East Africa is rather dry, caused by the prevailing monsoon systems; however, the highlands and high mountains cause some orographic precipitation. Precipitation in the East African Highlands peaks in the cloud belt between 1400 and 2200 m and decreases with further elevation, to about 850 mm on the summit of Mount Kenya and to approximately 200 mm on Mount Kilimanjaro (Hedberg 1964; Coe 1967).

In the tropical and subtropical Andes (between 9°N in northern Columbia and 23°S in north-west Argentina), the annual course of temperature amplitudes are lower (7–14°C) in the moist equatorial alpine areas or *páramo* (Venezuela, Colombia, Ecuador, northern Peru) than in the dry alpine areas called *puna* of southern Peru, Bolivia, north-west Argentina, and Chile (16–25°C). This is especially marked in the southern winter months of June and July, when there is a marked drop in minimum temperatures in the *puna*, all the more accentuated with increasing latitude (Fig. 4.13; see also figs 2.5 and 2.6 in Rundel 1994). There is a decrease in the precipitation along the Andes, from about 1080 mm year$^{-1}$ on Cotopaxi near Quito (0°60′N) to 330 mm on the Argentine–Bolivian border (22°10′S) on the Altiplano (Fig. 4.13). The decrease in annual precipitation is paralleled by an increase in the length of the dry season up to 7 months. Rainfall increases with elevation up to 1200–1500 m and decreases above. Eastern slopes receive more precipitation than west-facing ones as the prevailing wind is from the Amazon Basin.

# 4.6 Conclusions

We have shown the common features and processes related to energy in alpine environments. Despite the numerous common features, especially related to relief energy, alpine environments vary with respect to their atmospheric conditions, particularly air density, UVB radiation, and precipitation. Bioclimatic characterization of the environment provides a useful descriptive model of environmental conditions and allows modelling their changes. However, the relationship between the thus modelled abiotic environment and alpine habitats and communities on the ground is yet to be understood and made use of in modelling specific potential changes in the distribution of alpine organisms in response to climatic fluctuations or directional changes.

# 5 Habitat creating factors: landforms, hydrology, and soils

## 5.1 Introduction

Mountain building occurred in three major phases and as a result today we see two largely different types of high mountains: (1) worn-down mountains of Caledonic (Scottish Highlands, Scandes, Appalachians) and Variscan (Urals, Altai, Tianshan) origin, and (2) younger rugged and higher mountain chains of so-called alpine origin that formed relatively recently. Today's landforms are the result of tectonics, volcanic activity, weathering and mass wasting, water erosion, wind, and glacial and peri-glacial activity. The rate at which material is displaced towards lower elevations (denudation processes) is influenced by climate and rock type. The shaping of the alpine landscape in seasonally varying climates is by frost and snow (ice) impacts in the winter, followed by mass movement and fluvial activity during the summer. In arid areas, frost-shattered material is displaced downhill by gravity. Most aseasonal tropical high mountains have also been, at least in part, subject to glacial events in the past and today frequent daily freeze–thaw occurs in the alpine zone, increasingly so with elevation (e.g. Table 4.5).

## 5.2 Landforms

### 5.2.1 Tectonics, volcanism, and slope effects

From a plate tectonics point of view mountains have been formed by either collision or divergence. Depending on whether collision occurred between ocean and continental plates, ocean and ocean, or continental and continental plates, the resulting mountain chains are different (Veit 2004). Continental and oceanic plates built the Andes and the Cascades, both with high volcanic activity. Ocean plate collisions also resulted in volcanoes, e.g. Japanese Alps. Continental plate collisions formed the other major high mountain chains, such as the one that runs along the Atlas, Alps, Balkan Mountains, Caucasus, and Himalayas line. Rifting, or faulting is prevalent in East Africa, especially in the Ethiopian Highlands, where rift valleys had slipped down between neighbouring high-lying ground and later volcanic activity resulted in the formation of the Ethiopian Highlands. Some particularly active volcanic hot spots resulted in some of today's high mountains of pure volcanic origin, such as in Hawaii.

Volcanic landforms include volcanic cones and lava flows (see Figs 5.1 and 5.2). The number of active volcanoes in high mountains is relatively few today (e.g. Manua Loa, several in the Andes, Mount Etna, Ruapehu, Kamchatkan volcanoes); however, the number of extinct ones is large (e.g. several on the Andean Altiplano, Mount Kenya, Mount Kilimanjaro, Fujisan, Changbaishan, Taisetsusan of Hokkaido, Mount Teide of Tenerife), and numerous ancient volcanic mountains have undergone further orogenic uplift processes. Phenomena related to volcanic activity, such as regular lava flows (e.g. Manua Loa) and cinder deposits (e.g. Mount Etna), or both (Mount Cameroon) are important habitat factors. Mountain tops that receive regular ash deposits develop an undulating surface with broad, round or flat peaks with outcropping rock. Fumaroles

**Fig. 5.1**    The glaciated cones of Volcán Parinacota (left, 6132 m), Chile and Volcán Pomarapi (right, 6222 m), Bolivia in a hail storm in March. In the foreground, typical bunch-grass (*Festuca orthophylla, Stipa icchu*) high Andean semi-arid puna vegetation on volcanic ash soils of the lower slopes at c. 4600–4800 m of Nevado Sajama (6542 m), Bolivia, which receive a mean annual precipitation of about 350 mm. For detailed accounts on vegetation see Liebermann (1986) for Nevado Sajama and Luebert & Gajardo (2005) for Parinacota. (Photo: L. Nagy)

**Fig. 5.2**    Ancient lava flows on the slopes and foot of Volcán Socompa (6005 m), Argentina. *Stipa xanthophylla* bunch grass grows up to an elevation of about 4700 m. Observe the build-up of debris on the mountain slopes in this arid environment. (Photo: L. Nagy)

are gas-emitting holes and may create extra-climatic habitat conditions for plant and associated animal life by increasing temperature, such as in Iceland or Antarctica (see Chapter 6), or by increasing carbon-dioxide concentration and local air humidity. The latter is particularly important in arid environments, such as on Volcán Socompa, north-west Argentina, where fumaroles have facilitated rich bryophyte assemblages to thrive locally at elevations between 5700 and 6000 m, well above the prevailing upper limit of plant life (high desert puna), which is at approximately 4700 m (Halloy 1991).

Common to all mountains are their slope systems, which have been formed by one or more of the slope-forming processes, such as scree and rock glaciers, soil creep, earth flows, mudflows, debris flood, landslides, and rockslides. Characteristic differences between glacial and non-glacial landscapes are their valley systems, glacial valleys being broad U-shaped with flat bottoms and steep sides. Fluvial valleys of non-glacial landscapes are the primary result of water activity that results in V-shaped valleys. These two broad valley forms may co-occur in high rainfall alpine areas where the glacial landscape is modified by fluvial activity. In glacial landscapes with permafrost and in periglacial post-permafrost conditions solifluction occurs (Fig. 5.3), resulting in characteristic buried organic soil horizons (e.g. Mark 1994). In arid mountains, debris builds up from the

**Fig. 5.3**  Solifluction lobes in southern Iceland 2007; in the foreground and middle distance strips of turf pick out the crests of solifluction lobes that are moving down-slope from right to left. This area lies at about 450 m above sea level and has experienced extensive landscape modification since ninth century AD Norse colonisation. A once ubiquitous soil cover has been breached and aeolian erosion has resulted both the exposure of the underlying surfaces and in a thickening of surviving areas of soil (middle of photograph). Solifluction proceeds both on exposed gravel-mantled slopes and under the thinner soil cover. For further details of slope processes in Iceland see Kirkbride & Dugmore (2005). (Photo: A. Dugmore)

lower slopes, and as it is not transported away by water; it may accumulate to a point where such mountains appear to be 'drowning' in their own debris (see Fig. 5.2).

## 5.2.2 Glacial landforms and their habitats

Alpine glacial landforms include cirques or corries, ridges, horns, cols, glacial troughs, and hanging troughs, tarns or cirque and upper valley lakes, and moraines—for a detailed description of high mountain landscape forms and elements see Stahr and Hartmann (1999). All these landforms are important and complex habitat-forming structures. Corries host rock crevice- and scree-dwelling plant and animal assemblages, as well as snowbeds, often in the vicinity of tarns. Ridges are exposed to wind and have a thin snow cover that exposes them to winter soil freeze and thereby create a very harsh environment for plant growth and animal life. Glacial troughs can hold grasslands, snowbeds, or mires, while moraines provide well-drained, but nutrient-poor habitats (Fig. 5.4). Reisigl and Keller (1994) have given a general scheme for the European Alps that illustrates the contrasting landscapes of calcareous and siliceous ranges and the distribution of their habitats and plant communities (Figs 5.5 and 5.6).

As climates change, sometimes the only reminders of earlier colder epochs are remnant landform features (Fig. 5.7). For example, in the Bale Mountains Ethiopia, a host of glacial features bear witness to ancient glacial and periglacial activity: *roches moutonnées* (bedrock hills shaped by the passage of ice), ice striations, glacial cirques, and glacial tongue basins with recent swamps or lakes, moraines, screes from frost shattering, ancient patterned ground at the southern margin of the Sinetti Plateau, and boulder fields (Miehe and Miehe 1994). Today, all these monuments lie well below the snow line. Frost shattering only occurs above 4200 m; however, there is no stone sorting caused by freeze and thaw cycles, and only a weak solifluction occurs.

## 5.2.3 Landscape formed by water

Snow can actively shape the landscape, largely through its relief energy, or indirectly, through its effects on cryological processes. Snow flecks or late lying snowbeds can affect solifluction and the transport and accumulation

**Fig. 5.4**     Glacial landscape with corrie, tarn and moraines, Sierra de Urbion, Spain. (Photo: G. Grabherr)

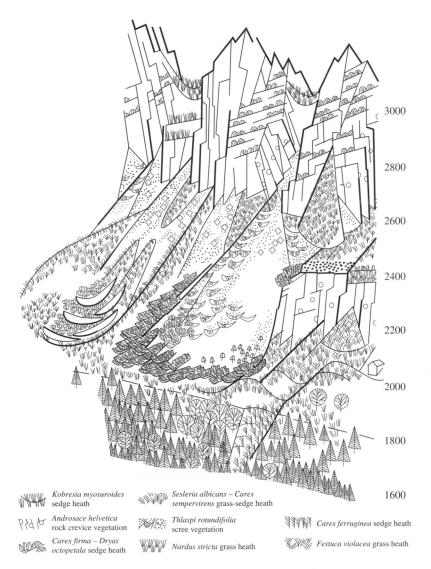

3000

2800

2600

2400

2200

2000

1800

1600

| | *Kobresia myosuroides* sedge heath | | *Sesleria albicans – Carex sempervirens* grass-sedge heath |
| | *Androsace helvetica* rock crevice vegetation | | *Thlaspi rotundifolia* scree vegetation | | *Carex ferruginea* sedge heath |
| | *Carex firma – Dryas octopetala* sedge heath | | *Nardus stricta* grass heath | | *Festuca violacea* grass heath |

**Fig. 5.5** Landscape and vegetation in a schematic representation in the calcareous Alps (after Reisigl & Keller 1994). The upper montane zone is with *Fagus sylvatica* and *Picea abies*, reaching the forest limit at around 1800 m, also pastured *Nardus stricta* grass heath, and rock outcrops with *Potentilla caulescens*; the treeline ecotone and lower alpine zone extends to 2300 m with scrub (*Pinus mugo, Rhododendron hirsutum*), sedge heath (*Carex ferruginea*), grass heaths (*Festuca violacea*) and scree (*Rumex scutatus*); in the mid alpine zone up to 2500 m grass-sedge heath (*Sesleria albicans, Carex sempervirens*), on ridges sedge heath (*Kobresia myosuroides*); the upper alpine zone up to 2800 m with *Carex firma* sedge heath, snowbeds (*Arabis coerulea*), grass heath with *Salix retusa*, scree (*Thlaspi rotundifolia*), rock (*Androsace helvetica*); in the sub-nival zone fragments of *Carex firma* sedge heath; above 3000 m scattered pockets with *Poa minor, Saxifraga aphylla*, and abundant cryptogams.

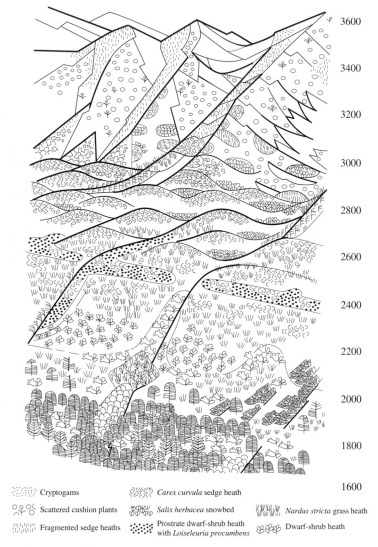

3600

3400

3200

3000

2800

2600

2400

2200

2000

1800

1600

- ░░░ Cryptogams
- ⌀⌀⌀ Scattered cushion plants
- ⌇⌇⌇ Fragmented sedge heaths
- ⌇⌇⌇ *Carex curvula* sedge heath
- ⌇⌇⌇ *Salix herbacea* snowbed
- ⣿⣿ Prostrate dwarf-shrub heath with *Loiseleuria procumbens*
- ⫶⫶⫶ *Nardus stricta* grass heath
- ⸙⸙ Dwarf-shrub heath

**Fig. 5.6**     Landscape and habitats in the siliceous massifs of the European Alps in a schematic representation (after Reisigl & Keller 1994). The upper montane forest grows up, on average, to 2000 m (1600–2400 m), in the east with *Pinus cembra*, *Larix decidua*, in the west with *Larix decidua*, *P. mugo*, *P. sylvestris*, *Alnus*, secondary grasslands with *Nardus stricta*, and *Agrostis schraderiana* grasslands; the lower alpine zone is up to 2400 m with dwarf-shrub heaths (*Juniperus*, *Arctostaphylus*, *Rhododendron*, *Vaccinium*), rocky outcrops (*Primula hirsuta*), scree (*Cryptogramma crispa*), pastured *Nardus* grass heath; the mid alpine zone (2600 m) is formed by *Carex curvula* sedge heath and *Loiseleuria procumbens* prostrate dwarf-shrub heath; in the upper alpine (2800 m) *Carex curvula* sedge heath mosaics with *Salix herbacea* snowbeds, scree (*Oxyria digyna*), and rock crevices (*Androsace vandellii*); the sub-nival zone with *Carex curvula*, *Kobresia myosuroides* sedge heaths, scree (*Androsace alpina*) and bryophytes in late snowbeds; above 3000 m *Poa laxa*, *Ranunculus glacialis*, cushions of *Saxifraga*, *Androsace*; above 3400 m cryptogams.

**Fig. 5.7**     Active stone sorting and stone polygons on the Taisetsusan, Hokkaido. Such polygons can survive long after periglacial activity has ceased. (Photo: G. Grabherr)

of fine soil material. Snow creep can affect vegetation and underlying soil, and may be the cause of erosion and soil slips (see chapter 7 in Stahr and Hartmann 1999). High-energy avalanche events can greatly modify the landscape, soil, and vegetation (Chapter 4).

Melt-water from glaciers or snow drives fluvio-glacial processes, which are major landscape forming features and habitat creating or modifying factors. Melt-water runnels in the upper reaches of the glacier foreland, gravel flats, and braided rivulets from glacier melt-water outflow on flat ground at the bottom of glacial valleys and *kames* or raised gravel flats are associated with water action and sediment deposition. Catastrophic flood and mudflow events can reshape fluvial landscapes overnight.

## 5.2.4 The role of animals in shaping alpine landscapes and habitats

Frost-driven abiotic processes, as described above, supplemented by wind and water action, and by mass movement are the primary drivers of shaping the characteristic landforms in alpine environments. However, animals may often modify directly or indirectly, mainly locally, the landscape. Their grazing, trampling, digging, burrowing, and mechanical erosive impacts can lead to soil compaction, removal of sediment, loading that leads to slope failure, and the introduction or removal of chemicals (nutrients). Indirectly, these activities may initiate sediment removal by abiotic processes (wind and water erosion), decrease slope stability by burrowing, increase surface wash and/or concentration of overland flow as a result of compaction, cause pedoturbation, and change soil-water chemistry by facilitating water penetration through burrowing (Hall and Lamont 2003). The degree of impact is a function of the intensity of animal activities, slope and parent material. For example, in the Canadian Rocky Mountains, burrowing (e.g. by marmots, *Marmota* spp., or ground squirrels, *Spermophilus* spp.) and digging (e.g. brown bear, *Ursus arctos horribilis*) activity is largely confined to easterly and southerly slopes, where there is less wind exposure, deeper but earlier melting snow, coarser and drier soil, higher temperature, and a higher and more species-rich vegetation cover than in northerly or westerly aspects.

Overuse of land for grazing in alpine areas can lead to significant landscape impacts (Hall *et al.* 1999). Loading of steep slopes by grazing animals (e.g. sheep, *Ovis aries* and cattle *Bos taurus* in European mountains, yak *Bos grunniens* in Tibet, caribou *Rangifer tarandus* in the boreal mountains) results in an extensive system of tracks (Fig. 3.7) and terracette development. The repeated trampling of these terracettes can cause the death of vegetation and bare, compacted terracettes (Fig. 5.8) can lead to water erosion by channel forming (Hall and Lamont 2003). These impacts can be greatly amplified if they expose permafrost, which, in turn, will degrade and contribute to further mass movement. In contrast, Tasser *et al.* (2003)

**Fig. 5.8**    Bare terracettes, caused by livestock trampling in the Sichuan Mountains, China. (Photo: G. Grabherr)

have argued in favour of continuous pasturing in the Alps, as in their view, the resulting terracettes were beneficial, by increasing surface roughness and reducing snow glide and abrasion, to preventing local small-scale slope failures (surface soil sliding, or '*Blaiken*') in the Alps. It appears that this argument for the maintenance of the terracettes bears a cultural analogy to the perceived need to maintain cultivation terraces to avoid large-scale soil erosion. However, the case for maintained trampling by grazing animals, whose presence resulted in the terracettes in the first place, does not appear to be convincing.

## 5.3 Hydrological characteristics of alpine landscapes

High mountain hydrology, in addition to landform sculpting, is a major habitat creating and maintaining factor. It is characterized by the presence of cryospheric elements and phenomena, such as ice caps and glaciers, seasonal snow cover, freshwater ice, and seasonally frozen ground, together with permafrost that play a part in defining habitats. Ice caps and glaciers are host to cryoconite dwelling organisms, largely found in solution cups; the source and upper parts of glacier melt-water streams are a specialist habitat for cold tolerant/cold requiring freshwater organisms; freshwater ice can, in extreme cases, manifest in the complete freezing up of lakes and water courses; seasonally frozen ground and snow cover are important determinants of vegetation growing season length, and animal and microbial activity; and permafrost can limit plant rooting depth and water percolation into the subsoil.

The fate and impacts of water in mountains are connected to topography, soils, and vegetation. The dissected mountain landscape results in numerous micro watersheds that channel their run-off to collect in the concaves of terrain (Fig. 5.9). Such a run-off network translates into habitats, such as runnels and rivulets with their characteristic vegetation and animal life. Glacial valleys, with the glacier snout in the upper valley, form well-defined large run-off channels, with numerous small non-glacial tributary channels from the valley sides. Following the melt of glaciers, tarns, or glacial lakes may form where outflow is dammed, small corrie (glacial cirque) lakes (Fig. 5.4) being a typical example. Precipitation patterns and soil permeability largely determine run-off and vegetation has an important role in stabilizing and fixing soil. Input from precipitation (rain and snow melt) may leave in the form of run-off on steep or little permeable surfaces and form permanent or transient water courses and water bodies, or alternatively, seep into the soil and add to underground reserves and contribute to underground flow. The annual course of the run-off in the alpine zone reflects the temporal pattern of snowmelt and summer rainfall, characteristic of seasonal climates (e.g. Fig. 5.10).

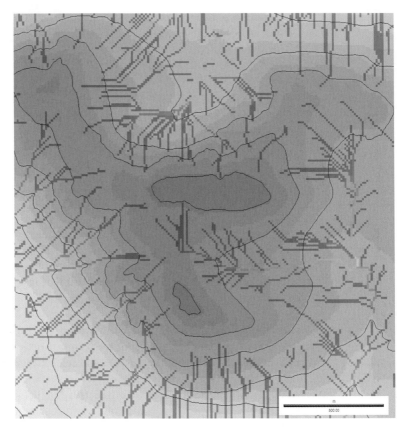

**Fig. 5.9**     Runoff pattern on Cul Mór (871 m), north-west Scotland. Black concentric lines are 100 m contours downwards from 800 m at the twin peaks; scale bar 500 m. Generated by the Runoff module in IDRISI ANDES, using a 10 m horizontal resolution digital elevation model (DEM, courtesy of Scottish Natural Heritage).

In seasonal climates, a large proportion of precipitation in the alpine zone is in the form of snow, largely in the winter (Körner *et al.* 1980). For example, in the Colorado Front Range of the Rocky Mountains about 80% of the annual precipitation falls as snow (Williams and Caine 2001). Much of the snow is redistributed by wind, causing denudation in exposed areas and vast accumulation in sheltered lees (see e.g. Jeník 1998). In addition to the consequences for soil freeze in winter and for soil moisture availability after snowmelt, snowdrift contributes to water loss by sublimation of as much as 30% of the total snow fall (Berg 1986). In aseasonal humid tropical and seasonal tropical–subtropical climates snowfall is possible at most nights in the alpine zone (Table 4.5); however, it only accumulates in the nival zone. As such, snow distribution has no ecological impacts on

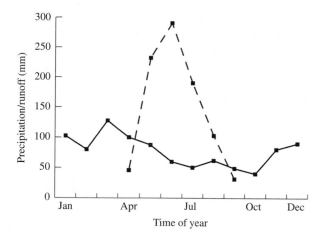

**Fig. 5.10** The annual course of mean monthly precipitation (solid line) and that of runoff (broken line) in the Green Lakes Valley, Niwot Ridge, Colorado. Runoff data refer to those measured at Green Lakes No. 4 station at 3550 m asl. Data from Williams & Caine (2001).

aseasonal alpine habitats. Conversely, the presence, duration, or absence of snow characterizes and defines habitats in seasonal alpine climates. In general terms, alpine habitats are: (1) snow protected, hosting chionophil (snow tolerating) species; (2) exposed, with little snow cover and home to a suite of chionophobe (snow avoiding and frost tolerant) species; or (3) intermediate (mesic) with snow protection in the winter and early melting in the spring–summer.

### 5.3.1 Alpine freshwater bodies as potential habitats

The major alpine freshwater habitats include glacial melt-water streams, alpine spring-fed streams and rivers, and water bodies such as tarns or glacial lakes. Water from these can mix as streams merge or feed into lakes. Glacier-fed streams are particularly cold—melting ice is 0°C, but further away from the glacier snout water temperature can reach 10°C in summer. It is during the summer high temperatures and rain events that glacial streams reach their peak water levels. Although at first sight, cold and highly turbid fast flowing water courses offer little in the way of a habitable environment, there is life in these troubled waters (e.g. Brittain and Milner 2001). As glacial streams carry and deposit a high amount of sediment, their course is often ill-defined, or braided, and can change during and after high water peaks, adding spatial instability to temporally fluctuating habitat conditions. The gravel and sand deposits serve as habitat for a specific assemblage of pioneer plants and animals (for examples in the Alps see Bressoud 1989).

Spring-fed streams receive their water from underground aquifers. They are warmer than glacial streams and have their peak water levels (or discharge) during snow melt and after high rainfall events, when they become rather turbid. They also have higher ionic (potential mineral nutrient) concentrations, having sometimes nearly twice as high a conductivity than that measured in glacial streams (Füreder 1999).

Tarns or glacial lakes are covered by ice, i.e. are cold and light limited, for 7–9 months a year in seasonal climates. As a result, they have a wide amplitude of annual temperatures, especially shallow lakes, which can freeze to the bottom in winter and can heat up considerably in summer. A high diurnal change in temperature is characteristic of shallow ponds, such as those reported from the Cordillera Vilcanota, central Andes, where at elevations of 4900–5300 m (comprising the upper limit of the alpine and the entire subnival zone) water temperature in the shallows of 1–2 cm depth of 18–22°C have been recorded at midday, after night ice (Seimon *et al.* 2007). Like other alpine freshwater environments, lakes are low in mineral nutrients and are exposed to high ultraviolet radiation. The structure, habitat differentiation, and biology of alpine freshwaters, related to gradients in water depth from littoral to deep water and benthos, are discussed in detail in specialist texts.

## 5.3.2 Water availability in terrestrial alpine habitats: climate, topography, and soils

Topography, snow duration, and alpine soil profile development are related (Fig. 5.11), as is vegetation, along slope systems. For defining terrestrial alpine habitats hydrological processes that affect soil moisture (e.g. precipitation, surface run-off and erosion, and snow melt) are particularly important. Too little or too much water in the soil can influence nutrient cycling, cause (bio)geographical exclusions, and, on a more local scale, it necessitates structural or physiological adaptations by all inhabiting organisms. A look at the vegetation types, the dominant life forms, and anatomical adaptations of plants across habitats ranging from high-altitude desert (see Fig. 5.2) to mires will be convincing. Animal species and soil microorganisms, apart from some widespread soil bacteria, are also different in their suite of taxa that have evolved/adapted to contrasting hydrological conditions. For example, as soil bacteria, nematodes, and protozoa live in water films in soil pores—and when available, in free water—they are likely to be little abundant in desert conditions. Soil-dwelling micro-arthropods are also sensitive to variation in soil moisture: too much of it restricts their living space in pores, and too little causes upsets in their body water balance (e.g. Bardgett 2005). Consequently, micro-arthropods can be predicted to have highest abundances in well-drained soils. Topographic position and relief energy have important effects on hydrology, and through it

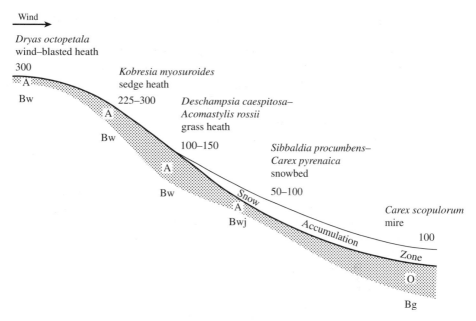

**Fig. 5.11**     Alpine soil profile development in relation to slope and snow duration in the Colorado Rocky Mts. Source: Birkenland *et al.* (1991). The thickness of A (mineral horizon with incorporated humified organic matter) and O (organic soil, peat) horizons shown is not based on measurements. Numbers are duration of snow-free period in days per year. Corresponding vegetation types are indicated after Walker *et al.* (2001b).

on soil physical properties (e.g. particle size distribution and pore volume). Soils on exposed ridges usually have higher sand, and lower silt and clay fractions than slope bottom soils, in part, as a result of the sorting of particles by the down slope flow of water. The result is well-aerated nutrient-poor sandy soils in ridge position and increasingly wetter and finer soils at the bottom of slopes. The ecological implications are twofold: higher water content results in better potential accessibility to mineral nutrients from solutes, and the capacity of soil to bind ions increases with increasing fine particle fraction, such as silt and clay. However, rates of decomposition of organic matter in wet soil are usually slower than those in well-aerated ones. Gravitational energy results in a washout of minerals along slopes and their accumulation on level ground, such as has been shown by Makarov *et al.* (1997) in the Caucasus (Fig. 5.12), a pattern repeated at the landscape scale (e.g. Mount Trikora, Papua, Indonesia; Mangen 1993). The example from the Caucasus illustrates well the hydrological impact of snow accumulation—from very little at the ridge to about 6 m at the valley bottom—and subsurface water movement from melt-water causing

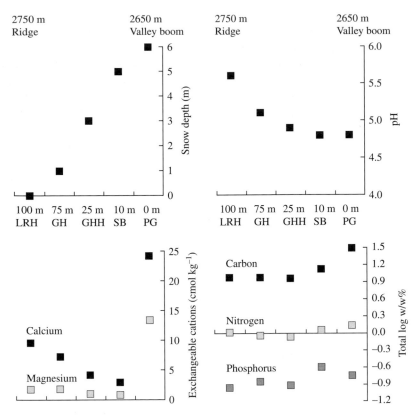

**Fig. 5.12** Changes in snow depth, topsoil (0–10 cm) pH, exchangeable Ca and Mg, and total C, N, and P concentration along a ridge—slope—valley bottom system at Malaya Khatipara, Teberda, northern Caucasus. Along the slope (all Umbric Leptosol; ranker) there is a decrease in pH, and cation concentrations, presumably caused by washout. At the valley bottom (histosol; peat), there is a water surplus during the c. two-month snow free period and there is a large build-up of carbon (peat) and cation concentrations. LRH, lichen-rich dwarf-shrub heath; GH, grass heath with *Festuca varia*; GHH, herb-rich grass heath with *Geranium gymnocaulon*; SB, snowbed with *Sibbaldia procumbens*; PG, *Nardus stricta* snowbed. (Data from Makarov *et al.* 1997). Note that the value axis for C, N, P concentrations is log-scale.

nutrient leaching along the slope. Excess water/impeded drainage in the valley bottom results in peat formation (carbon accumulation) and the accumulation and binding by soil organic matter of nutrients from the slope washout.

At the landscape scale, surface run-off can cause soil erosion, and in extreme cases the total removal of soil. Vegetation cover is paramount in this respect. On islands with an oceanic high rainfall climate, such as Taiwan, the rate of soil erosion can be higher than the rate of chemical weathering of the soil on very steep slopes, especially in the montane zone. Taiwan alpine watersheds (>3500 m) have a relatively more gentle topography than the montane slopes and the soils of the lower alpine zone have a well-developed vegetation cover (*Juniperus squmata* var. *morrisonicola* scrub and prostrate dwarf shrub, cushion-forming *Rhododendron* spp.; and *Yushania niitakayamensis* bamboo thicket in the treeline ecotone). This, in turn, helps retain the soil and allows chemical weathering to produce labile cations ($K^+$, $Na^+$, $Ca^{2+}$) (Selvaraj and Chen 2006). In another example, in the French southern Alps it has been estimated that soil loss rates by erosion can be 100–1000 times higher from bare soil surfaces than from areas under closed grassland; in addition, northern, humid slopes can lose two to four times more soil than southern, dry ones (Descroix and Mathys 2003). The kind of precipitation and its intensity is another important factor for soil loss. For example, an extreme heavy hail storm event caused an erosion of 60 kg soil 100 $m^{-2}$ on bare soils at a site in the Central Alps (2100 m), 50 times more than heavy rain without hail but of much longer duration (1.2 kg soil $m^{-2}$) (Kurz 1987). Freeze–thaw cycles and burrowing animals can substantially add to potential erosion rates by bringing unconsolidated soil to the surface and exposing it to rainfall and run-off.

# 5.4 Soils

## 5.4.1 Landscape-scale patterns of soil types and vegetation

Soil types, described in terms of their characteristic structure and physico-chemical properties, are not only useful categories for classification of soils but they are telltale indicators of the dynamics of soil development at the large-scale. The development of soil types reflects the interplay of soil forming factors (topography, climate, biota, the frequency of disturbance, and parent material) at work over time. Climate, parent material, and their combinations are general factors that affect high mountains, as well as lowlands. More characteristic of alpine landscapes are the impacts of topography and specific types of disturbance.

Typical spatial habitat and soil sequences form along topography. The steep rocky summit areas are often devoid of soil and contain shattered fragments and dust only, as the steepness of slopes causes slope instability, which precludes *in situ* soil forming processes. Typically, on such mountains large expanses of bare rock and scree habitats dominate, such as in the Corsican high ranges, or on the granitic dome of Mount Kinabalu Borneo. At high altitudes physical weathering through frost shattering dominates. Slope exposure is important as high insolation in summer may result in very high temperatures on insolated slopes and that can lead to heat shattering of rock. In contrast, non-insolated slopes may have a coarser structure, maintain block fields much longer, and have less developed soils than insolated ones. The fine scree habitats are much different from block fields. Fine screes are colonized by plants that contribute to stabilizing it and to soil formation; block fields offer a fundamentally different habitat, whose microclimate is more balanced than that of screes and favours tall herbs that are protected from large herbivores. Gentle slopes or flat areas accumulate the finest fraction of weathered mineral material and wind-blown dust. Mineral nutrient and water availability combined with fractal properties of the physical space make such areas amenable for early colonization by microbes, plants, and animals, which, in turn accelerates the biological soil-forming processes.

The alpine life zone has experienced a combination of perturbations that have had a large impact on its soils and their biota. At the historic time-scale, glaciations, where present, profoundly reshaped the landscape and in the process removed all previously formed soils and reset biological soil-forming processes to zero, except on nunataks, where old soils survived (e.g. the so-called *terra* soils of the Alps; Anonymous 1998). Ongoing periglacial processes, such as cryoturbation and solifluction are characteristic not only of post-glacial landscapes, but of all sufficiently cold and humid high mountain environment where mass movement can further decrease the stability of soils. Cryoturbation is an important alpine and arctic periglacial phenomenon, related to hydrology and soil type. It is important for nutrient cycling by mixing the soil whereby organic material is translocated downwards and weathered mineral materials are moved upwards and laterally. Cryoturbation affects especially those alpine soil types that have a high moisture retention capacity (e.g. volcanic ash soils) or soils in general under moist conditions, such as particularly well exemplified in Iceland (Wookey *et al.* 2002; Arnalds 2004).

In addition to *in situ* weathering and soil-forming processes, the role of wind transported fine material may contribute to the character and nutrient relations of alpine soils. In the northern calcareous Alps, deposits of wind-blown loess of late glacial origin with a mean thickness of 30 cm are found over large areas in the alpine zone (Küfmann 2003); even more striking is the scale of loess deposition found in the Himalayas (Baumler 2001; Baillie *et al.* 2004). Contemporary aeolian deposition is widespread. For example, snow patch soils in the Australian Snowy Mountains have a particle size and mineralogy consistent with dust enrichment (Walker and Costin 1971; Johnston 2001). Mean contemporaneous aeolian deposition rates have been estimated at about 100 kg ha$^{-1}$ in the Alps (Gruber 1980) and in the Pacific Ranges of British Columbia (Owens and Slaymaker 1997). The best studied site in the Alps is the Gamsgrube (covered largely by wind-blasted *Kobresia* heath and eroded patches at elevations of 2500–2650 m, see monograph by Friedel (1956)), where, in some areas Gruber (1980) has found aeolian deposition of about 9000 kg ha$^{-1}$ year$^{-1}$, which is equivalent to an increase of soil depth of >1 mm year$^{-1}$.

In some regions it is precipitation that has contributed so much to re-allocating soil nutrients. The New Guinean Highlands have particularly high rainfalls, which has caused nutrient leaching from ridge and mountain slope soils and colluvium forming on lower slopes from transported soil; high water tables on the flattish bottoms of glacial valleys have resulted in the development of hydromorphic, or water-impacted soils (e.g. peats and gleys; Mangen 1993). Such patterns are repeated in, but not restricted to, highly oceanic climates all around the globe from the Scottish Highlands to the Fjordland in New Zealand.

As mountain climate changes along a vertical succession of bioclimatic zones, mountain biota reflect this not only in the arrangement of vegetation belts (and associated animal assemblages), but also in the succession of soil types. The distribution of soil types in mountain landscapes, such as exemplified by the Rocky Mountains Colorado Front Ranges (Birkenland *et al.* 2003; Fig. 5.13), the Himalayas (Smith *et al.* 1999), the Andes and the European Alps (Fig. 5.14; Table 5.1), or the Arctic (Fitzpatrick 1997) provides a snapshot of how the climate-driven distribution of natural vegetation belts or zones relates to soil types.

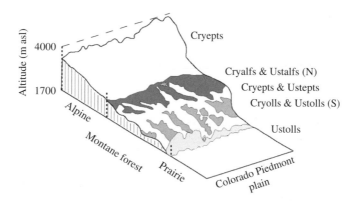

**Fig. 5.13**     The sequence of soil types along an altitude and climate gradient from the Colorado Peidmont plains (c. 1700 m asl) to the Continental Divide (east–west watershed) at 3600–4000 m in the Rocky Mts after Birkenland *et al.* (2003). Soil profiles reflect the climatic transect from the dry plains to the mesic alpine: A/Bw or Bt/Bk or K (grassland) changes into A/E/Bw or Bt/C (forest) and then A/Bw/C (alpine vegetation). A, mineral surface layer with incorporated humified organic matter; B, mineral subsoil with accumulation of clays; C, unconsolidated mineral horizon with rock structure; E, heavily leached (podsolised) layer; K (Bk), carbonate horizon, typical of arid soils (for further details see Birkenland 1999). For FAO soil classification equivalents see legend to Fig 5.17.

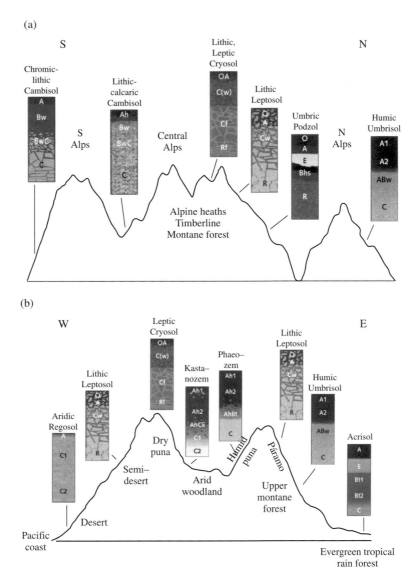

**Fig. 5.14**     A comparison of soil types along altitude, parent material and moisture gradients in the European Alps and in the Andes (modified from Zech & Hintermaier-Erhard 2002). (a) a north–south cross section of the Alps with bioclimatic vegetation belts and corresponding soil types (FAO), note that the northern and southern Alps are predominantly calcareous and the central Alps siliceous; (b) a schematic west to east cross section of the tropical and subtropical Andes with vegetation belts and soil types (FAO). Note that páramo occurs in the northern Andes and puna in the central tropical-subtropical Andes. The eastern side of the Andes is humid, whilst the western Pacific side is arid.

**Table 5.1** The vegetation and soils of bioclimatic zones or belts in the Alps (from Veit 2002)

| Altitude belt | Vegetation | Soil on calcareous substratum | Soil on siliceous substratum | FAO Soil category |
|---|---|---|---|---|
| Nival | Scattered individuals | Skeletal soil on frost shattered rock debris | Skeletal soil on frost shattered rock debris | Lithic leptosols |
| Subnival | Open patchy, with abundant bare substratum | Skeletal soil on frost shattered rock debris; patches of rendzina | Skeletal soil on frost shattered rock debris; alpine rankers and brown earths | Lithic and other leptosols; cambisols |
| Alpine | Sedge heaths | Rendzinas, pararendzinas brown earths | Alpine ranker; Regosol; Alpine pseudogley; Alpine pararendzina; Podsol (nanopodsol); | Leptosols, arenosols, regosols, mollic and eutric gleysol, podsol |
| Treeline ecotone (subalpine) | Mosaic of trees, dwarf-shrub heath and grassland | Pararendzinas; Brown earths | Humus, iron and brown earth podsols; Brown earths | Podsols; cambisols |
| Montane | Montane forest | Rendzinas; *Terra fuscas*; Brown earths; Chernozems in Inner Alps valleys | Brown earths Podsolic brown earth; podsols | Cambisols, podsols |
| Lowland and hill | Lowland and hill forest | Rendzinas; *Terra fuscas*; Pararendzinas; Chernozem in Inner Alps valleys | Brown earths | Cambisols |

A typical soil profile is described in the form of a succession of organic and mineral horizons. In non-water-saturated conditions the organic horizon may be formed by undecomposed litter (L), partially decomposed fine litter (F), and humus or decomposed plant litter (H); under water-saturated conditions peat (O) forms. The mineral horizons may contain a surface mineral horizon, rich in humus (A), leached subsurface horizon (E) in podsols, mineral subsurface soil (B), and unconsolidated mineral horizon with rock structure (C). The typically relatively shallow alpine lithosols have an A–C structure, and the resulting soil type on acidic base rock is called ranker (<2% carbonate content), and on calcareous bedrock rendzina (>75% carbonate, or pararendzina (2–75% carbonate). In humid conditions (e.g. Scottish Highlands, Scandinavian west coast mountains), where precipitation excess in relation to evapotranspiration leads to the leaching of iron and aluminum and their accumulation at depth, podsols develop with an A–E–C structure. Brown earths have an A–B–C structure.

In the alpine zone, lithosols prevail in all high mountains that are of relatively young age, or have until recently been glaciated, such as for example parts of the European Alps, the Pyrenees, or the Himalayas and the Andes (see Fig. 5.15 for the distribution of lithosols in arctic and boreal alpine ranges). Lithosols form a large group of soils (e.g. leptosols, regosols), characterized by an incomplete development, or have no clearly expressed soil morphology and consist of freshly and imperfectly weathered rock or rock

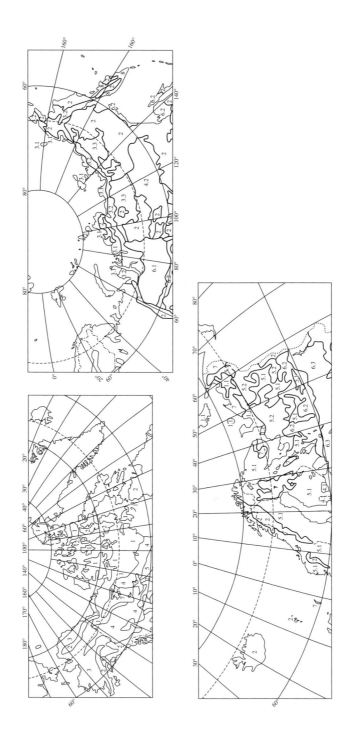

**Fig. 5.15** Circum-Arctic soil types in North America (top left), Asia (top right), and Europe (bottom). Note that at such a large scale all soils in all the alpine mountain ranges are classified as lithosols (redrawn from Marion et al. 1997).

| Code | FAO name | USSC name |
|------|----------|-----------|
| 1 | Regosols | |
| 1.1 | Gelic regosols | Pergelic Cryorthents |
| 2 | Lithosols | |
| 3 | Gleysols and gleyic sols | |
| 3.1 | Gelic Gleysols and Gelic Regosols | Pergelic Cryaquepts, Histic Pergelic Cryaquepts, Pergelic Cryorthents |
| 3.2 | Gelic Gleysols | Pergelic Cryaquepts, Histic Pergelic Cryaquepts |
| 3.3 | Gelic Gleysols and Gelic Cambisols | Pergelic Cryaquepts, Histic Pergelic Cryaquepts, Pergelic Cryochrepts, Pergelic Cryumbrepts |
| 4 | Cambisols | |
| 4.1 | Vertic Cambisols | Pergelic Cryochrepts |
| 4.2 | Gelic Cambisols | Pergelic Cryochrepts, Pergelic Cryumbrepts |
| 5 | Podsols | |
| 5.1 | Orthic Podsols | Pergelic Haplocryods |
| 5.2 | Gleyic Podsols | Pergelic Cryaquods |
| 6 | Podsoluvisols | |
| 6.1 | Gleyic | Pergelic Cryoboralfs |
| 6.2 | Dystric | Pergelic Cryoboralfs |
| 6.3 | Eutric | Pergelic Cryoboralfs |
| 7 | Histosols | Pergelic Cryfibrists, Pergelic Sphagnofibrists, Pergelic Cryohemists, Pergelic Cryosaprists, Lithic cryofolists |
| 8 | Andosols | Typic Gelicryands |

fragments. Many other soil types also occur in the alpine zone world-wide, reflecting climatic variability and volcanic activity (see Section 5.3.3). For example, the FAO-UNESCO (1992) world soil map has 22 additional types that occur above 4000 m elevation; however, they account for about 10% in total, the rest being lithosols.

## 5.4.2 Temporal aspects of alpine soil development

Temporal soil development in glaciated mountains may be observed along altitude sequences, where the higher areas often bear younger soils than lower lying areas, which have had a longer ice-free history (Table 5.2). Typical elevation sequences are sometimes inversed. Where glaciations scoured mountain sides and valley bottoms but left mountain tops with their established soils ice free, such as at the Niwot Ridge in the Colorado Rocky Mountains, the ridge top areas comprise the oldest and deepest soil, the valley sides are largely bare rock or have thin young soils, as well as do valley bottoms on glacial till (Burns 1980).

Soil time series can also be observed after the retreat of valley glaciers. Short- to medium-term changes can be followed along deglaciated valley moraines that contrast with the few locations at similar altitudes that were unglaciated (e.g. Egli *et al.* 2001). Such series allow the study of changes with soil age in chemical and physical properties. In general, during soil development soil organic matter (C, N) accumulates and main mineral elements (Ca, Mg, K, Na, Fe, Al, and Si) are lost through leaching,

**Table 5.2** The general elevation scheme of soil types and soil-forming processes on Mount Jaya, Papua, Indonesia after Bleeker (1980) and Mangen (1993)

| Altitude (m) | Soil type | Characteristics | Vegetation |
|---|---|---|---|
| 4600–4800 | No soil, bare rock | | Nival zone |
| 4000–4600 | Lithosols and bare rock | Soil formation limited because of low temperature, sparse vegetation cover, and steep terrain | Alpine zone grasses; above 4300 m dwarf-shrubs, herbs, mosses, and lichens |
| 3700–4000 | Alpine peat and alpine humus soils on gentle slopes and valley bottoms; lithosols, including colluvia on steep slopes | Acid organic matter accumulation by vegetation; slow mineralization rates | Treeline ecotone with patches of forest and extensive grasslands |
| 3000–3700 | Alpine peat (under montane forest) and alpine humus soils, podzolic and podzolic gley soils (under grassland) with rankers (rendzinas on limestone), humic brown clay soils at lower altitudes | Relatively flat areas in glacial valley bottoms, high water table | Upper montane forest, tree ferns, grassland |

especially during the early stages of soil formation (e.g. up to 3000–4000 years in the Alps; Egli *et al.*, 2001), after which depletion levels off. In addition to the accumulation of organic carbon, the build-up of clay minerals, followed by a later decline through podsolization (the disintegration of clay minerals and release of minerals) is characteristic. The rate at which these processes operate is controlled by the weathering properties of the parent material, climate, and to a large extent, by biological activity (e.g. Bardgett 2005). Humus accumulates under vegetation as a result of low mineralization rates (= low biological activity) in cool and humid environments, while in drier areas the rate of decomposition of organic matter keeps up with litter production. The speed of the process at which soil–vegetation sequences develop on morainic deposits in glacier forefields after the melting of glacier ice can be observed in many parts of the world (e.g. Fig. 5.16). For example, since the end of the Little Ice Age in the mid-nineteenth century, the plant cover at Rotmoos, North Tyrol, Austria has built up on bare mineral deposit to 15–30% after 25 years, 40–70% after 40 years, and 50–100% after approximately 140 years. The corresponding

**Fig. 5.16**     Glacial deposits, exposed after the retreat of ice at the end of the Little Ice Age show different stages of colonisation by vegetation at the north-west shores of Sibinacocha (c. 4900 m), Cordillera Vilcanota, Peru. The vegetation reflects the development of soil: the top of the undulating morainic hillocks are largely devoid of vegetation and are of mineral nature, the lower lying areas in between the hillocks have developed shallow soils under prostate shrubs such as *Baccharis* spp.; the lake shore has closed hard cushion wetland (*bofedal*) vegetation with *Distichia muscoides* (Cyperaceae) on waterlogged soil. The mountain sides, on either side of the moraine, were not glaciated during the Little Ice Age. (Photo: L. Nagy)

changes in soil development entailed a gradual increase of organic matter, and a decrease in pH and in sand fraction (Fig. 5.17). An example of soil formation and development along a longer chronosequence (150–10,000 years) in the Swiss Alps is provided by Egli *et al.* (2001) and shows that lithosols develop into other types approximately 450 years after the retreat of ice (Table 5.3). Glacier forefields are a particular case, to some extent, as they contain a large amount of relatively fine mineral matter, and when exposed after the ice retreats the fragments are further weathered by frost shattering. On fine mineral material, the build up of vegetation cover and

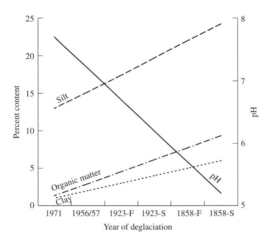

**Fig. 5.17** Temporal trends in soil particle size, organic matter content and pH along a chronosequence on Rotmoosferners, Ötztaler Alps, Austria (Erschbamer *et al.* 1999). The initial skeletal soil phase (1971 and 1956/57) is characterised by a few millimetre of organic matter rich mineral A layer; alpine pararendzina with a c. 4 cm A horizon is found in areas which were deglaciated in 1923 and 1858. The most dramatic change over a period of c. 120 years is the drop in pH from 7.7 to an acid 4.8. F, flat at the bottom of moraines; S, moraine slopes.

**Table 5.3** The temporal development of soils after the retreat of valley glaciers (up to 700 years) in contrast with old soils in the Swiss Alps (source; Egli *et al.* 2001). Note the change from lithosols (leptosols) to brown earths (cambisols, with A and B horizons) appear after approximately 450 years

| Soil type (FAO–UNESCO 1992) | Age (years, dated by $^{14}C$) |
| --- | --- |
| Lithic leptosol | up to 150 |
| Umbric leptosol | 260–300 |
| Umbric leptosols to dystric cambisol | 450 |
| Dystric cambisol | 700 |
| Haplic podsol | 10,000 |

the initiation of (soil) biological processes (including biological weathering) is relatively rapid, as has also been also shown by Costin *et al.* (1969) in heavily wind-eroded areas in the Kosciuszko Mountains, Australia, where the first 30 years saw a steep increase in soil depth, cation exchange capacity and nitrogen content in sheltered locations. The vegetation of such sheltered areas changed from open prostrate wind-clipped *Epacris petrophila* dwarf shrub to closed grass–herb vegetation over approximately 50 years. The role of plants in the early stages of colonization in building up and cycling nutrients in alpine soils is illustrated by the work of Gigon (1999): the root systems of pioneer plants transport mineral nutrients from the depth they reach to the plant, and the litter shed by the plant will lead to the gradual building up of a humus layer. Such mechanisms can counteract to some extent leaching in humid climates.

## 5.4.3 Bedrock type, soils, and vegetation

Different parent material types, formed in geological times, or transposed during orogenesis, give rise to different soils (Table 5.1). At a local to regional scale, acidity and the presence of $Ca^{2+}$ ($Mg^{2+}$) ions are some of the most important factors for plant species and vegetation type distribution. While the physiognomy of the vegetation of habitats on contrasting bedrock may appear similar, a rather different and richer suite of plant species is found in calcareous habitats (Fig. 5.18) than in habitats on acid bedrock (e.g. Europe-wide, Virtanen *et al.* 2003; Alps, Ozenda and Borel 2003; Carpathians, Coldea 2003; Pyrenees, Gómez *et al.* 2003). Extreme soil conditions, such as those found on ultramafic soils—soils developed on serpentinite bedrock and having high concentrations of magnesium, iron, and often nickel—(see e.g. Egger 1994) may lead to the evolution of edaphic endemism, such as in the Pindos Mountains (Strid *et al.* 2003) and have striking difference in plant cover and species composition across sharp boundaries (Fig. 5.19).

The differences in soils across the treeline ecotone, from forest to alpine heath are well marked. However, in the alpine zone proper, especially on largely uniform bedrock, sometimes 'it is difficult to establish a clear correlation between vegetation and soils since the occurrence of certain vegetation types is governed by climatic, topographic and drainage factors, as well as soil, and all these factors are interrelated' (Bleeker 1980). This is supported by the example of the toposequence shown in Fig. 5.11, where soil chemical properties showed directional trends and could be linked to distinct habitats (from exposed to snow-protected) and their plant communities. An ecologically diverse range of plant communities (lichen-rich dwarf-shrub heath, grass-sedge heaths, *Sibbaldia* snowbed) occurred on the same main soil type (leptosol or ranker) and only the *Nardus* snowbed in a valley bottom position grew on a different soil type (histosol or peat) (Makarov *et al.* 1997). Similarly, sampling across the mountains of the

| Vegetation type | Sesleria albicans - Carex sempervirens grass-sedge heath | Kobresia myosuroides sedge heath | Carex curvula sedge heath | Carex firma sedge heath |
|---|---|---|---|---|
| Altitude range (m) | 2000–2900 | 2200–2800 | 2300–2700 (1750–3300) | 2000–3000 |
| Mean number of species | 50 (25) | 40 (20) | c. 30 (15) | 25 (20) |
| Habitat characteristic | steep insolated slopes | wind exposed and dry | more or less dry, warm, with medium wind exposure | exposed, rocky |
| Winter snow cover | early snow melt | little | full | sometimes early melting |
| Snow free period (months) | 6–7 | up to 12 | 4–7 | 4–7 |
| Growing season length | 5 | 5 | 5 | 5 |
| Plant biomass   above-ground   below-ground | 1200<br>1600 | n.d. | 775<br>1275 | 1260<br>750 |
| Depth of soil (cm) | deep | shallow to deep | c. 30 cm | shallow |
| pH of surface soil | 6.2–7.1 (5.6–7.4) | 6–7 (3.5–8) | 4.5–5.5 (3.2–6.5) | 6.2–7.2 (5.6–7.8) |
| Humus (wt%) in surface soil | c. 10 | c. 10 | 5–15 | 30 (10–40) |
| Most frequent soil type | mull rendzina/leptosol | brown earth; pararendzina/eutric cambisol | brown earth/dystric cambisol | rendzina/leptosol |

**Fig. 5.18** Characteristics of alpine habitats on calcareous (columns 1–2, 4) and siliceous bedrock (column 3). Source of information in table: Reisigl & Keller (1994). Plant biomass is given in g m$^{-2}$. (Photographs from left to right: L. Nagy, H. Pauli, G. Grabherr, L. Nagy)

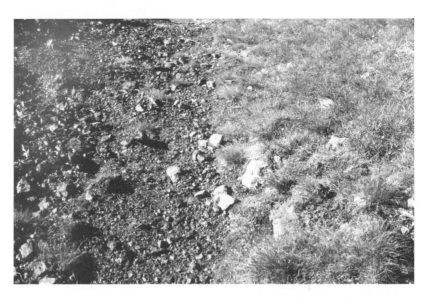

**Fig. 5.19**     Sharp contrast, caused by soil and bedrock, in vegetation cover at the boundary between serpentinite (left, low plant cover) and amphibolite (right, closed vegetation), Montafon Alps, at c. 2300 m elevation (Photo: G. Grabherr)

Scottish Highlands by McVean and Ratcliffe (1962) showed only a broad pattern between habitats and their vegetation, and soil type (Table 5.4). Contrary to the single slope study by Makarov *et al.* (1971), measured soil chemical characteristics and corresponding vegetation across habitats and mountain ranges in the Scottish Highlands has shown no clear correlation between vegetation and soils that was attributable to topographic differences and resulting snow cover length. Similarly, in the New Guinean Highlands, variation in soil fertility within the same vegetation type was found to be equal or larger than between vegetation types; there was an altitude effect, presumably owing to an increase in precipitation and leaching (total cation exchange capacity, exchangeable $Ca^{2+}$ and $Mg^{2+}$ and base saturation in the 0–10 cm soil samples decreased with altitude) (Wade and McVean 1969).

Soils derived from recent volcanic deposits (volcanic ash soils or andosols) are found on active or recently extinct volcanoes. Although andosols cover <1% of the Earth's land surface, their contribution to mountain soils is much higher, as many high mountains are of volcanic origin or are in areas of volcanic activity (Table 5.5). Alpine soils formed from recent (<10,000 years) volcanic ash are high in organic carbon content (e.g. 6.1–20.4% in the top 20 cm of soil in the páramos of Ecuador, Ortiz *et al.* (2003); Table 5.6) and are of a coarse sandy texture (e.g. Poulenard *et al.* 2001), yet have a high water retention capacity, which, in moist alpine environments can

**Table 5.4** Soil characteristics of acid alpine vegetation types from the Scottish Highlands after McVean and Ratcliffe (1962). Soil chemical analyses refer to top 0–22.5 cm soil; outstandingly high values of peat samples are in parentheses

| Habitat/vegetation type | Soil type(s) | n | pH | LOI (%) | Ca (g/kg) | $P_2O_5$ (g/kg) |
|---|---|---|---|---|---|---|
| **Prostrate dwarf-shrub heaths** | | | | | | |
| Wind-blasted | | | | | | |
| *Loiseleuria procumbens– Empetrum nigrum* ssp. *hermaphroditum* | Alpine ranker | 1 | 4.4 | 11.4 | 21.5 | 3.9 |
| *Cladonia* spp.–*Calluna vulgaris* | Shallow alpine podsol (2) Alpine ranker | 2 1 | 4.3–4.4 4.3 | 12.2–14.5 19.1 | 7.6–8.9 24 | 0.21–0.43 4.82 |
| Snow-protected | | | | | | |
| *Vaccinium myrtillus– Empetrum nigrum* ssp. *hermaphroditum* | Podsolic brown loam with mor humus | 2 | 4.4–4.7 | 9.9–17.5 | 5.7–6.6 | 0.46–0.6 |
| *Cladonia* spp.– *Vaccinium myrtillus* | Shallow alpine podsol Alpine ranker | 1 1 | 3.7 3.9 | 17.4 (87.8) | 40.3 28.3 | 3.25 7.03 |
| *Festuca ovina– Vaccinium* spp. | Brown loam with mor or moder humus, skeletal or slightly podsolized | 5 | 4.1–4.9 | 12.7–27.3 (58.9) | 5.6–28.4 | 0.45–2.45 |
| **Grass-sedge heaths** | | | | | | |
| *Alchemilla alpina– Sibbaldia procumbens* dwarf herb | Periodically irrigated loam with moder humus horizon: sometimes podsolic | 3 | | 12.8–26.2 (66.4) | 5.9–27.2 | 0.91–3.6 |
| *Nardus stricta* | Alpine podsol | 1 | 5.2 | 12.2 | 10.2 | 6.23 |
| *Dicranum fuscescens– Carex bigelowii* | Alpine sod podsol with variable amount of gleying | 6 | 3.7–4.2 | 6.1–50.3 (83.3) | 7.6–41.0 | 0.20–3.04 (19.67) |
| *Cladonia* spp.–*Juncus trifidus* | Shallow alpine podsol | 1 | 4.3 | 3.5 | 2.0 | 0.64 |
| *Juncus trifidus–Festuca ovina* | Brown loam, truncated alpine podsol, subject to solifluction | 4 | 4.7–4.9 | 3.9–6.5 | 3.9–6.6 | 0.43–3.12 |
| **Moss heath** | | | | | | |
| *Carex bigelowii– Racomitrium lanuginosum* | Well developed but shallow alpine podsol | 4 | 4.2–4.8 | 5.3–10.6 | 2.1–17.8 | 0.13–2.5 |
| **Late snowbeds** | | | | | | |
| *Polytrichum alpinum– Carex bigelowii* | Alpine podsol with variable amounts of humus and gleying | 5 | 4.5–4.9 | 6.1–12.4 | 5.1–18.5 | 0.16–1.43 |
| *Deschampsia cespitosa– Rhytidiadelphus* spp. | Alpine podsol | 3 | 4.3–4.9 | 6.4–12.9 | 5.0–53.9 | 0.54–0.8 |
| *Polytrichum norvegicum–Kiaeria starkii* | Shallow alpine podsol, humus rich and slightly gleyed | 3 | 4.3–5.4 | 4.3–22.2 | 2.0–5.4 | 0.11–4.81 |
| *Gymnomitrium concinnatum–Salix herbacea* | Amorphous solifluction soil | 2 | 4.4–4.9 | 3.6–7.6 | 1.6–12.4 | 0.12–0.2 |

**Table 5.5** The distribution of volcanic ash soils (andosols) in high mountains world-wide

| Continent | State/country | High mountain volcano examples (elevation, m asl) |
| --- | --- | --- |
| Europe | Iceland, Sicily, Canary Islands | Sneafellsjökull (1448), Etna (3350), Teide (3715) |
| Africa | Tanzania, Rwanda, Cameroon | Meru (4565), Karisimbi [Virunga] (4507), Cameroon (4095), |
| Americas | Alaska, British Columbia, Washington, Oregon, California, Mexico, Guatemala, Costa Rica, Colombia, Ecuador, Peru, Bolivia, Argentina, Chile | Churchill (5005); Garibaldi (2678), Glacier Peak (3213), Hood (3426), Shasta (4317), Popocatepetl (5426), Atitlan (3535), Turrialba (3340), Nevado del Ruiz (5321), Cotopaxi (5911), El Misti (5822), Parinacota (6348), Nevados Ojos del Salado (6887) |
| Asia and the Pacific | Kamchatka, Japan, Korea, New Zealand, Hawaii | Kliuchevskoy (4835), Fuji (3776), Changbaishan (2744), Taranaki (2518), Manua Loa (4170) |

lead to intense cryoturbation. In many areas, andosols are important agricultural soils, such as in Iceland, where most of the island is under andosols, in Japan, where, 27% of the soils are of volcanic ash origin (Takahashi and Shoji 2003), and in the tropical Andes (e.g. Ecuador, Ortiz *et al.* 2003), where they are the basis for high-altitude potato cultivation. Andosols are readily erodible after perturbation, be it natural, such as by cryoturbation and solifluction, or after trampling by grazing animals and cultivation. Erosion may be followed by aeolian redistribution and wind-blown volcanic ash deposits may modify or over-ride bedrock influences on soil development such as in the Wallowa Mountains, Cascades Range, Oregon, USA where they have led to the formation of andosols with a near neutral pH on acidic granodiorite bedrock (Allen and Burns 2000).

## 5.5 Nutrient budgets: the atmosphere–plant–soil system

The general principles of element cycling in the plant–soil system are usually formulated in terms of fluxes among compartments or pools (soil organic matter, undecomposed plant residues, living plants, and soil microbes). The plant–soil system is not a closed system and can receive or lose nutrients; external input can be in the form of atmospheric deposition, nitrogen fixing from free air, or by inflow through the hydrological cycle; leaching, hydrological outflow, and the formation of gases causes losses from an ecosystem (Ågren and Bosatta 1996). The majority of elements can be accessed from their mineral reservoirs by plants and the organic or elemental forms of many (e.g. carbon, nitrogen, phosphorus, and sulphur) usually do not enter the biological cycle. The availability of these elements is largely through the mineralization of accumulated organic matter.

**Table 5.6** Climate, vegetation, and soil characteristics in various alpine environments

| Range | Altitude (m) | Annual mean precipitation (mm) | Annual mean temperature (°C) | Topography | Parent material | Vegetation |
|---|---|---|---|---|---|---|
| Mount Giluwe* | 3570 | | | Terminal moraine ridge | Andesitic glacial till | Alpine tussock grassland |
| Mount Trikora† | 4000–4250 | | | Summit ridge | Limestone scree | *Deschampsia klossii* |
| | 4000 | | | Slope (45°) | Limestone scree | *Deschampsia klossii* tussock grassland |
| | 4100 | | | Flat | Limestone rock | *Astelia alpina* herb field |
| | 4000 | | | Slope 45° | Sandstone with calcareous matrix | Alpine dwarf shrub heath |
| El Angel,‡ north Ecuador | 3500 | 1150 | 9 | – | Volcanic ash (<3000 years) | *Espeletia picnophylla, Stipa icchu* |
| Cordillera Central,§ Colombia | 4000 | 725 | 4 | Northern slope 35° | Volcanic ash | *Espeletia hartwegiana, Calamagrostis* spp. |
| Australia:¶ Snowy Mountains, Bogong High Plains, Tyndall Ridge, Central Tasmanian Plateau | – | 1780–2660 | 4–5.6 | Ridges, crests, saddles | – | Fjaeldmark (>50% open ground) |
| | – | | | Upper slopes | – | Tall alpine herbfield |
| | | | | Mid- and mid- to lower slopes | | Heath, Short alpine herbfield, Grassland |
| | – | | | Lower slopes, 7–10° | – | Cushion-dominated heath, Alpine sedgeland |
| Pyrenees,** France | 2340–2390 | | | | Granite | *Loiseleuria procumbens, Vaccinium uliginosum, Rhododendron ferrugineum, Calluna vulgaris* |
| Caucasus (Malaya Khatipara, Teberda, Russia)†† | 2750 | 1400 | −1.2 | Ridges, upper slopes | Granite, schist | *Carex umbrosa, Vaccinium vitis-idea* lichen-rich sedge heath |
| | | | | Middle slopes | Granite, schist | *Festuca varia, Nardus stricta* grass heath |

| Depth of soil (cm) | Sample depth (cm) | pH | C% | N% | C/N | Clay (%) |
|---|---|---|---|---|---|---|
| 30 | 5–23 (23–30) | 5.3 (5) | 9 (51.3) | 0.65 (1.21) | 13.8 (42.2) | 3.0 (7.0) |
| 10 | 0–10 | 4.2 | 8.0 | 0.69 | 11.6 | ND (organic) |
| 30 | 0–20 (20–30) | 6.1 (4.3) | 3.1 (7.9) | 0.25 (0.62) | 12.0 (12.6) | 24.0 (23.8) |
| 15 | 0–5 (5–15) | 5.6 (4.2) | 5.1 (1.9) | 0.27 (0.15) | 18.8 (12.4) | 13.9 (20.6) |
| 15 | 0–15 (>15) | 5.1 (4.9) | 1.1 (0.6) | 0.06 (0.06) | 17.1 (9.4) | 4.5 (3.5) |
| 165 | 0–30 | 4.3 | 21.2 | 1.12 | 18.9 | 26.1 |
| 40 | 0–10 | 4.2–5.5 | 6.6–26.7 | 0.55–1.18 | 16.2–19.3 | |
| 10 | 0–5 | 5.2 | 11 | 0.43 | | 4.0 |
| 18 | 0–5 | 4.9 | 12 | 0.67 | | 9.4 |
| 19–25 | 0–5 | 4.9–5.3 | 13–17 | 0.66–0.78 | | 7.1–9.5 |
| 26–33 | 0–5 | 4.8–5.1 | 22–26 | 0.76–0.92 | | 6.4 |
| 20/40 | 2–20 | 4.4–4.7 | 11.3–22.7 | 0.54–1.80 | 17–23 | |
| 30 | 0–10 | 5.5 | 8.3 | 0.67 | 12.2 | |
| 40 | | 5.2 | 9.4 | 0.72 | 13.1 | |

**Table 5.6** *Continued*

| Range | Altitude (m) | Annual mean precipitation (mm) | Annual mean temperature (°C) | Topography | Parent material | Vegetation |
|---|---|---|---|---|---|---|
| | | | | Lower slopes | Granite, schist | *Nardus stricta, Geranium gymnocaulon, Hedysarum caucasicum* herb meadow |
| | | | | Slope foot | Granite, schist | *Sibbaldia procumbens, Phleum alpinum* snowbed |
| Changbaishan[‡‡] | 1950–2650 | 900–1300 | | | Volcanic | Rocky, stony, typical, meadow and swamp tundra (*Dryas octopetala, Vaccinium* spp., *Rhododendron* spp., *Phyllodoce caerula*) |
| (10–20) | | | 0.32–0.54 (0.18–0.34) | 15.2–19.3 | – | |
| Sikkim Himalaya[§§] | 3800–4800 | 2300 | | | Gneiss | |
| West Greenland[¶¶] | 935 | Approx. 300 | Approx. –10 | Northern slope | Gneiss | *Carex bigelowii, Pedicularis* sp. (algae) |

[*]Rutherford (1968).
[†]Mangen (1993).
[‡]Poulenard *et al.* (2003).
[§]Hofstede (1995).
[¶]Kirkpatrick and Bridle (1999), texture scored on a scale from 1 (sand) to 15 (clay); not percentage values.
[**]Cassagne *et al.* (2000).
[††]Makarov *et al.* (2003).
[‡‡]Wu *et al.* (2005) peaty Alpine tundra soil, meadow Alpine tundra soil, grey Alpine tundra soil, lithic Alpine tundra soil and cold desert Alpine tundra soil.
[§§]Singh and Sundriyal (2005).
[¶¶]Sieg *et al.* (2006).

Ecosystems can be described by their stocks of essential elements stored in their soils, vegetation, and animal assemblages. A large number of studies have quantified alpine vegetation (above- and below-ground biomass, necromass, and production) (e.g. Larcher 1977; Körner 2003, pp. 155–160) and invertebrate mass (see e.g. Chapter 6). These studies are being revisited today with the advent of interest in ecosystem carbon

| Depth of soil (cm) | Sample depth (cm) | pH | C% | N% | C/N | Clay (%) |
|---|---|---|---|---|---|---|
| 40 | | 4.7 | 8.8 | 0.7 | 12.5 | |
| 30 | | 4.4 | 15.1 | 1.1 | 13.7 | |
| | 0–10 | | | | | |
| | 0–15 (15–30) | 5.4 (4.9) | 4.5 (4.7) | 0.4 (0.3) | 11.3 (15.9) | 10.9 (14) |
| 10(5) | 0–10(5) | 4.9 (5.9) | 9.8 (1.3) | 0.75 (0.10) | 13.1 (12.6) | |

dynamics and possible changes in it with global change. Vegetation carbon content and the ratio of carbon to nitrogen are relatively conservative. For example, live vegetation carbon concentration (structural and non-structural carbohydrates) is about 50% on a dry mass basis and nitrogen concentration is about 1%; these values decrease through the reallocation of carbohydrates and proteins at senescence, before the death of plant parts. Interestingly, ericaceous plants store high amounts of fat, *Loiseleuria* in particular (>10% dry weight). As most of this fat passes into the litter compartment without being reallocated into living tissues before litter fall, ericaceous species appear to 'waste' high amounts of assimilates gained from photosynthesis (see also Larcher 1977). Dead vegetation in cool humid environments, becomes partially decomposed and forms organic matter, which is a more stable form than plant litter and can accumulate in large quantities. The soil carbon/nitrogen quotient in alpine soils is typically about 15 (Table 5.6) as opposed to 5, estimated globally (Ågren and Bosatta 1996). As a result of the transformation and accumulation of organic matter a 'nutrient bank' forms that links between soils and plant mineral nutrition. Until organic matter is broken down into mineral components by micro-organisms it forms a stock, hence the importance for carbon sequestration of soils in cool humid environments, such as in the alpine, and especially in the arctic.

There have been a number of studies to describe the pools or stocks of essential elements (primarily that of nitrogen, and in some cases phosphorus and potassium) in alpine vegetation components (e.g. chapters 10–13 in Bowman and Seastedt 2001; chapter 10 in Körner 2003; Monsoon *et al.* 2006); however, they usually contain no information about external inputs and outputs. While most mineral nutrients are derived from the soil by weathering and enzymatic processes, there can be major inputs from the atmosphere, especially as a result of volcanic activity, wind-blown dust, or pollution. Water movement through the soil (vertical or horizontal) can cause ecosystems to lose nutrients, which can be measured in stream discharge at the landscape (watershed) scale. Such a study of the entire atmosphere–plant–soil system was undertaken in a tropical alpine ecosystem by Hofstede (1995), who described the nutrient budget of a Colombian páramo by analysing input from precipitation, element stocks locked up in vegetation and soil, and stream outflow from the watershed studied. The rather large annual input rates found (e.g. approximately 5 kg N ha$^{-1}$, see Table 5.7), presumably emanating from nearby volcanic activity or received from wind-blown dust, are comparable with those reported from pollution in industrial countries (see Chapter 9). The results suggest that nutrient deposition, affecting alpine ecosystems in some regions may be more frequent than commonly perceived and may have been a long-term feature of many alpine ecosystems. The comparison of areas that have been affected by volcanic activity, with those without such known recent activity could be instructive in identifying long-term impacts of external mineral inputs on vegetation dynamics, composition, and productivity.

The study by Hofstede (1995) also showed that half the total nitrogen and phosphorus was locked up in dead plant material, with the other half more or less equally shared in above- and below-ground live plant material; the extractable, or plant available (as opposed to total acid digestible) soil stocks of the two elements were about 5% of the combined total for plant and soil. A variable but higher proportion of exchangeable potassium, calcium and magnesium were stored in the soil (Table 5.7; Hofstede 1995). Where estimates of total major element stocks in the soil are available for comparison (last column in Table 5.7) it becomes apparent just how small a fraction of the major elements is present in a readily available ionic form in the soil (0.4–1.2 %) and that the amount of major elements

**Table 5.7** Nutrient ledger for a páramo ecosystem in the Colombian Andes (Agua Leche, Cordillera Central, 4000 m asl) in 1991. The input is by way of precipitation and the loss is measured in stream discharge. The total storage comprises soil, above- and below-ground biomass, and dead plant material. The high input is likely to be from a nearby active volcano, Nevado del Ruiz or from wind-blown dust.

| Nutrient | Input (kg ha$^{-1}$) | Loss (kg ha$^{-1}$) | Total storage (of which soil)* (kg ha$^{-1}$) | Total soil vs. extractable soil nutrient stocks† (kg ha$^{-1}$) |
|---|---|---|---|---|
| $NH_4^+$ and $NO_3^-$ | 5.3 | 3.0 | 410 (30) | 5110 (19.8) |
| $PO_4^{2-}$ | 0.8 | 0.0 | 28 (2) | 550 (2.4) |
| $K^+$ | 10.7 | 3.0 | 290 (110) | 5260 (61.1) |
| $Ca^{2+}$ | 26.0 | 8.0 | 400 (290) | ND |
| $Mg^{2+}$ | 4.6 | 2.0 | 160 (130) | ND |
| $SO_4^{2-}$ | 60.2 | ND | ND | ND |
| $Na^+$ | 16.6 | ND | ND | ND |
| C | | | | 91960 |

*The soil stocks refer to extractable amounts of ions and not total digestible amounts—for comparison of total versus extractable amounts of nutrients see last column of table;
†Values are from a study 1 year after burning of páramo.
Source: Hofstede (1995).

held in the vegetation (live, dead, and roots) is only 3–7% of that locked up in the soil. The fraction of loss of elements from the watershed, estimated by measuring stream discharge, was smaller than the atmospheric input. Although the study best represents an approach to full nutrient budgeting, the figures are not generalizable. Bunch grass *páramo* with the giant rosette *Espeletia hartwegiana*, is not representative of all alpine habitats along climatic and hydrological gradients. Although total (above- and below-ground) mass of vegetation and litter (approximately 4750 g m$^{-2}$) was higher than found in most similar studies and probably did not capture the range of variation in vegetation mass along altitude and topo-gradients shown for example in the Changbaishan (600–2960 g m$^{-2}$, Wu *et al.* 2006), the alpine dwarf-shrubs of the Alps (2610–6057 g m$^{-2}$ (Pümpel 1977), and 800–3300 g m$^{-2}$ along a snow gradient—see Fig. 6.15), Colorado Rockies (approximately 1700–6000 g m$^{-2}$) and about 1250 g m$^{-2}$ from grazed alpine pastures in the Himalayas (Rikhari *et al.* 1992; Singh and Sundriyal 2005), vegetation nitrogen pools did not appear to differ too much (Table 5.8).

Importantly, the input of mineral nutrients from volcanic activity, or wind-borne dust might, to some extent, decouple availability of nutrients for plant uptake from microbial mineralization. How this may work in reality is difficult to estimate, as there are large temporal uncertainties with regard to input. Airborne input from precipitation depends on the concentration of elements in the atmosphere and the amount of precipitation,

**Table 5.8** Nitrogen pools (kg ha$^{-1}$) in different alpine ecosystems of the world

| | Soil | Vegetation above-ground | | Vegetation below-ground | | Microbial | Total vegetation |
|---|---|---|---|---|---|---|---|
| | | Live | Dead | Live | Dead | | |
| Páramo* | 5110 | 62 | 254 | 74 | ND | ND | 220–380 |
| Alpine meadows Rocky Mountains[†] | 6570–6960 [4380–4640][‡] | 24–64 | 66–84 | 53–120 | 268–740 | 31–78 | 411–993 |
| Changbaishan[§] | 2050–4100 | 0.5–17.4 | 5.3–11.3 | – | – | ND | 7.4–28.5 |
| Tundra types, Brooks Range, Alaska[††] | ND | 20–88.5 | 31–54 | ND** | ND | ND | |

*Hofstede (1995), 0–10 cm soil depth

[†]Fisk et al. (2001), vegetation measured at peak biomass in early August 1992–93.

[‡]Soil depth was 0–15 cm, i.e. assuming homogeneous distribution of nitrogen in that depth, soil values need to be reduced by one-third, as given in square brackets, to be comparable with those given for the páramo.

[§]Wu et al. (2006), soil 0–10 cm; values for vegetation look 10 times less than expected–may be a problem with decimal points in paper?

Litter in Wu et al.; below-ground included in above-ground values.

**Hahn et al. (1996).

[††]Based on studies in arctic ecosystems the below-ground:above-ground ratio is about 10:1. Accordingly, the below-ground live and dead vegetation nitrogen stocks are expected to be similar to those in the Colorado Rockies by Fisk et al.

both of which can vary considerably from year to year. These are unknowns that are important not only concerning volcanic input, but also for anticipating the impacts of climate variability on alpine ecosystems.

# 5.6 Climatic seasonality and plant available soil nutrients

At first sight, biological soil-forming processes are likely to be active all year round in tropical high mountains. However, it is not known to what extent observed large fluctuations in soil arthropod densities (Diaz et al. 1997) might be indicative of the periodicity of activity of other soil organisms. In seasonal climates, the growing season for vascular plants and the season of activity for micro-organisms and animals is restricted, largely as a function of temperature and snow cover. Less so for microbial activity and organic matter decomposition, which are surprisingly high in frost-free soil in the winter, even under snow (Fisk et al. 2001; Bardgett et al. 2005). Mediterranean climates impose a double seasonal restriction on mineralization because, in addition to winter freeze and snow, drought and high temperatures in the summer also affect weathering and soil biological activity. Plant available soil nutrients in an aseasonal humid tropical alpine environment might be thought to be higher than in seasonal alpine areas because mineralization of organic matter is not limited by lasting periods of frozen soil, or drought. However, the issue appears to be more complex, as a comparison of litter decomposition in aseasonal humid tropical (páramo) and seasonal dry subtropical (puna) has shown (Pansu *et al.* 2007). In the humid páramo, although the mineralization of experimentally added litter continued all year round, its rate was much slower than in the puna, and as a result, about five times more organic matter accumulated in the páramo after 2 years. In contrast, there was a seasonality in the puna and the dry cold winters (and equally, the dry spells that occurred outside the winter) led to a near cessation of mineralization. The fundamental difference between the two ecosystems (both cultivated and fallowed periodically) was in the rate of soil respiration and the soil microbial biomass—about three times higher in the puna than in the páramo.

The importance of biological activity in alpine soils lies in the fact that ionic nutrient concentrations are usually low; this is due to several factors that may change from place to place, such as low weathering rates, or high leaching. Unless physico-chemical weathering replenishes plant available inorganic nutrients, and frost action or ground-dwelling organisms rejuvenate the soil by mixing, the accumulating undecomposed organic matter in theory can limit major nutrients such as nitrogen or phosphorus. That biological activity is low in alpine soils is evident from the accumulation of undecomposed litter and organic matter (see e.g. Larcher 1977, p. 327 who reported 10–25% higher amounts of litter and dead standing material than above-ground biomass for alpine *Vaccinium* and *Loiseleuria* heaths in the Alps). However slowly, decomposition of organic matter does happen, the dynamics of which is dependent on, apart from litter quality, two climate-related factors, temperature and humidity. Seasonal variation in temperature is compensated, to a certain degree, by a seasonal change in the activity of different micro-organism types, however, soil freezing brings a halt to decomposition.

The humidity optimum for decomposing organisms is a balance of about 20–40% air and 60–80% water in the soil pore space (Flanagan and Veum 1974). This condition in an alpine landscape is not always met as warm insolated slopes can be too dry, while glacial troughs and valley bottoms can be too wet. It is important to differentiate between surface litter decomposition and organic matter decomposition in the soil. Litter decomposition is largely controlled by hydrology, related to snow cover and duration, a process not confirmed for soil organic matter mineralization (Seastedt *et al.* 2001). Pulses of nitrogen and phosphorus release occur after snowmelt and additional smaller ones later in the growing season

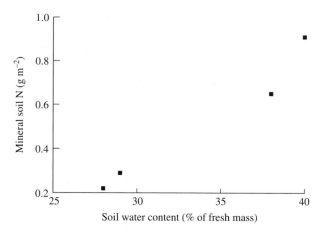

**Fig. 5.20**     Top soil (0–15 cm) nitrogen mineralization rates as a function of soil moisture content in a field incubation experiment on a well-drained hill slope, Mt Kenya, 4160 m asl (source: Rehder 1994). The values of N mineralization in increasing magnitude are from the end of dry season, beginning of dry season, end of rainy season and beginning of rainy season.

(e.g. Fig. 6.19); the dynamics of other nutrients is poorly documented as they are assumed to be present in sufficient quantities and not to be limiting to plant growth. The available evidence (Mount Kenya, Rehder 1994; Andes, Pansu *et al.* 2007) is insufficient to enable making generalizations about the temporal dynamics of nitrogen mineralization in tropical alpine soils. A hydrological control appears to be at work, with the moister part of the year resulting higher soil nitrogen mineralization (Fig. 5.20) both on Mount Kenya and in dry puna soil (approximately 350 mm precipitation year$^{-1}$) on the Bolivian Altiplano.

## 5.7  Nutrient limitation in alpine ecosystems?

There are some clearly limited marginal environments in the nival zone of high mountains where unweathered substratum or arid climate minimizes the functioning of biogeochemical cycles towards organic matter production and a high elevation aeolian life zone is present. In aeolian ecosystems (see Chapter 6), organic matter input is via wind transport, adaptations to which have developed highly specialized local faunas (e.g. Loope and Medeiros 1994).

For terrestrial alpine ecosystems from the arctic to temperate latitudes, the notion of nutrient limitation, presented in the literature largely as nitrogen limitation (e.g. Jonasson *et al.* 2001; Bowmann *et al.* 2001), is somewhat controversial. Generally, it is stated that these seasonally cold-limited ecosystems are nutrient limited because the experimental addition of mineral nitrogen increases plant biomass, mainly by satiating the needs of the microbial decomposers, and thereby increasing the amount of nitrogen available to plants for biomass growth. The fact that arctic and alpine plants have also been observed to access low chain organic compounds (e.g. amino acids, see, e.g. Bowman and Seastedt 2001, p. 215), in addition to mineral forms, has reinforced this notion of nitrogen limitation. It is easy to demonstrate that plants either increase their nitrogen concentration or grow more tissue without changes in concentration when they have unlimited access to nitrogen. All living organisms respond to a supply of resources; however, in the end, all this has to be offset by an adaptation to best survive in any given environment. Therefore, it may be argued that alpine plants are in a dynamic equilibrium with their atmospheric and soil environment, biotic and abiotic. This is illustrated by the distribution of physiognomically different vegetation types in the mosaic of microclimatically different sites within the alpine zone and their corresponding characteristic uptake of nitrogen (as demonstrated by Fisk *et al.* 1998). For example, there would be little advantage for plants to grow additional tissue that was less winter hardy, unless there was a corresponding amelioration in the harshness of winter.

Nutrient availability has consequences not only for growth alone, but for reproduction as well. Variation in the weather from year to year does have perceptible consequences for the vegetation (Walker *et al.* 1995) and involves temporary changes in below-ground processes and nutrient availability. Ameliorated atmospheric conditions in exceptionally warm years may indeed produce higher seed maturation than usual; however, flowering will have been determined in the preceding growing season when most flower buds will have formed. In rosette plants, flower bud formation requires a minimum rosette size, which in turn is connected to nutrient availability. It is possible to prompt the flowering within a year of sowing in the field of long-lived perennials, such as *Lychnis alpina*, if nutrients (primarily phosphorus and nitrogen) are supplied in surplus quantities (L. Nagy unpublished). Does this mean that there is indeed a nutrient limitation? Yes, one could argue, because clearly, the addition of nutrients has made dramatic changes to the growth and reproduction of the plants. However, in natural situations such a high nutrient supply is likely to have historically changed the distribution of the species and have excluded it from the alpine zone. The *Lychnis* example highlights *ad extremis* how nutrient supply can change the life cycles of plants (in this case a long-lived perennial turned into uniparous biennial that completed its life cycle within a single calendar year). Such a life cycle is clearly not selected for alpine environments; however, the plasticity shown by the species is remarkable, none the less.

In contrast to the ever-present discussions on nitrogen availability and ecosystem biomass production (size), phosphorus availability has been rarely investigated. However, temporal changes in the availability of phosphorus alone can have dramatic effects on the reproduction and population dynamics of non-clonal species (e.g. Nagy and Proctor 1997). The consequences of ameliorated availability to plants of nitrogen and phosphorus with sustained soil and atmospheric warming may have different ecological consequences on different alpine habitats: nitrogen-driven larger plant stature may change plant to plant interactions, while phosphorus could drive reproduction, and resulting colonization through dispersal (see Chapter 9 on Global change impacts on alpine habitats for further discussion).

# 5.8 Conclusions

Landforms determine the microclimatic, hydrological and soil environments of a wide range of contrasting alpine habitats. Glacial landforms generally maximize habitat diversity, both terrestrial and aquatic. Convex and concave forms result in differences in the main habitat characteristics with regard to wind exposure, soil moisture, and snow accumulation in seasonal climates. Slope and substratum (e.g. rock versus till) further affect soil moisture availability and solute movement. Bedrock is important for soil development, and soil physical and chemical properties, and thus selecting for certain species and excluding others. The diversity of landform-related habitat creating factors greatly increases biological diversity and influences the functioning of alpine ecosystems. There are habitat-linked gradients in growing season length in seasonal extra-tropical climates. Along these gradients, resource availability and biomass production varies. Mineral nutrient availability and potential responses to changed mineral cycling can reshape alpine ecosystems, where in a warmer climate, species responses to changed nutrient availability are likely to differentiate among species that have enhanced growth and those that can increase their reproduction and dispersal potential.

# 6  Alpine terrestrial habitats and community types/assemblages

## 6.1  Introduction

The habitat diversity of high mountains is well developed along topo-sequences, where ecological conditions such as wind exposure, snow cover and duration, soil availability, soil depth, nutrients and moisture, and disturbance regimes vary. The variation in habitat factors is largely predictable along an idealized slope (e.g. Fig. 6.1) and is reflected in a sequence of habitats/vegetation types. Such habitats may be divided into two main types: zonal, or habitats that reflect the average growth conditions (climate, soil, and water availability) that are prevalent in a latitudinal zone at a given altitude; the existence of azonal habitats is linked to locally deviating growth conditions, largely to topographical extremes. In relation to a slope topo-sequence (e.g. Figs 5.5 and 5.6), zonal habitats are analogous to mid-slope areas and azonal ones to those near to either end. For example, at the top end, gravity-related processes and phenomena exacerbate the scarcity of soil in rocky habitats and disturbance from frost shattering and rock fall, or avalanches create habitats that are exploited by specialist organisms, adapted or resilient to recurrent disturbance. At the lower end of the slope, materials accumulate, such as talus or scree, a surplus of snow (snowbeds), water (mires), or nutrients (colluvia). While it may appear from the above that azonal habitats are characteristic of the extremities of a mountain slope, in reality, they form a patchwork within the expanse of zonal habitats, in topographically fragmented localities (see for example Figs 5.5 and 5.6).

For a generalized classification of alpine habitats into zonal and azonal types see Table 6.1. Such a classification is most applicable to Holarctic humid mountains, although ecologists have reported zonal patterns in almost all mountains in the world (Burga *et al.* 2004). Zonal habitats and

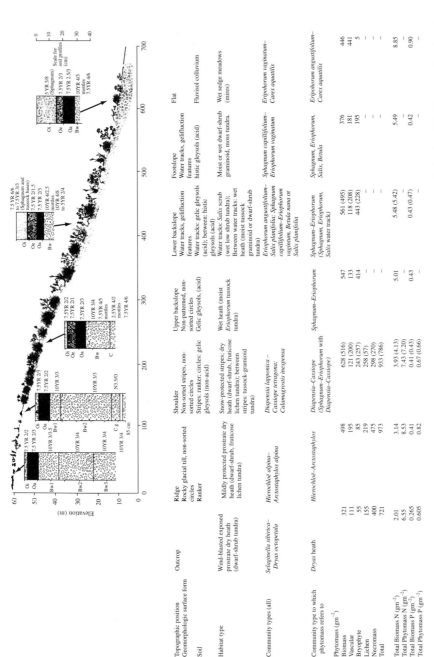

**Fig. 6.1** Idealised slope and corresponding soils and vegetation from the foothills of the Brooks Range, northern Alaska (modified from Walker & Walker 1996). The toposequence along the slope illustrates a gradient along which, from top to bottom, exposure to wind diminishes and snow accumulation increases; alongside with topography soil water and nutrients also change. See text for explanations on analogies to zonal and azonal plant communities. The corresponding phytomass for the vegetation in each topographic feature is shown after Hahn et al. (1996).

**Table 6.1** Alpine terrestrial habitat types, their geographic distribution and principal life forms and taxa.

| Habitat type | Geographic distribution | Principal taxa or life forms |
|---|---|---|
| Aeolian (rock, ice, 'soil') | Highest mountain ranges (e.g. Himalayas, Pamirs, Andes); locally in others (e.g. Alps) | see Table 6.2 |
| Rock fissures, ledges | Ubiquitous | Epi-, chasmo-, and endolithic cryptogams (crustose and foliose lichens; mosses), chasmophytic phanerogams (ferns, rosettes); mites, collembola |
| Screes | Ubiquitous | Epilithic lichens, hypolithic cyanobacteria, green algae; phanerogams (forbs, small graminoids and dwarf shrubs) |
| Springs and rills | Restricted by parent rock and arid climate | Cyanobacteria, algae, liverworts, mosses, phanerogams (graminoids and forbs) with a requirement for flowing water |
| Mires (ombro-, topo-, and soligenous) | Not in arid climate | Ombrogenous oligotrophic mires (peat bogs) in humid cold environments: fruticose lichens, bryophytes (e.g. *Sphagnum* spp.), graminoids and dwarf shrubs<br>Topogenous mires in low-lying areas, usually with a watertight layer in the subsoil<br>Soligenous mires (flushes, fens) surface irrigated by ground water: algae, bryophytes, graminoids, forbs |
| Salt flats | In arid climates | Salt-tolerant phanerogams |
| Snowbeds | In seasonal climates with late melting snow | Cyanobacterial crusts, abundant and numerous liverworts, bryophytes, few phanerogams |
| Tall herb meadows | Local, excess nutrients from animal input or of soil origin/disturbance | Tall forbs and graminoids |
| Extrazonal fumaroles | Local, supply of heat, water vapour and $CO_2$ | Lichens, bryophytes |
| Zonal habitats of mountain slopes | Ubiquitous | Grasslands (grass-, sedge-, and moss-heaths); dwarf-shrub heath, giant rosette formations |
| Treeline ecotone scrub, including *Krummholz* | Ubiquitous, except very dry | Trees, scrub, and shrub species, graminoids, forbs |

their vegetation are a distinctive physiognomic feature of alpine landscapes and regional descriptions are usually based around this characteristic; for example, dwarf-shrub heaths, in the boreal zone alpine grasslands (sedge heaths) in temperate alpine areas, and giant rosette formations in African or north Andean tropical alpine environments. Azonal formations are usually less conspicuous in the landscape; however, there are exceptions. The impressive bare limestone rocks or sweeping screes can locally be co-dominant features. The frozen world above the limit of plant life (nival and aeolian habitats, Table 6.2) is sometimes referred to as extrazonal. To aid the treatment of zonal habitat variation at the global scale (table 1 in Grabherr *et al.* 2000; Breckle 2002) they are discussed along latitude life zones, introduced in Chapter 2. Some of the azonal habitats are also specific to geographies—precipitation and its distribution during the year can give rise to specific habitats (e.g. ombrogenous mires in a cold and humid climate, or salt pans in an arid and hot climate), or largely influence the structure and composition of plant and animal assemblages. Other azonal habitats (e.g. rock, scree) are ubiquitous and do not require a geographic treatment.

## 6.2 Arctic alpine habitats and community types

Arctic landscapes have distinctive landforms that have developed as a result of permafrost (patterned ground) and freeze–thaw cycles (solifluction lobes). Patterned ground includes the ice-cored pyramids, or pingos; hummocks; peat mounds, palsas and plateaux; polygons, and smaller patterned

**Table 6.2** Azonal or extrazonal habitats that occur globally in mountains above the permanent snow-line (nival/aeolian)

| Habitat type | Principal taxa or life forms | Notes/references |
|---|---|---|
| Glacial ice | Micro-organisms, predatory invertebrates | Swan (1992); Thaler (1999) |
| Snow | Bacteria, fungi, green algae | e.g. Thomas and Duval (1995); Novis (2002) |
| Ice–rock interfaces: cryoconite | Cyanobacteria, green algae, diatoms | Fogg (1998, p. 57) |
| Cryoconite | Protozoans, rotifers, tardigrades | See references in Thaler (1999) |
| Rock surfaces and fissures | Epi-, chasmo-, and endolithic cryptogams; mites, collembola | |
| Screes | Epilithic lichens, hypolithic cyanobacteria, green algae | |
| Habitats overlying permafrost | Various | Largely waterlogged habitats in the Arctic; some alpine mountains |
| Glacial till (recently de-glaciated) | Cyanobacteria, beta-proteobacteria, diatoms, algae | Nemergut *et al.* (2007); Schmidt *et al.* (2008) |
| Nival zone soils | Cyanobacteria, heterotrophic bacteria, algae, chytrids, fungi | Freeman *et al.* (2009); Ley et al. (2004) |

ground, made obvious by stone sorting (e.g. Fitzpatrick 1997). In a thawing landscape solifluction lobes develop as the melting active layer (soil) gravitates downslope (Fig. 5.3); slumping of peat is produced after the deep thaw of permafrost. Habitat diversity is linked primarily to topographic variation (Webber 1978; Fig. 6.2); species diversity is related to habitat diversity (gamma diversity) on the one hand (high diversity in exposure and water availability provides for more species with different ecological needs), and to latitude on the other hand, as overall species richness (species pool) decreases from south to north. For example, in the Canadian Arctic a total of about 400 plant species is found in the southernmost tundra, reducing to about 50 in the northern islands (Pielou 1994); the same pattern is found for animals (see examples in Chernov 1995).

Arctic tundra has been defined to cover all areas north of the latitudinal treeline (Payette *et al.* 2002). Usually, a latitudinal, temperature- and vegetation-based subdivision of the tundra distinguishes cold (parts of the high arctic, polar desert), cool (most of the high arctic, arctic), and mild (subarctic, low arctic) regions (Longton 1997). These latitude zones have their analogues in mountains both south and north (e.g. Sieg *et al.* 2006) of the northern tree limit.

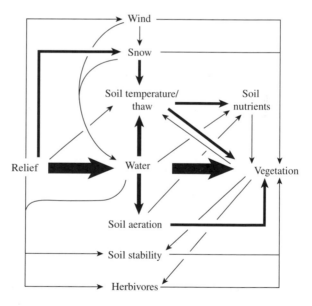

| Fig. 6.2 | A conceptual model of vegetation control by habitat factors in the arctic. The environmental control is largely manifested through topography and related hydrology. Hydrology, in turn, controls pore air space in the soil and selects species for aeration vs. anoxia at the two extremes, with further variation with laterally flowing water (soligenous mire) that is richer in oxygen than stagnant water and carries a supply of water-soluble mineral nutrients (flushing). Source: Webber (1978). |

Latitudinal cold deserts are found where summer mean monthly temperatures reach up to 2°C (northern Svalbard, northern Greenland, Franz Josef Land, and Novaya Zemlya). Their equivalent is found in the mountains of mid and southern Greenland, Iceland, northern Scandes, Polar Urals, and the north Siberian mountains, where they form the nival zone. The sparse ground cover of plants (<2%) is largely of lichens and mosses, with crustose and foliose lichens on rocks; in moist conditions closed bryophyte, lichen or alga cover occurs in extensive patches. The few scattered vascular plants include *Papaver radicatum*, *Silene acaulis*, *Draba* spp., and *Saxifraga* spp. The invertebrates include protozoa, tardigrades, nematodes, enchytraeid worms, mites, and collembola, which feed on bacteria, algae, and fungi. Occasional vertebrates are herbivores such as the arctic lemming (*Dicrostonyx groenlandicus*) and arctic hare (*Lepus arcticus*), and their predator, the snowy owl (*Bubo scandiacus*).

In the high arctic or cool tundra closed, vascular plant dominated vegetation is restricted to favourable areas and mountains are mostly of open fell-field type. It is typical of most of the Canadian and Alaskan Arctic, where the mean summer monthly maxima range from 3 to 7°C and the length of the growing season is about 3 months (Reynolds and Tenhunen 1996; Fogg 1998).

The soil invertebrate fauna of the tundra includes protozoa, nematodes, rotifers, enchytraeid worms, tardigrades, crustaceans, collembola, and dipteran larvae, feeding on bacteria, algae, and fungi. The above-ground invertebrates include moths, butterflies, and diptera (tipulids). There are a number of herbivores such as the arctic and brown lemming (*Lemmus sibiricus*), arctic hare, snow geese (*Anser caerulescens*), musk ox (*Ovibos moschatus*) feeding on sedges, and reindeer or caribou (*Rangifer tarandus*) on lichens in winter and on wet meadow plants in summer. The main carnivores are the arctic fox (*Alopex lagopus*), wolf (*Canis lupus*), snowy owl, and skuas (*Stercorarius pomarinus*); the main insectivores are buntings (*Calcarius lapponicus*, *Plectrophenax nivalis*) and sandpipers (*Calidris* spp.).

The mild or low arctic is characterized by mean monthly temperatures of about 6–10°C in the warmest month. The temperature ranges recorded 10 cm below the soil surface on the highest peaks at Latnjajaure, north Scandes (5–8°C) and the Polar Urals (6–9°C) and the peaks in the treeline ecotone (9–10°C and 10–11°C) (GLORIA, unpublished data) place these mountain areas in the mild to cool arctic equivalent latitude classes. Extensive areas are under the cover of grass and dwarf-shrub heaths and mires; *Sphagnum* is abundant in many mire types; scrub or woodland occurs in sheltered locations. Exposed areas have fell-fields (open vegetation on skeletal soil, with surface stone cover).

Midges (Culicidae) are a feature of these habitats. Nematodes, aphids, moths, and butterflies are notable invertebrate herbivores; vertebrates include the Norwegian lemming (*Lemmus lemmus*), voles (*Clethrionomys* spp.), mountain hare (*Lepus timidus*), grouse (*Lagopus lagopus*), and the reindeer. Their carnivores are the occasional brown bear (*Ursus arctos*), wolf, wolverine (*Gulo gulo*), red fox (*Vulpes vulpes*), and the short-eared owl (*Asio flammeus*). Insectivores include the willow warbler (*Phylloscopus trochilus*) and meadow pipit (*Anthus pratensis*).

## 6.2.1 Functioning

Arctic tundra is a collective term for all the vegetation types it is made up of: (1) exposed fell-field with lichens (*Cetraria, Rhizocarpon*); (2) dry heaths (wind-blasted outcrops: *Dryas*), medium exposed, light snow cover (*Arctostaphylos, Betula*); snow-protected (*Diapensia, Cassiope*) and mosses and lichens; (3) tussock tundra with *Eriophorum*, dwarf-shrubs, *Sphagnum* mosses and lichens; (4) vegetation of areas with and bordering flowing water; (5) mires (*Eriophorum angustifolium, Sphagnum* spp.) (e.g. Walker and Walker 1996; Fig. 6.1). As to how habitats and vegetation types are controlled by environmental factors in arctic alpine environments, a conceptual model that was proposed by Webber (1978) for arctic tundra at Barrow (Alaska) may be used as a pointer (Fig. 6.2). Topography and, in turn, water (its availability or excess) are important in determining the suitability of a site for what kind of plants can grow. This conceptual model well captures the relative constancy of the existence of habitats and their vegetation in a landscape. While a gamut of biogeochemical, hydrological, biophysical, and biochemical processes are at play and provide control via various feedback mechanisms, any visual change in habitat distribution is primarily related to an irreversible change in water availability. For example, the thawing of permafrost as a result of climate warming causes the disappearance of a watertight layer and can lead to the draining of previously water logged areas, changing the vegetation in the process from a wet meadow (soligeneous mire, or fen) type to dwarf-shrub.

Habitats support different vegetation types and their production and standing biomass or phytomass show wide ranges across habitats, as well as across latitude within the arctic (Longton 1997). In the cold arctic, living phytomass is only about 10 g m$^{-2}$ on average, but can be a great many times higher in favourable locations. In the cool arctic the total annual production is about 100–300 g m$^{-2}$ and total plant biomass is in the range of 300 to 3000 g m$^{-2}$, the former characteristic of fell-fields and the latter of mires (Table 6.3); see also Svoboda and Freedman (1994) for locally occurring warm areas or polar oases. In the mild arctic, the corresponding figures for production range from 45 g m$^{-2}$ to 1000 g m$^{-2}$ and for phytomass from 1470 to 5800 g m$^{-2}$.

Functional linkages of ecosystem components and the energy and element flows among them are essentially similar in all latitude zones from

**Table 6.3** A sample of the range of productivity and phytomass recorded in the high Arctic (from Longton 1977)

| Habitat type | Annual net production (g m$^{-2}$) | | | Phytomass (g m$^{-2}$) | | | | |
| --- | --- | --- | --- | --- | --- | --- | --- | --- |
| | Vascular plants | Cryptogams | | Vascular plants | | | Cryptogams | |
| | Above-ground (below-ground) | Bryophytes (Lichens) | Total | Above-ground | Below-ground | Standing dead | Bryophytes (Lichens) | Total |
| Fell-field | 15–27 (3–5) | 2–20 (2–3) | 23–54 | 89–126 | 50–57 | 192–298 | 15–600 (23–49) | 508–991 |
| Dwarf-shrub heath | 18 (90) | 20 (4) | 132 | 159 | 1041 | 228 | 423 (48) | 1899 |
| Graminoid heath (dry) | 4–13 (4–13) | 1–32 (<1) | 9–58 | 4–13 | 88–519 | 74–120 | 76–2128 (9–20) | 728–2323 |
| Wet meadow (mire) | 45–58 (104–130) | 15–103 (0) | 182–279 | 78–112 | 1295–2023 | 120–202 | 908–1100 (0) | 2592–3208 |

the arctic to the tropics; for an arctic example see fig. 4.18 in Fogg (1998); a boreal sedge-moss heath example is shown in Fig. 6.3. In other latitudes some components are present in different proportions, or are missing. The differences primarily lie in the availability of water that affects both the components and the processes; for example, humid and cold environments favour the growth of cryptogams, such as land algae, lichens, and mosses, while these organism are less abundant in semi-arid and arid regions. Organic matter decomposition is slow and results in the accumulation of soil organic matter under cool and humid conditions and the opposite is true in drier areas (see e.g. soil section in Chapter 5; Fig. 6.1). The presence or absence of herbivores, the relative magnitude of their impact, and their availability as a prey for carnivores drives some regulatory feedback processes. The interplay between large herbivores and below- and above-ground processes has been well illustrated on Svalbard tundra (van der Wal *et al.* 2004; van der Wal 2006): increased trampling by reindeer reduces moss depth, which leads to increased soil temperatures. These, in turn, stimulate grass growth via increased mineralization rates and plant available nutrients; herbivore input of faeces and urine further drives these cycles.

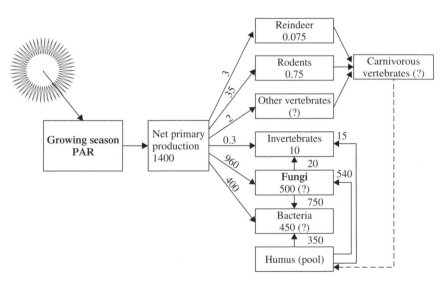

**Fig. 6.3**   The flow of energy among and accumulation in ecosystem components constructed for a low-alpine ecosystem at Hardangervidda, southern Norway in a peak year of rodents. Values are means (kcal m$^{-2}$, 1 kcal = 4.26 kJ), redrawn after Wielgolaski (1997b). A list of plant and animal groups for the compartments is given in the text. Mosses have an insulation effect and generally lower soil temperatures in the summer (e.g. Longton 1997).

### 6.2.2 Fumaroles

Geothermal activity causes specific local habitat conditions by raising temperatures and, often, contributing to water flow in a frozen environment. In addition to heat, fumaroles emit vaporized water that condensates, and various gases, including carbon dioxide. Polar examples of geothermal activity related vegetation are found in the Svalbard, Iceland, Greenland (Halliday et al. 1974), and in Antarctica (Longton and Holdgate 1967). The latter is of special interest as the bryophyte and algal colonies of Mount Melbourne, Antarctica (74.4°S, 164.7°E) at 2733 m altitude grow in an isolated situation, far away from potential colonization sources, comparable with the bryophytes and other cryptogams found on the high Andean desert of Volcán Socompa, Andes (24.3°S, 68.3°W) between 5700 and 6000 m (Halloy 1991). The origin of such colonies of plants is difficult to explain. A recent study of the moss *Campylopus pyriformis* growing in two localities on Mount Melbourne showed genetic diversity within the population. Genetic evidence for dispersal between the two localities, together with some genetic variation within individual colonies, indicates that a single colonization event has probably occurred at this extremely isolated location followed by multiple mutations. Interestingly, material from near the summit of another volcano, Mount Erebus (3794 m; on Ross Island, 300 km south of Mount Melbourne) was found to be genetically closely related to the mosses on Mount Melbourne (Skotnicki et al. 2001). This example is worth bearing in mind for the role of long distance dispersal in the evolution of alpine floras in Chapter 7.

## 6.3 Boreal zone mountains: dwarf-shrub heaths

Arctic habitat and community types penetrate south of the arctic treeline, at increasingly higher elevations, into the mountains of northern Europe (Caledonids-Scottish Highlands, Scandes, and the Urals, which were all formed during the Caledonian folding—the Chibiny Mountains of the Kola Peninsula and the volcanic mountains of Iceland), the Siberian mountains, with the exception of the extreme far eastern ones, the Alaskan ranges, and the Canadian Rockies (Ozenda 2002). The southern limit of the boreal zone mountains is about 56°N in Scotland, 48°N in the Altai, 50°N in the Baikal Mountains, and about 49°N in Canada (Figs 2.1 and 2.7). Recent accounts on the habitats and vegetation of the Scandes (Virtanen 2003) and the Scottish Highlands (Nagy 2003) have summarized community types, and Gorchakovsky (1989) has analysed the vertical zonation in the Urals along the entire range from the arctic to the boreal-temperate zone. An overview of the Siberian mountain vegetation is given by Shahgedanova *et al.* (2002a,b), for the Altai by Revyakina and Revyakin (2004), and for North America by Billings (2000); Mark *et al.* (2001) have reported the vegetation patterns on Tierra del Fuego, for the Southern Hemisphere boreal zone.

The topography–habitat relationship and the importance of water are evident in the boreal mountains, similarly to that in the arctic (e.g. Billings 2000). At the small-scale, this can be seen along gradients on terrace slopes in drier or more mesic conditions, such as in the Cairngorms, Scotland as illustrated by Watt and Jones (1948). The grading of habitats is a function of soil moisture, exposure to wind, and snow cover duration, factors all related to topography. Some northern boreal alpine habitats are relatively poor in vascular species, and plant communities, formed by a limited pool of species can often distinguished only by the differing proportions of the same constituent species. However, the stature of vascular species and the accompanying, or often dominant cryptogams are important distinguishing features, closely related to the above habitat factors. The habitat gradients can result in variation in the growth of the character giving species within a plant community, and when the gradient is strong enough, contrasting community types will result (Fig. 6.4). Fire, of natural and sometimes of management origin, can reshape boreal treeline ecotones and alpine landscapes (e.g. Vajda and Venelainen 2005). Forest re-growth after fire may be checked in marginal areas, where open dwarf-shrub heath may become the dominant vegetation type. Uncontrolled moorland fires spreading on to alpine prostrate dwarf-shrub heath may cause the loss of thin soil cover and result in bare rock or scree, such as observed in the Scottish Highlands. On the windswept Tierra del Fuego gales can flatten *Nothofagus* forest stands and browsing by guanaco (*Lama guanaco*) then can prevent the regrowth of trees, converting the smaller blown down patches into alpine heath (Rebertus *et al.* 1997).

## 6.3.1 Boreal alpine treelines

The treeline in Iceland, Scandes, the Chibiny Mountains, and on the western slopes of the northern Urals is formed by *Betula pubescens* ssp. *tortuosa*, and in Scotland by *Pinus sylvestris* and *Betula pubescens* (Fig. 6.5; Table 6.4). On the eastern slopes of the northern Urals there is *Picea abies* ssp. *obovata* and *Larix sibirica*; *L. sibirica* and *Betula ermannii* in the Putorana, Verkhoyansk, Tchersky, Kolyma, and Kamchatka Mountains; *Larix dahurica* in the Anadyr, and *Pinus pumila* in the far east of Siberia. *Pinus sibirica* and *Larix sibirica* are the treeline-forming species in the southerly chain of Siberian mountain ranges between 50°N and 60°N, including the Djugdjur, Baikal Mountains, Sayan, and the Altai. The

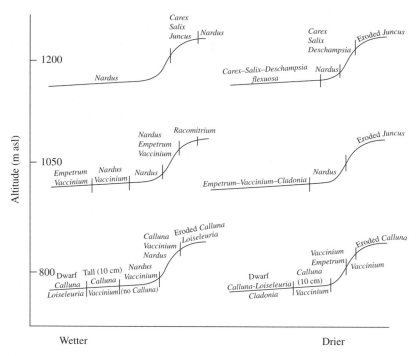

**Fig. 6.4**  The variation of the vegetation with altitude, position along a slope and soil water status in the Cairngorms, Scottish Highlands (from Watt & Jones 1948)

treeline in the North American boreal mountains is formed by *Picea glauca* and *P. mariana*. In the southernmost Andes, *Nothofagus pumilio, N. antarctica* and *N. betuloides* form the treeline at about 600 m asl in Tierra del Fuego (54°S).

The upper limit of the treeline ecotone is at about 1300 m in the southern Scandes (Fig. 6.6), 600 m in the northern Scandes, 400–500 m in the northern Urals, 600–800 m in the northern Siberian ranges and 2000 m in the Sayan and Altai. The treeline ecotone in the Scandes is characterized by scattered *Betula pubescens* ssp. *tortuosa* trees and *B. nana* heaths, willow scrub (*Salix lapponum, S. glauca,* and *S. lanata*) and extensive *Vaccinium myrtillus* heath (Virtanen 2003). In the Urals, a mosaic of trees, *B. nana* heaths, which also occur in the north-eastern and southern Siberian ranges, secondary grasslands, and snowbed patches form the treeline ecotone (Gorchakovsky 1989). There is a notable contrast between the western and eastern slopes in the polar and northern Urals (69–59°N). On the wetter and milder western slopes *Betula* is the main tree species in the treeline ecotone, while in the drier and more continental east scattered

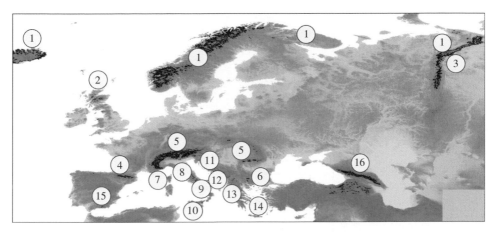

**Fig. 6.5**    Treeline-forming tree species and replacement scrub vegetation in European boreal, temperate and Mediterranean, and Pontic high mountains (after Nagy 2006; Grabherr *et al.* 2003; Ozenda 2002). The map was compiled using public data provided the U.S. National Geophysical Data Center (Global Ecosystems Database and The Global Land One-kilometer Base Elevation (GLOBE) Digital Elevation Model). Alpine areas are indicated in black; the numbers refer to the following species and mountain areas:

1. *Betula pubescens* ssp. *tortuosa* (Iceland, Scandes, Chibiny Mts of the Kola Peninsula, eastern Urals)
2. *Betula pubescens*, *Pinus sylvestris* (Scottish Highlands)
3. *Picea abies* ssp. *obovata*, *Larix sibirica* (western Urals)
4. *Pinus uncinata* (Pyrenees)
5. *Larix decidua*, *Picea abies*, *Pinus cembra*, *P. mugo*, *Alnus viridis* (Alps, Carpathians)
6. *Picea abies*, *Pinus mugo* (Rila)
7. *Alnus viridis* ssp. *suaveolens* (Corsican mountains)
8. *Fagus sylvatica*, *Pinus mugo* (central Apennines)
9. *Pinus leucodermis*, *Juniperus communis* ssp. *hemisphaerica* (southern Apennines)
10. *Genista aetnensis* [*Abies nebrodensis*] (Mt Etna)
11. *Pinus mugo* (northern Dinarids)
12. *Pinus peuce*, *P. heldreichii* (southern Dinarids)
13. *Pinus heldreichii* (Hellenids)
14. *Cupressus sempervirens* (Crete)
15. *Juniperus communis* ssp. *hemisphaerica*, *J. sabina*, *Genista baetica* (Sierra Nevada)
16. *Picea orientalis*, *Pinus kochiana* (dry slopes); *Betula litwinowii*, *Acer trautvetteri*, *Sorbus caucasigena* (humid); *Quercus macranthera*, *Q. pontica*, *Fagus orientalis* (Caucasus)

*Larix* is characteristic. In addition to *Salix* spp. and *Betula nana*, the subarctic dwarf-shrubs (*Cassiope tetragona*, *Diapensia lapponica*) are found from the Scandes to the Anadyr in northern Siberia. In the far eastern Siberian ranges *Pinus pumila* forms a scrub in the treeline ecotone (Fig. 6.7). In the Altai tall-herb meadows appear in the treeline ecotone along with dwarf-shrub heath and grasslands (Suslov 1961).

**Table 6.4** Main treeline tree taxa, the mountain regions where they form the treeline, types of the particular treeline ecotones (narrow, broad, prostrate woody plant formations), and human impact (historic fire, cleared forests, historic and recent pasturing, special recent impacts)

| Region | Important genera or families | Type of treeline ecotone | Human impact |
|---|---|---|---|
| Northern Hemisphere (Holarctic realm) | Gymnosperms: *Pinus, Picea, Abies, Larix, Juniperus, Tsuga, Chamaecyparis* Angiosperms: *Betula, Alnus, Fagus, Quercus* | Relatively narrow: Alps, Pyrenees, Scandes; Broad, patchy: Rocky Mountains–*Krummholz*\* type, Prostrate woody formations:[†] Alps, Dinarids, Urals, Russian far eastern mountains, Japan (Taisetsusan) | Pristine or very natural: e.g. Rocky Mountains, Urals, mountains of Japan Moderately altered: e.g. Alps; Altered—almost treeless: mountains of Tibet |
| Equatorial (Neotropical, Paleotropical realm) | Gymnosperms: *Podocarpus* Angiosperms: Ericaceae (*Philippia, Rhododendron, Vaccinium*) Myrtaceae, Asteraceae, *Quercus* (evergreen; Central America) | Broad, gradual change or patchy: giant rosette formations; no *Krummholz* or prostrate woody plants formations | Pristine to moderately altered: East Africa, Mount Kinabalu, mountains of New Guinea, equatorial Andes Altered: Highlands of Ethiopia |
| Southern Hemisphere (Australian, Antarctic realm, Capensis) | | | |
| Andes | Angiosperms: *Polylepis*, deciduous *Nothofagus* (south of 36°), *Alnus* | Narrow: Patagonian Andes Broad; patchy: Andes north of 36°: isolated *Polylepis* up to >5000m; | Pristine to moderately altered: Patagonian Andes altered: Altiplano—almost treeless |
| Australia | Angiosperms: *Eucalyptus niphophila* | Broad, patchy; *Krummholz* formations | Very natural; pasturing in historic times; affected by fires |
| New Zealand | Gymnosperms: *Podocarpus, Dacrydium* Angiosperms: evergreen *Nothofagus*, Myrtaceae, Malvaceae | Very narrow: *Nothofagus* treeline; broad, gradual: *Nothofagus* gap in the west | Pristine to natural; in some regions severely impacted by introduced mammals (opossums, red deer) |

| Islands and volcanoes (examples) | | | |
|---|---|---|---|
| Mount Haleakala (Hawaii) | Angiosperms: *Metrosideros* (Myrtaceae), *Sophora* (Fabaceae) | Broad, patchy | Natural to moderately altered |
| Yushan (Taiwan) | Gymnosperms: *Abies, Juniperus*; Angiosperms: *Rhododendron* | Narrow | Natural (influenced by fire) |
| Mount Teide (Tenerife) | *Pinus canariensis* | Relatively narrow | Natural |
| Mount Pico (Azores) | *Erica acorica* | Broad, gradual | Altered |
| Mount Fuji (Japan) | Gymnosperms: *Abies, Larix*; Angiosperms: *Betula, Alnus* (prostrate) | Broad, gradual | Natural |
| Levka Ori (Crete) | Gymnosperms: *Pinus, Cupressus*; Angiosperms: *Acer, Quercus* | Relatively narrow | Moderately altered to altered |
| Mount Etna (Sicily) | *Genista aethnensis* (Fabaceae) | Broad, gradual to patchy | Natural; fires by volcanic activities |

*Krummholz: wind- and snow-shaped groups of gnarled trees (often clonal)—the prostrate form not genetically fixed (e.g. *Picea engelmannii* in the Rocky Mountains).

†Prostrate woody formations: dense shrublands of genetically fixed prostrate woody—often clonal—plants (*Pinus mugo, P. pumilio, Alnus viridis* s.l.) (modified after Holtmeier 1989).

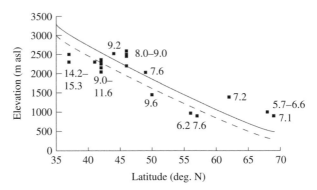

**Fig. 6.6**     Generalised potential (solid line) and actual (broken line) limits of the European alpine zone. The solid line was fitted to data from Körner & Paulsen (2004) and Ozenda (2002); the broken line was fitted to data supplied by experts (see Nagy 2006). Figures are growing season means, from Körner *et al.* (2003) for sites (indicated by solid squares), selected by experts to be above the potential treeline by about 250 m. Sites from the Scandes (69°N) to the Alps (41°N) show good correspondence between modelled treeline elevation and expert selection of alpine sites at c. 250 m above the potential treeline. Outliers are the two sites in Scotland, where the alpine temperatures were in the range obtained for the Scandinavian sites, however, at much lower elevations, owing to the limited insolation during summer in the prevailing oceanic climate. The site at 50°N is from the Giant Mts, which do not reach the predicted treeline elevation, but locally bear alpine-like vegetation. South of 42°N, the elevation of the alpine zone, appears to have been determined too low by experts (also confirmed by mean growing season temperature data cited in Körner *et al.* 2003).

In eastern North America, the Labrador, Laurentian, and Newfoundland mountain ranges, which were formerly all heavily glaciated, form a transition from the Canadian arctic mountains to the temperate northern Appalachians. Apart from their major treeline-forming species (*Pinus banksiana, Picea mariana*), their flora and vegetation have a high similarity with those in the arctic and in the boreal mountains of Asia and Europe, the dominant alpine vegetation being low dwarf-shrub heath. The western North American boreal ranges extend from Alaska to the Canadian Rockies. In its northernmost position, the Brooks Range (approximately 68–69°N) is prominent in having a treeline (*Picea mariana, P. glauca*, Figs 6.8–6.10) at about 700 m on the north-facing slopes and at about 1000 m on those facing south. (The treeline at corresponding latitudes near Abisko, in the Scandes is about 700 m.) A succession of *Alnus viridis* ssp. *sinuata* and *Salix* spp. scrub, dwarf-shrub heath, closed sedge heath, and dwarf-herb meadow follows from the treeline, until the open subnival vegetation largely disappears above 1400 m.

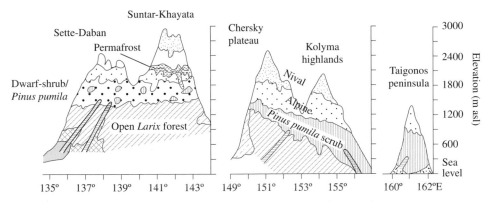

**Fig. 6.7**     The altitudinal zonation of vegetation between longitudes 135° E and 162° E along 60° 30′ N latitude in northern Siberia, Russia (redrawn from Parmuzin 1979, in Shahgedanova *et al.* 2002). Note the appearance and dominance of extensive closed *Pinus pumila* scrub in the Far East.

**Fig. 6.8**     *Picea glauca* treeline with *Alnus* and *Salix* scrub, giving way to snow-protected dwarf-shrub heath and *Dryas octopetala* wind-blasted heath in the Hockley Hills, Selawik, Alaska (c. 66°N, 160°W). Note the trees and shrubs in the sheltered concaves of the landscape and the barren looking ridges and shoulders (convex landforms). (Photo: L. Nagy)

**Fig. 6.9**     Open *Picea glauca* treeline ecotone stand with *Alnus viridis* ssp. *sinuata* and *Salix* spp. scrub understorey and with low dwarf-shrub heath (*Arctostaphylos alpinus*, *Betula nana*, *Vaccinium uliginosum*, *Cassiope tetragona*, *Ledum palustre*) in a sheltered valley strongly contrasts with the wind-blasted hillside in the foreground (*Dryas octopetala* heath). (Photo: L. Nagy).

In Tierra del Fuego (54°S), *Nothofagus antarctica Krummholz* [crook-stemmed scrub] (Table 6.4) interdigitates with *Berberis buxifolia* dwarf-shrub and alpine grassland (Mark *et al.* 2001).

## 6.3.2 The boreal alpine zone proper

A growing season length of 106–144 days approximately 250 m above the treeline in the Scandes (62–69°N) in 1999–2000, with a mean soil temperature 10 cm below the surface of 5.7–7.2°C, and with the mean of the warmest months ranging from 7.3 to 9.3°C across four sites (see Körner *et al.* 2003 for details) places the boreal alpine sites of the Scandes range at the colder end of the alpine spectrum in Europe (Fig. 6.6). Temperature values are similar for Scotland; however, the length of the growing season there is 175–188 days, reflecting the prevailing oceanic climate. The activity of plant species is largely confined to the growing season (Körner 2003), which, in turn, is largely determined by suitably high temperatures in the absence of late-lying snow cover. The combination of snow cover (a topographically determined factor) and thawing degree-days, for example, have been related to the onset of flowering in 144 species studied in a

**Fig. 6.10**    Struggle for survival—local site conditions had provided shelter for the seedling and early growth stages of this *Picea glauca* tree in a small patch of snow protected dry heath, however, since it had grown above the snow protection its leader has died back several times. (Photo: L. Nagy).

subarctic–boreal alpine area in the Scandes, Sweden (Molau *et al.* 2005). The authors have found a large degree of conservatism in the sequence of flowering from year to year for most species. In exposed habitats (e.g. cliffs and ridges) plants flower more or less at the same time every year–as a function of temperature alone–as the length of snow cover varies relatively little from year to year (plants are free of snow from mid-May at the study site). In contrast, habitats with snow accumulation may experience large inter-annual variation in snow cover, snow-melt date, and consequently in the timing of flowering. A similar pattern has also been confirmed for temperate alpine snowbed sites (Hülber *et al.* 2006).

In boreal mountains dwarf-shrub heaths are the predominant feature of the landscape above the treeline ecotone. In the Scandes, well-drained areas provide habitats for *Empetrum* heaths. In the upper part of the alpine zone, *Empetrum* heaths become fragmented (in the north, *Cassiope tetragona*, an arctic dwarf-shrub species occurs commonly); snowbeds are characterized by dwarf willows (*Salix herbacea* and *S. polaris*), and grass and rush heaths (*Festuca ovina* and *Juncus trifidus*) are typical where snow duration is not limiting (Dierssen 1996, pp. 486–607; Virtanen 2003).

**Table 6.5** Alpine dwarf-shrub heath types in boreal mountains

| Topography | Scotland | Scandes | North Siberia | Alaska |
|---|---|---|---|---|
| Exposed | *Arctostaphylos alpina; Calluna vulgaris; Dryas octopetala; Juniperus communis* ssp. *alpina* | *Arctostaphylos alpina; Dryas octopetala; Empetrum nigrum* ssp. *hermaphroditum; Loiseleuria procumbens* | | *Arctostaphylos alpina; Dryas octopetala; Salix* spp.; *Vaccinium uliginosum* |
| Snow protected | *Empetrum nigrum* ssp. *hermaphroditum; Vaccinium myrtillus* | *Cassiope hypnoides; Phyllodoce caerulea; Salix* spp.; *Vaccinium myrtillus* | *Cassiope tetragona; Diapensia lapponica; Salix* spp. | *Cassiope tetragona; Salix* spp. |

Data from McVean and Ratcliffe (1962); Viereck *et al.* (1992); Påhlsson (1994).

Largely the same dwarf-shrub species occur from the Scandes across to the Far East and in the Alaskan ranges (Table 6.5). In the Alaskan ranges, *Dryas octopetala–Selaginella sibirica* dwarf-shrub dominates exposed habitats and in snow-protected alpine habitats mostly ericaceous dwarf-shrub heath is found (*Hierochloë alpina–Arctostaphylos alpina, Diapensia lapponica–Cassiope tetragona* types) (Viereck *et al.* 1992).

In Tierra del Fuego, the altitude sequence of alpine vegetation is of a cushion heath, or grassy cushion heath on well-drained lower slopes with snow-protected grasslands in snowbeds, followed by a high alpine type, lichen-rich cushion vegetation on the sparsely vegetated upper slopes (Mark *et al.* 2001).

Fell-fields with few scattered vascular plants, such as *Luzula arcuata* ssp. *confusa* and *Ranunculus glacialis* and stone sorting, causing polygon fields are common in the Scandes. Snow lies until July–August, or may stay year round in patches. Lichen-dominated block fields in exposed locations form the subnival zone in the high Urals and the Siberian mountain ranges (Fig. 6.11).

Many landscape elements in the alpine zone of the boreal mountains bear vegetation similar to that found in the arctic tundras. For example, the barrens of the cold arctic are similar to the *goltsy* in north Asia, the dwarf-shrub heaths are a dominant feature in the boreal alpine zone, and the dominant mires of the mild polar regions are often found in boreal mountains on flat topography. For all the physiognomic similarities of vegetation the underlying causes sometimes can be very different. For example, arctic mild-polar mires lie over permafrost and have a vegetative period of about 5 months. Their vegetation is similar to that found on high-altitude ombrogenic peat bogs further south, such as the *Vaccinium–Empetrum* type in the oceanic Scottish Highlands between approximately 600–900 m, or in the western Scandes. The dominant dwarf-shrubs include the

**Fig. 6.11**     Lichen-covered boulder fields (*goltsy*) cover large areas in the Urals and the northern Siberian mountains. (Photo: H. Pauli).

dwarf birch (*Betula nana*), crowberry (*Empetrum nigrum* ssp. *hermaphroditum*), *Vaccinium* spp., with *Rubus chamaemorus* and *Eriophorum vaginatum* over a closed moss carpet of *Sphagnum* spp. The high-altitude peat bogs in Scotland are the result of excess precipitation in relation to evapotranspiration and there is no permafrost underlying them.

### 6.3.3  The role of animals in boreal alpine ecosystem processes and dynamics

Reindeer is the only large herbivore and is common in the Polar Urals, but once common, has been absent from other parts of the Urals since the end of the nineteenth century. Apart from domestic grazers that largely include cattle and sheep, semi-domestic reindeer or wild caribou in North America are the main herbivores that mostly impact vegetation, locally or at the landscape-scale. The impacts are many-fold: at low to intermediate levels grazing and associated trampling causes increased species richness compared with areas not grazed. However, at a high level, foraging and trampling at first cause changes in vegetation (growth forms and species richness, whereby sensitive species, such as dwarf-shrubs are lost and are partially replaced by a small number of grazing tolerant graminoids). Yet higher levels of herbivores can lead to changes in hydrology, loss of vegetation cover, and eventually soil erosion. An extreme example is Iceland (Fig. 5.3) where it is estimated that half the total original vegetation cover on about 65% of the total land area (that

**Table 6.6** The impacts of herbivores on boreal alpine vegetation in the Scandes

| Habitat/vegetation type | Herbivore | Reference |
|---|---|---|
| *Betula pubescens* ssp. *tortuosa* Krummholz | Reindeer: browsing of shoots, checking regeneration | Lehtonen and Heikkinen (1995) |
| | Autumnal moth (*Epirrita autumnalis*) outbreaks: defoliation, patchy or extensive dieback of trees | Kallio and Lehtonen (1975) |
| Snowbed | Norwegian lemming (*Lemmus lemmus*): outbreaks in approx. 30-year cycles causing periodic removal of mosses | Henttonen and Kaikusalo (1993) |
| Dwarf-shrub heath | Grey-sided vole (*Clethrionomys rufocanus*) and other voles: outbreaks every 4–5 years preferentially grazing *Empetrum* cover | Emanuelsson (1984) |
| Lichen heath | Reindeer: annually recurring winter grazing on foliose lichens | |

of the alpine zone not quantified separately) has been lost during the history of human occupation of about 1100 years (Arnalds 1987).

The cyclic nature of rodent populations in northern European boreal mountains has a characteristic impact on vegetation. For example, in the Scandes, in peak years lemmings invade snowbeds and graze on bryophytes and graminoids. It has been demonstrated that in the absence of recurrent grazing, snowbeds would be dominated by bryophytes and would support a much higher biomass (approximately threefold more *Polytrichum* mosses and up to sixfold for more graminoids; Virtanen 2000). Besides lemmings, there are a number of other rodents whose overall impact depends on their combined population sizes. The impact of an individual species depends on its hierarchy in relation to the other rodents present and on the landscape structure (Table 6.6). Generally, larger and stronger species dominate smaller ones; however, the habitat range of a dominant species is usually narrower than that of an inferior one, which enables the survival of the latter one. Abrupt cyclic changes in densities highly influence the degree of competition; in addition, the peaks vary in magnitude. In the Scandes, *Lemmus lemmus* and *Microtus oeconomus* are largely superior to *M. agrestis*, while *Clethrionomys* species (*C. rufocanus*, *C. glareolus*, and *C. rutilus*) are inferior to *M. agrestis* (Henttonen *et al.* 1977). None the less, *C. rufocanus* is known to periodically drastically reduce the cover of *Empetrum nigrum* ssp. *hermaphroditum* and thereby initiate the expansion of other dwarf-shrubs such as *Vaccinium myrtillus* and *V. vitis-idaea*. Another cyclic event is the mass defoliation of *Betula pubescens* ssp. *tortuosa* by the autumnal moth (*Epirrita autumnata*). The effects of such defoliations can be dramatic and result in long-term vegetation change over large areas. For example, the 1964–1966 outbreak transformed about 5000 km$^2$ of upper montane birch forest/scrub into alpine heath (Kallio

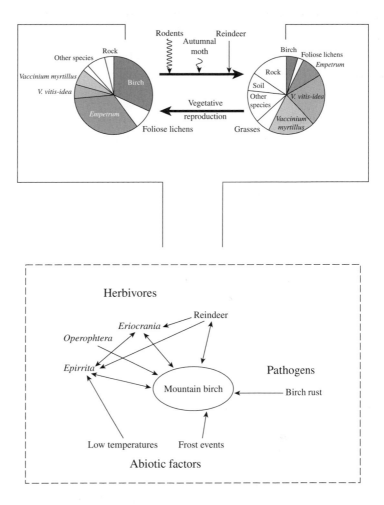

**Fig. 6.12**     Alpine vegetation dynamics in the boreal Scandes mountains (modified from Väisänen 1998; original sources: Callaghan & Emanuelsson (1985), top; Tenow (1996), bottom). Top panel, conceptual model of the effect of grazing on alpine heath vegetation in Swedish Lapland. Vegetative growth increases the cover of the dwarf-shrub *Empetrum nigrum* ssp. *hermaphroditum*, foliose and racemose lichens (*Cladonia*) and that of the mountain birch *Betula pubescens* ssp. *tortuosa*. Herbivore impacts are cyclical (rodents), incidental recurring (autumnal moth), or relatively stable (reindeer). Rodents peak every 4 to 5 years—their winter grazing decreases the cover of *Empetrum* and the faster growing *Vaccinium vitis-idaea* and *V. myrtillus* increase at the expense of *Empetrum*. The reindeer consume largely foliose lichens and through trampling it may increase bare ground. Larvae of the autumnal moth defoliate the mountain birch about every 10 years and break up birch stands. See also Lindgren (2007).

Bottom panel, details of the interactions among herbivores and between herbivores and mountain birch at Abisko, Swedish Lapland. Extreme winter cold kills the eggs of *Epirrita*; arrows from reindeer to insects indicate unintentional predation of insects by reindeer that feed on sprouts.

and Lehtonen 1975). The overall dynamics of herbivores and vegetation is shown in Fig. 6.12 after Tenow (1996) and Väisänen (1998).

Burrowing animals such as marmots (*Marmota* spp.) and ground squirrels (*Spermophilus* spp.) in North America and Siberia have direct local impact by their earth moving, feeding, and manuring within their range. In addition, in the spring ground squirrels attract brown bears, which dig the burrows up, causing large bare areas devoid of vegetation.

Predators, such as red fox, brown bear, common stoat (*Mustela erminea*) occur in small numbers throughout the Urals; wolverine occurs in the Polar Urals the north Siberian ranges and in Alaska. Several species of shrews (*Sorex* spp.) are also found.

The use of different habitats by breeding birds in the alpine zone of the Scottish Highlands during incubation and chick rearing has been shown by Thompson *et al.* (2003). Densities of birds were low on mires (skylark *Alauda arvensis*; golden plover *Pluvialis apricaria*; dunlin *Calidris alpina*) and *Nardus*-dominated snowbeds (dunlin for nesting), and high on dwarf-shrub heath (meadow pipit *Anthus partensis*), sedge and moss heaths (dotterel *Charadrius morinellus*). This was specific to species, and to their food preferences; e.g. ptarmigan had no preferences for nesting, but occupied dwarf-shrub heath during chick rearing. Waders (dotterel, dunlin, golden plovers) had marked habitat preferences, related to vegetation structure and nest site requirements, and some habitat partitioning: dotterel nested on and used *Racomitrium lanuginosum* dominated moss heath as a feeding ground. Snow buntings and wheatears are associated with scree and rock strewn habitats. Such differential use of habitats by insectivorous birds is related to their food, such as the presence and density of certain invertebrates (in this case the crane fly *Tipula montana*).

Soil invertebrates of the alpine zone in the Polar and northern Urals include protozoa, nematodes, rotifers, earthworms, tardigrades, and chilopods (for full details see Mikhailov and Olschwang 2003). Soil invertebrates are concentrated under moss and low-stature grassy vegetation in the top 5-cm soil and their biomass is dominated by the earthworm *Eisenia nordenskioldi* (Lumbricidae), found in the alpine zone throughout the Urals. It can contribute over 90% to the biomass in the Polar Urals (4–7 g m$^{-2}$), or 85–95% in northern Urals (6–30 g m$^{-2}$). The rest is made up by 10–15 species of ground-dwelling arthropods: crane fly larvae in the Polar Urals (150 individual m$^{-2}$, about 0.5 g m$^{-2}$) are complemented by ground beetles and millipedes in the northern Urals (up to a total of 400 m$^{-2}$, 0.5–2.0 g m$^{-2}$). Diptera and Lepidoptera make up about half the species of the insect fauna, which interestingly is about 2.5 times more taxon-rich in the Polar than in the northern Urals.

**Fig. 6.13**   North Cirque, Mt Poroshiri, Hokkaido (42° 43′ N, 142° 41′ E; 2052 m; 25 July 1971), the highest mountain in the Hidaka Range of Hokkaido. In this area tree limit is at c. 1650 m asl, above which *Pinus pumila* scrub grows, interspersed with scree and alpine heath. The Cirque has numerous tarns and is a favoured habitat of the brown bear (*Ursus arctos* ssp. *yesoensis*). Insert: Far Eastern high mountain ranges: Changbaishan (1), Sikhote-Alin (2), and the Japanese mountains of Honshu and Hokkaido. (Photo: M. Ohsawa).

## 6.4 Temperate mountains: sedge heaths

The most distinguishing feature of temperate alpine mountains is the expanse of sedge heaths that defines the physiognomy of the vegetation of the large alpine massifs. Northern temperate mountains form a large and somewhat heterogeneous group that lie between the arctic-boreal and

Mediterranean zones in Europe, in addition, they include the Caucasus and the north Anatolian Mountains (Fig. 6.5), the Far Eastern ranges of the Changbaishan, Sikhote Alin and the Japanese Alps (Fig. 6.13), and the Cascades Range, Rocky Mountains, and the northern Appalachians in North America (Fig. 2.1).

The European mid-latitude temperate high mountains include the Alps, Carpathians, Dinarids, Rila and Pirin of the Rhodopean system and the north Balkan Mountains, and the northern Apennines; the Pyrenees together with the Cantabrian Mountains, central Apennines, and the Corsican massifs are at the interface between temperate and Mediterranean climates. The limestone massifs of the Picos de Europa, the highest part of Cantabrian Mountains, Spain show a north–south duality in their vegetation, the humid northern slopes bearing a moist temperate vegetation, while the heavily insolated southern slopes have a Mediterranean character. This situation repeats itself in other mountain ranges that form climatic barriers, or are at the transition zone between major climate zones, such in the Pyrenees and Corsica and in many other parts of the world, the Himalayas being the most prominent example, on the border between subtropical and temperate climates.

The few examples of temperate mountains in the Southern Hemisphere include the Andes of temperate Chile (north Patagonia) and mainly the South Island Ranges in New Zealand (Fig. 2.1).

## 6.4.1 Temperate treelines

The treeline in mid-latitude European mountains is formed by conifers (*Larix decidua, Picea abies, Pinus cembra*), with scattered broadleaves and abutting *Pinus mugo* scrub (not west of the Alps), or *Alnus viridis* in sheltered locations with late-lying snow (Fig. 6.5). The exceptions are the central Apennines, where *Fagus sylvatica* forms the treeline and where the *Pinus mugo* scrub has largely been lost through human land use, and Corsica, where *Alnus viridis* ssp. *suaveolens* scrub has altogether replaced trees. The Caucasus lies at the intersection of a variety of climates meeting (and is sometimes considered as a subtropical chain), which is expressed by the variation in the elevation of treelines and the species that form them (Fig. 6.5). In addition, its vegetation of the treeline ecotone (scrub, tall-herb meadow, feather grass steppe, and anthropogenic grasslands) is especially varied because of a humidity gradient from the west (Colchis, Western Caucasus) to the dry to arid east (Eastern and Minor Caucasus). An extension of the Southern or Little Caucasus to the west is the north Anatolian (or Pontic) ranges, which have *Picea orientalis* and *Pinus sylvestris* as the highest elevation trees. In the Far Eastern Changbaishan, separating south-eastern China from northern Korea the treeline-forming species are *Larix olgensis* and *Betula ermannii* at approximately 2000 m (Qian *et al.* 1999); in the Sikhote-Alin *Betula ermannii* and *Pinus pumila* at about 1600 m, although thought be 200–250 m lower than climate would allow (Grishin *et al.* 1996), and in the Japanese Alps *Alnus maximowiczii*, *Betula ermannii*, *Larix leptolepis*, and *Pinus pumila* at approximately 2500 m (e.g. Gansert 2004).

The dotted North Island alpine areas are found above 1500 m at a latitude similar to that in south-eastern Australia. The treeline in New Zealand is formed by evergreen *Nothofagus* species as opposed to *Eucalyptus* in Australia. Most of the alpine areas in the temperate oceanic climate are found on South Island, where, the treeline descends to 900 m in the south. In the temperate Andes in Chile deciduous *Nothofagus* trees (*N. antarctica*, *N. betuloides*, *N. pumilio*) reach elevations of approximately 1900 m at 36°S and the treeline decreases southwards to 500 m at 56°S, in the southern boreal mountains. In North America, in the Cascades, the treeline is at about 2000 m, formed by *Picea engelmannii* and *Abies lasiocarpa*; additionally, species of *Pinus*, *Larix lyallii*, and *Tsuga mertensiana* contribute to the Rocky Mountains treelines, which are at approximately 3400–3600 m on the Niwot Ridge, Colorado Front Range (Walker *et al.* 2001b). In the north Appalachians *Picea rubens*, *Abies balsamea*, *Betula papyrifera*, and *Alnus crispa* give way to the alpine zone at approximately 1300 m.

The treeline ecotone is characterized by dwarf-shrub heaths intermixed with trees in European mountains; ericaceous dwarf shrubs are conspicuously absent in western North America, but well developed in the northern Appalachians. In the Caucasus, a distinctive feature of the treeline ecotone is the large extent of megaforbs (Apiaceae, Asteraceae, Dipsacaceae). The most notorious of its constituent species is the giant hogweed (*Heracleum mantegazzianum*) that has become an invasive weed after its introduction as an ornamental the world over. The treeline ecotone has been heavily influenced by human activities that have led to the development of a largely secondary vegetation of dwarf-shrub heaths and grasslands in many parts in the European and Caucasian mountains (e.g. see Fig. 6.17 below).

## 6.4.2 Temperate alpine habitats

Temperate mountains have a strongly seasonal climate that causes seasonality of growth and related physiological activities, similarly to that in the boreal zone. The range of growing season length at an estimated 250 m above the treeline has been estimated to range from 126 days to 184 (203) days in the European temperate alpine zone, with a growing season mean soil temperature of 7.6–9.2 (11.6) °C (Fig. 6.6; Körner *et al.* 2003). The calculated growing season length in a separate study that measured summit soil temperatures in all four main compass directions (see Pauli *et al.* 2009) decreased from the treeline ecotone (mean approximately 200 days) to about 60 days on the highest summit recorded about 1000 m above the treeline ecotone in the Alps. The corresponding decrease in mean growing season soil temperatures from 9.8°C at the treeline ecotone, was about 2°C, as measured over 1 year (GLORIA, unpublished). Seasonality is fine tuned by local topographic and microclimatic differences: relative to north facing slopes, all other slopes are warmer (mean annual temperature): N < W (+0.5°C); < E (+1.1°C); < S (+1.8°C) (GLORIA unpublished).

European alpine habitats, community types, and species diversity have recently been reviewed (Pyrenees, Gómez *et al.* 2003; Alps, Ozenda and Borel 2003; Carpathians, Coldea 2003; Apennines, Pedrotti and Gafta 2003; Corsica, Gamisans 2003). A general pattern between alpine species richness on the one hand and geological substratum, latitude, and snow gradient on the other, is that: (1) vascular species richness is highest in the temperate mountains, while cryptogams are most numerous in the boreal mountains; (2) calcareous mountains have more species than siliceous ones; and (3) wind-exposed and late snow-lie habitats are less species rich than habitats with an intermediate length of snow cover (Virtanen *et al.* 2003).

Habitat types vary in their extent and proportion they occupy in an alpine landscape. At one extreme, rock and scree can predominate and shelter scant vegetation, confined to crevices and ledges, such as in the Picos de Europa and in the Corsican high ranges, leaving most of the slopes bare. In such cases, the equation of alpine habitat types with the small pockets of vegetation that they bear may appear to be forced. In stark contrast, most massifs of the major high ranges such as the Alps, Pyrenees, or the Carpathians display a visually distinctive zonation of vegetation with elevation, and typical sequences along soil moisture gradients. The dwarf-shrub, tall-herb meadows, and secondary grasslands of the treeline ecotone are replaced by zonal sedge heath, the predominance of which is only interrupted by pockets of azonal vegetation types, such as rock and scree formations, snowbeds, and mires. There is a clear contrast between acid bedrock and limestone landscapes, well illustrated in Figs 5.5 and 5.6. Limestone landscapes are much dissected, be it by karstic features or by the steep and bare summits and abundant scree as, for example, in the Dolomites in the Alps (Stahr and Hartmann 1999).

The Caucasus has a variety of environmental conditions that result in a high diversity of plant communities. Volodicheva (2002) have presented a description of the geography of the Caucasus, including numerous altitude distribution schemas for the vegetation by region, a detailed description of which is available in Russian (Guliashvili *et al.* 1975). The habitats and vegetation of the alpine zone of the central Caucasus (Georgian Greater Caucasus) were recently reviewed by Nakhutsrishvili (2003), and on the northern side by Onipchenko (2005). Alpine vegetation (2400–3000 m) is found almost throughout the Greater and Minor Caucasus. It is dominated by sedge heaths, dwarf-herb meadows, alternating with lichen-rich dwarf-shrub heath in exposed locations, snowbeds and rock and scree vegetation, largely similar to that found in the Alps (Nakhoutsrichvili and Ozenda 1998); above 3000 m the open patchy vegetation of the subnival zone is found up to 3750 m. The alpine vegetation of north-east Anatolia is poorly known.

In central Japan, the alpine zone occurs above 2500 m. One characteristic that shapes habitats and their distribution in the mountains of Japan is the contrasting climate between the continental and the Pacific side of the mountains. This results in differences in depth and duration of snow cover, the number of days with clear versus wet days, sunshine hours, and prevailing wind directions (Yoshino 1978). The main alpine habitats and their vegetation (Table 6.7) bear similarities to those found in the boreal and temperate regions of Eurasia (Ishizuka 1974; Holzner and Huebl 1988).

In North America, the climatic gradient (oceanic to continental) from the west coast to inland results in a high contrast in the alpine vegetation in the North Cascades Range. Most of the inland community types in the east are similar to those found in the Rocky Mountains, southern Alaska and the southern Yukon, whereas communities in the western North Cascades are more similar to communities in other west coast ranges (Douglas and Bliss 1977). In the Colorado Rockies, the principal habitat types are similar to those in the European temperate mountains; however, plant community types differ in that ericaceous dwarf-shrub types, characteristic of the altitude sequence of arctic and boreal mountains, and to a lesser extent of the European temperate ones, are not represented in the Colorado Rocky Mountains where large expanses of the landscape is dominated by bare rock and scree (Komárková and Webber 1978). The best studied part of the Rocky Mountains, the Niwot Ridge in the Indian Peaks area is a local exception in that it has a range of zonal plant community types in habitats arranged along a topo-hydro gradient. They include extremely wind-exposed open wind-clipped heath or fell-field (*Dryas octopetala, Paronychia pulvinalis, Carex rupestris*), wind-blown dry sedge heath (*Kobresia myosuroides, Carex rupestris*), mesic grassland in early melting snowbeds (*Acomastylis rossii, Deschampsia caespitosa*), late (*Sibbaldia procumbens, Carex pyrenaica*) to very late snowbed (*Polytrichum alpinum, Pohlia obtusifolia*), soligenous mires on slope bottoms (*Carex scopulorum, Caltha leptosepala*), and low willow scrub (*Salix villosa, S. planifolia*) (Fig. 6.14). Additionally, azonal habitats such as rock crevices, boulder fields, scree, and springs and rills complement zonal habitats (Walker *et al.* 2001b). The northern Appalachian Mountains have traditionally been classified as temperate; however, their alpine vegetation is rather more akin to that found in boreal and arctic alpine mountains than to temperate ones, with the ericaceous dwarf-shrubs being dominant. Exposed and snow-protected heath and fell-field with open cryptogamic vegetation dominates the landscape

**Table 6.7** The main habitats / vegetation types and their characteristic species above the treeline in the Japanese alpine mountains. Source: M. Ohsawa pers. comm., Ishizuka (1974).

| Habitat / community type | Typical species |
|---|---|
| Open cryptogamic communities | *Cladonia* spp. |
| Rock communities | *Sedum japonicum* var. *senanense, Saxifraga cherlerioides* var. *rebunshirensis, Campanula lasiocarpa, Rhododendron tschonoskii* var. *tetramerum* |
| Scree communities | *Deschampsia flexuosa, Carex stenantha* var. *stenantha, Cardamine nipponica, Oxyria digyna, Aconogonum weyrichii* var. *alpinum, Dicentra peregrine, Lagotis glauca* |
| Snowbed communities | *Moliniopsis japonica, Carex blepharicarpa, Fauria crista-galli, Primula nipponica* |
| Mild snowbeds/snow-protected grasslands | *Phyllodoce aleutica, Geum pentapetalum* |
| Late snowbeds | *Carex pyrenaica*, bryophytes |
| Mires | |
| Topogenous | *Moliniopsis japonica, Fauria crista-galli* |
| Ombrogenous | *Sphagnum* spp., *Rhynochospora yasudana* |
| Springs and rills | *Juncus beringensis, Deshampsia caespitosa* var. *festucaefolia, Epilobium foncaudianum* |
| Tall-herb meadows | *Trollius riederianus* var. *japonicus, Trautvetteria japonica* |
| Grass, sedge and moss heaths | Wind-exposed: *Kobresia – Oxytropus japonica* Snow-protected: *Calamagrostis* spp. |
| Dwarf-shrub heath | *Empetrum nigrum* var. *japonicum, Arcteria nana, Loiseleuria procumbens, Vaccinium uliginosum, Diapensia lapponica* var. *obovata, Geum pentapetalum* |
| Scrub | *Pinus pumila, Alnus maximowiczii, Sorbus commixta, Salix reinii, Acer tschonoskii, Sasa kurilensis, Juniperus communis* var. *montana, Ilex sugeroki* var. *brevipedunculata* |
| Treeline ecotone | *Abies veitchii, A. mariesii, Betula ermanii, Larix leptolepis, Pinus pumila* |

and *Carex bigelowii* heath accounts for about 5% cover (Kimball and Weihbrauch 2000).

The southern latitude alpine vegetation is rather different from that of the Holarctic one. Comparisons have indicated potential common origin of southern latitude floras (e.g. Wardle *et al.* 2001). In the temperate Andes in north-west Patagonia (41°S), Ferreya *et al.* (1998) differentiated four broad zonal alpine vegetation types in three mountains along a precipitation gradient from west to the east. Based on their environmental preferences these were: (1) warm-loving *Gaultheria pumila–Senecio argyreus–Perezia*

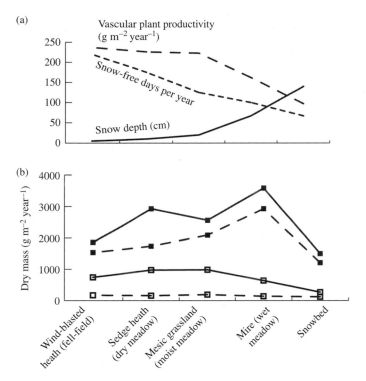

**Fig. 6.14**  Annual mean productivity (a) and phytomass (b) of alpine vegetation in habitats along a snow-duration sequence. Solid squares in (b) refer to below-ground, empty squares to above-ground total (solid line) and live (broken line) phytomass. (Data from Walker *et al.* (1994) and May & Webber (1982))

*bellidifolia* community on the warmest north- and west-facing slopes, immediately above the treeline ecotone (mainly in the arid east); (2) *Gaultheria pumila–Baccharis magellanica* community at the same altitude but on easterly slopes (mainly in the humid west); the highest elevations were occupied by (3) *Senecio portalesianus* (west) and (4) *Nassauvia pygmaea–N. revoluta* (east) (also found at lower altitudes in cold south-facing slopes or in areas of late-lying snow).

In New Zealand, the zonal vegetation in the lower part of the alpine zone is tussock grassland (with dwarf-shrubs in the treeline ecotone), dominated by a number of *Chionochloa* species, a grass genus largely confined to New Zealand (Mark and Adams 1995; Mark and Dickinson 1997). The tall tussocks form a series of grasslands above the treeline and the spatial distribution patterns of dominant individual species have been related to parent material (*C. rubra* occurs on volcanic soils on North Island), land form and soils, altitude and topography related snow-lie. Ombrotrophic mires (bogs) developed in depressions or flat surfaces in the grassland zone are dominated by cushion plants (e.g. *Celmisia argentea–Coprosma perpusilla*) and *Sphagnum* species. Screes represent another azonal habitat type, with their characteristic flora, crossing the tussock grasslands from the upper mountain slopes into the montane forest zone. Above the *Chionochloa* tussock zone the vegetation is of an open nature. Stable block fields and fell-fields may represent zonal habitats with their unmistakable character given by the extensive cushions of *Raoulia* spp. and *Haastia* spp. in the dry eastern mountains that lie in a rain shadow. In the oceanic west, such habitats are usually more species-rich and have a higher cover of herbs, in addition to cushions. At the extremes of the snow gradient lie prostrate dwarf-shrub (e.g. *Dracophyllum muscoides*) and cushion-dominated vegetation in wind-beaten exposed plateau summits, and snowbed vegetation in locally favourable locations for snow accumulation and late melt. Species of *Celmisia* characterize communities of different snow duration. The role of snow protection versus frost resistance in exposed habitats as a means of adaptation and the potential cause of plant distribution has been verified by Bannister *et al.* (2005), who found that the overall frost resistance of the principal species of snowbed habitats (*Celmisia haastii* and *C. prorepens*) was lower when compared with species from exposed habitats (*Dracophyllum muscoides*, *Poa colensoi*, *Celmisia viscosa*).

### 6.4.3 The functioning of temperate alpine ecosystems

Climate links the elements of the environment and alpine habitats, and their assemblages to ecosystem processes from plot to region scales. As has been emphasized throughout, geomorphology- and wind-determined snow distribution along elevation is the primary cause that differentiates alpine habitats with respect to growing season length and hydrology, and in turn ecosystem structure and processes. This works both at the landscape scale (Fig. 6.14) and at the scale of a few metres (Fig. 6.15). However, while climate is the main cause of differentiation, it can also be the source of uniformity at the landscape-scale. For example, in the Colorado Rocky Mountains prevailing westerly winds are so influential that they have an overriding effect that diminishes differences among habitats on the west side of the Continental Divide and causes a high degree of uniformity (Walker et al. 2001b).

The importance of snow cover and snow lie as habitat-creating factors and as drivers of alpine biological diversity and phytomass and below-ground processes is clearly linked to land forms and topographic sequences (Fig. 6.14). It is important to recognize that the observed snow gradients, while correlated with topography, are by far from being determined by it alone. It is the redistribution of snow by wind year after year that leads to the (re)distribution of certain species in the long-term. This applies to small stature plants (Fig. 6.15), as well as to some scrub species, such as *Alnus* spp. and *Rhododendron* spp., which are chionophilous while others, e.g. *Pinus pumila* are controlled by deep snow (Okitsu and Ito 1984). The

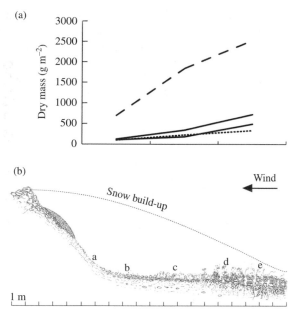

**Fig. 6.15**  Small-scale impacts of wind-blown snow accumulation on vegetation structure and phytomass (Pümpel 1977) in the European Alps: (a) Growing season length difference along the snow gradient causes a large reduction in phytomass (below-ground, broken line; above-ground, top unbroken line), biomass (lower unbroken line), and litter (dotted line); (b) Winter snow build-up in a north-east facing position. The resulting gradient in snow melt from (a) to (e) is c. 1 month and causes the co-existence of a range of snowbed types and sedge-heath within a short distance: (a) *Polytrichum sexangulare* late snowbed; (b)–(d) three varieties of *Salix herbacea* snowbed, (b) typical, (c) with *Omalotheca supina*, and (d) with *Ligusticum mutellina*; and (e) *Carex curvula* sedge-heath. Modified after Braun-Blanquet (1951).

result of the existence of such snow gradients is that (1) protection from frost by snow cover will differ with snow depth, and (2) snow-melt patterns may much deviate from what could be expected from solar irradiation in different slope exposures. As a further consequence, soil water conditions, and in turn, soil mineralization process will be affected. All these factors have led to the development and maintenance of contrasting seasonal alpine habitats and plant community types.

Alpine vegetation types contribute to stabilizing their habitats and their biological functioning by fixing soil and influencing hydrology and geochemistry. Hydrology has been highlighted as particularly important in areas where tall bunch grasses intercept fog and aerosols, such as has been suggested in New Zealand (Ingraham and Mark 2000), where *Chionochloa* species grow up to a height of 0.8–1.2 m and their above-ground phytomass can reach over 8000 g m$^{-2}$. It is debated, however, if it is fog interception or low transpiration by *Chionochloa* that leads to a positive hydrological balance (Davie *et al.* 2006). The soil retention by the enmeshing of the soil by plant roots and the ecosystem 'healing' afforded by species whose roots are capable of rapid elongation (e.g. *Agrostis schraderiana* in the European Alps) are fundamental to the maintenance of alpine ecosystems.

Productivity, can much vary among habitats (Fig. 6.16), as well as among years (e.g. Walker *et al.* 1994); however, in general the mean is in the region of 100–300 g m$^{-2}$ year$^{-1}$ for above-ground parts, reaching over 500 g m$^{-2}$ year$^{-1}$ in tussock grasses. Standing biomass and litter also show much variation among habitats; however, overall, the range is comparable with that described from the Arctic (see Table 6.3, Figs 6.14 and 6.15, 6.17 and 6.18). On the other hand, structure, related to life forms and species composition, is rather conservative. The best studied vegetation types are sedge heaths and some dwarf-shrub heaths of the European Alps (Larcher 1977; Grabherr *et al.* 1978; Grabherr 1997), the main vegetation types of the Niwot Ridge, Colorado Rockies (Bowman and Seastedt 2001), and the *Chionochloa* tussock grasslands, cushion and herb meadows of the New Zealand Alps (Mark *et al.* 1997). *Loiseleuria procumbens*, a prostrate

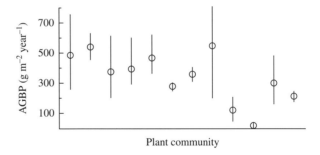

*Bromus erecta* (1600–1900 m)
*Bromus - Nardus stricta*
*Nardus stricta* (1800–2200 m)
*Rumex pseudoalpinus* (1600–2000 m)
*Festuca paniculata* (1800–2100 m)
*Cynosurus cristatus* (1600–1900 m)
*Carex nigra* (1800–2200)
*Festuca eskia* (1800–2300 m)
*Festuca gauteri* (1800–2300 m)
*Iberidion spathulata* (1800–2400 m)
*Primula intricata* (2000–2500 m)
*Kobresia myosuroides* (2200–2700 m)

**Fig. 6.16**   Mean (circle) and range (bar) of above-ground biomass production (AGBP) values in anthropogenic and natural upper montane and alpine plant communities in the Pyrenees (D. Gómez, pers. comm.; García-González and Marinas (2008)). The values are means of 2–37 samples per each community). The treeline is at 2200–2400 m. The list of plant communities and their altitude ranges correspond to values from left to right in the graph.

dwarf-shrub of exposed ridges, already cited for its high rates of loss of stored lipids, is also remarkable in that its root to shoot ratio is about unity, in sharp contrast to that found in other, more sheltered dwarf-shrub heaths (Fig. 6.17). The below-ground/above-ground ratio of phyto-mass is between approximately 2.5 for live and 4 for total phytomass in the sheltered *Vaccinium* heath and both become about 1 in the exposed *Loiseleuria* heath. Interestingly, the ratios found in New Zealand commu-nity types, ranging from 0.3 in *Chionochloa* tussocks to 0.8 in high alpine short herb meadows do not support the often-promoted adaptation by a below-ground storage mechanism in alpine vascular plants. It seems that root to shoot ratios of ≥1 occur in Holarctic alpine areas, as distinct from Southern Hemisphere alpines, and it may be of interest to test if and how the origin of floras (presumed Asiatic in the Holarctic, and of Antarctic in the Southern Hemisphere) bears on adaptation to alpine conditions. The vertical structure of the biomass/necromass distribution in the zonal sedge heath and associated early and late snowbeds of the European Alps (Fig. 6.18) all appear to be concentrated near the ground surface. The noticeably high cryptogam component in the *Carex curvula* sedge heath is held as an example of structural facilitation afforded by the sedge to the mainly lichen-dominated cryptogams.

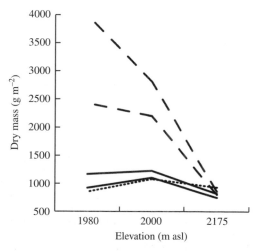

**Fig. 6.17**     The phytomass of three dwarf-shrub communities (*Vaccinium*, *Loiseleuria*, *Loiseleuria procumbens—Alectoria ochroleuca*) along an exposure gradient with altitude in the European Alps: total above-ground phytomass (upper solid line), above-ground bio-mass (lower solid line), litter (dotted line), total under-ground phytomass (upper broken line), under-ground biomass (lower broken line). Data from Larcher (1977).

The decomposition of litter (accession to the organic matter component of the soil) can much vary, depending on plant species and plant part, and rates can be as little as 15% (e.g. the roots of sedges) in a year and may reach approximately 50% in leaves. In general, the warmer the environment the more rapid the rate of decomposition. However, this is relative in the alpine. In an alpine landscape, sites that are warmest in the summer (insolated and dry) may readily become the coldest in winter (wind-blown with no snow cover and frozen). As with soil moisture alone, both lack and excess of water can retard decomposition and therefore, as the warmest sites are the driest, decomposition will peak at intermediate moisture and temperature combinations (e.g. see Seastedt *et al.* 2001). In addition to the abiotic environment, the availability and activity of decomposers is crucial for the functioning of the plant–soil complex.

The below-ground processes that link plant growth with the above-ground habitat conditions have been extensively studied in North American alpine ecosystems (for a synthesis see Schmidt *et al.* 2007). Decomposition of litter and mineralization of organic matter both follow major seasonal dynamics, accompanied by turnover of microbial communities (Fig. 6.19). It appears that generations and successional suites of microbes keep a tight hold of soil nitrogen. Leakage of dissolved mineral and organic nitrogen occur at each turnover. The most remarkable of these leakages is that which follows the crash of cold-adapted microbes after snowmelt. Very few plants are able to benefit from the available nitrogen at that time, which is at the very beginning of the growing season. Later in the season, the turnover of growing season microbial cycles provides sufficient leaked dissolved nitrogen for vascular plant uptake. This latter, has been estimated to amount to no more than 5% of the total microbial nitrogen (Fisk *et al.* 1998). There is a seasonality in the activity of the microbes that can be measured by enzyme assays. For example, there is a peak decomposition of plant residue under winter snow (cellulase activity), while the released organic nitrogen from the collapse of the winter microbes is followed by an intense protease activity, signalling a new set of microbes in action from the beginning of the vascular plant growing season (see Schmidt *et al.* 2007 for details). It is not known, however, how other, non-seasonal systems (aseasonal tropical, or seasonal, but not snow-impacted tropical-subtropical alpine) work in terms of nutrient dynamics.

## 6.5.4 Animals and their impact on the functioning of temperate alpine ecosystems

Alpine habitats can support a rich animal life, such as has been reported by Reichel (1986), who identified 47 species of mammals from alpine areas of the Pacific Northwest, 12 of which were resident, some habitat specialists, while others occurred in a variety of habitats (Table 6.8). Of the habitat

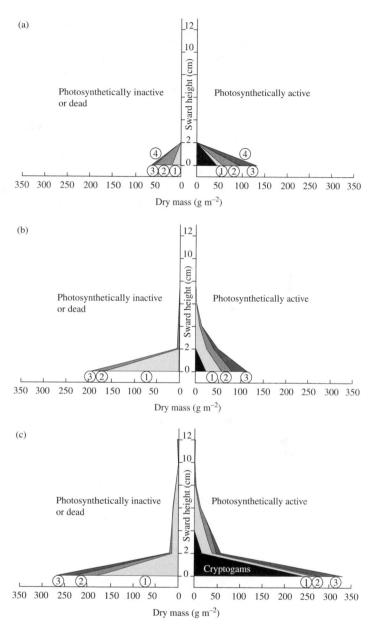

**Fig. 6.18**    The vertical distribution of phytomass in a *Salix herbacea–Omalotheca supina* snowbed (a), *Deschampsia caespitosa* early snowbed grassland (b), and *Carex curvula* sedge heath (c) at 2-cm height intervals at 2300 m asl in the Austrian Alps, August 1976. The contributing fractions in the snowbed: 1, *Omalotheca*; 2, *Salix*; 3, herbs; 4, grasses; in the *Deschampsia* grassland: 1, *Deschampsia flexuosa*; 2, other grasses; 3, herbs; in the *Carex* heath: 1, *Carex curvula*; 2, grasses; and 3, herbs. All cryptogams were deemed as live and fully contributing to the assimilating fraction. Data from (Pümpel 1977).

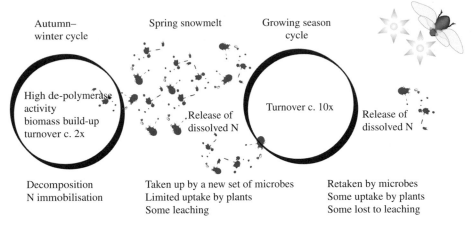

Autumn–winter cycle | Spring snowmelt | Growing season cycle

High de-polymerase activity biomass build-up turnover c. 2x | Release of dissolved N | Turnover c. 10x | Release of dissolved N

Decomposition
N immobilisation

Taken up by a new set of microbes
Limited uptake by plants
Some leaching

Retaken by microbes
Some uptake by plants
Some lost to leaching

**Fig. 6.19**    A conceptual model, adapted from Schmidt *et al.* (2007), of the dynamics of within-year succession of suites of microbes and microbial N cycles. Under snow, there is a slow turnover and a large build-up of psychrophilic microbial biomass (with all N locked up), primarily concerned with decomposition of litter. On snowmelt, the population of this suite of microbes crashes that leads to a massive release of N (largely in the form of proteins, but also as mineralised N) that is rapidly taken up by proteolytic microbes that will undergo an approximate 10 cycles in the growing season. At each turnover they release N, some of which is available for plant uptake.

the types, *Krummholz* (22), mire (wet meadow: 21), and rock (20) had the greatest number of small mammal species, and dwarf-shrub heath (heather) the least (three)—corresponding figures for Niwot Ridge, Colorado Rocky Mountains are available in Armstrong *et al.* (2001) and a study on ground squirrel habitat use in the Altai by Ricankova *et al.* (2006). Figures from the North American alpine much contrast with the perception of low animal diversity in the alpine zone in Europe, where the only true rodents are the marmot (*Marmota marmota*) and the snow vole (*Chionomys nivalis*) (Allainé and Yoccoz 2003).

Animals, especially large ungulates, contribute to the shaping and functioning of alpine environments at many levels from being consumers and facilitators of element cycling to being the engineers of long-term changes in vegetation structure or landscapes (Dearing 2001; Erschbamer *et al.* 2003; Loison *et al.* 2003). Impacts that change landscapes are usually attributable to domestic stock and are considered in Chapter 10. The most impact associated with ungulates that inhabit temperate alpine areas (e.g. mouflon *Ovis musimon*; chamois *Rupicapra rupicapra* and ibex *Capra ibex* in Europe; or the bighorn sheep *Ovis canadensis*, bison *Bison bison*, mule deer *Odocoileus hemionus*, and elk *Cervus canadensis* in North America) are those in sheltering areas where they cause local nutrient enrichment. They are selective grazers and favour graminoids; however, the extent of this selective grazing on vegetation dynamics is not well quantified.

**Table 6.8** Resident alpine small mammals in the Pacific Northwest, North America

| Species | Common name | Habitat |
| --- | --- | --- |
| *Peromyscus maniculatus* | Deer mouse | All alpine habitats |
| *Phenacomys intermedius* | Heather vole | A wide variety of habitats |
| *Marmota* sp. | A marmot | A wide variety of habitats, typically near large rocks |
| *Thomomys* sp. | A pocket gopher | Grasslands, herbfields |
| *Ochotona princeps* | Pika | Rocks |
| *Spermophilus lateralis, S. saturatus* | Mantled ground squirrels | *Krummholz*, rocks |
| *Sorex monticolus* | Dusky shrews | Mires (wet meadows), *Krummholz* |
| *Clethrionomys gapperi, Eutamias* sp. | Southern red-backed vole, a chipmunk | *Krummholz* |
| *Microtus richardsoni, Sorex palustris, Zapus trinotatus* | Water vole, water shrew, Pacific jumping mouse | Willow scrub |

After Reichel (1986).

Small mammalian herbivores, similarly to that in the boreal alpine (e.g. Lindgren 2007), can be important in the North American and Asian temperate mountains. Depending on their life cycle, rodents use different habitats. For example, gophers (*Thomomys talpoides*) are most abundant in wet meadows where they find a deep enough snow cover in the winter when they burrow and forage under snow. Pikas (*Ochotona princeps*) have an interesting duality with respect to summer and winter diets. In the summer they are generalist grazers; however, for the winter cache they collect leaves of an average of 28 kg fresh weight, of which about two-thirds is made up of *Acomastylis rossii*. The impact of small mammals on the vegetation is related (in a yet to be quantified way) to their habit of foraging around their burrows, and thereby depleting their favoured food species and to the seemingly specialist winter food preference of pikas, for example. In addition to removing plant tissue, earth moving by gophers and associated accelerated mineralization of soil organic matter, and occasionally erosion contribute to the patch-scale dynamics of temperate alpine habitats (Sherrod and Seastedt 2001). For example, on the arid temperate Tibetan Plateau, Zhang *et al.* (2003) have found half the species richness and above-ground biomass, and four times less below-ground biomass and vegetation cover in areas with plateau zokor (*Myospalax fontanierii*) activity, in comparison with areas where the animals were absent from.

Invertebrates have been much studied, especially in the European Alps (e.g. Meyer and Thaler 1995); however, most of this has been devoted to the taxonomic richness of the alpine fauna (see Table 3.3). Functional aspects, both above- and below-ground, have been less studied. One exception is the study by Blumer and Diemer (1996) that provides a rare estimate of the rate of herbage removal by grasshoppers (*Aeropus sibiricus, Melanoplus frigidus*) in two *Carex*-dominated habitats in the Swiss Alps. Values of 19% and 30% biomass removal in zonal *C. curvula* sedge heath and adjacent snowbed (*C. foetida*) perhaps appear high; however, in the absence of longer-term observations on the dynamics of host–herbivore interaction little can be said about potential impacts grasshoppers may contribute to the replacement of these sedge species by other, less palatable species. The below-ground/above-ground cycles of herbivores and vegetation are likely to operate in a similar way to that described for the arctic tundra in Section 6.2.1.

# 6.5 The interface between temperate and subtropical mountains: the Himalayas

The Himalayas extend from approximately 74°E 36°N to 95°E 29°N, reaching as far south as 27°N in Bhutan. The elevation of the treeline and tree species composition much varies: in the north-west it is at approximately 3800 m and is formed by *Pinus wallachiana, Salix* spp., and *Betula utilis*; on insolated slopes, *Juniperus excelsa* forms stands. In the central southern Himalayas, *Abies spectabilis* with *Betula utilis* and *Juniperus recurva* form the treeline at approximately 3900 m. In the inner ranges, away from the monsoon, the potential treeline rises to 4500 m (see fig. 9.2 in Miehe 1997).

Plant communities of zonal alpine habitats in the Central Himalayas include scrub of *Juniperus* to 4200 m, of *Rhododendron* to approximately 4500 m, and *Rhododendron–Kobresia pygmaea* heath 4500–4800 m (Miehe 1997), but *Kobresia* heath can reach up to 5600 m e.g. in the Khumbu, Nepal, where *Rhododendron nivalis* can form closed stands up to 5300 m. Azonal habitats include snowbeds in the north-western Himalayas with *Sibbaldia* and *Salix*; boulder fields, overgrown with *Rhododendron* scrub in sheltered locations, and crustose lichens in exposed places. Numerous moraines provide well-drained habitats. The nival zone is heavily impacted by gelifluction and plants occur as scattered individuals (for the floristic structure of habitats near the snowline see Rawat (1987)). The nival/aeolian zone is extensive and it is discussed in Section 6.5.1.

Vertebrates of the montane forest and treeline ecotone in the eastern Himalayas in Nepal include the musk deer (*Moschus chrysogaster*), Himalayan tahr (*Hemitragus jemlahicus*), weasel (*Mustela sibirica*), mice, voles and shrews, with birds such the bearded vulture (*Gypaetus barbatus*), pheasants (*Ithaginis cruentus, Lophophorus impejanus*), blue-fronted redstart (*Phoenicurus frontalis*), or the fire-tailed sunbird (*Aethopyga ignicauda*). In the alpine zone the snow leopard (*Panthera uncia*) and its main prey the bharal or blue sheep (*Pseudois nayaur*) are rare, endangered and protected. Other typical species include the grey wolf, red fox, marmot (*Marmota bobak*), pika (*Ochotona roylei*), snowcock (*Tetraogallus himalayensis*), snow pigeon (*Columba leuconota*), pipits and accentors (Klatzel 2001).

The vegetation of arid temperate zone mountains north of the Himalayas have been treated by, for example, Chang and Gauch (1986) along environmental gradients of the Tibetan Plateau, or in the Kunlunshan (Dickoré 1991; Kürschner *et al.* 2005), by Rawat and Adhikari (2005) in the Ladakh; Agakhanyantz and Lopatin (1978) and Breckle and Wucherer (2006) have treated the ecosystems of the Pamirs, Zlotin (1978, 1997) the Tianshan, and Dickoré and Miehe (2002) the Karakoram Mountains.

Grazing by domestic livestock is a major ecological factor in the Himalayas and much overrides the effect of small burrowing animals on vegetation dynamics (e.g. Bagchi *et al.* 2006). Estimates of phytomass and productivity of alpine ecosystems in the Himalayas have shown large variation, mainly due to variation in delineating the potential versus actual treeline and the sampling of pastures versus meadows. Singh and Sundriyal (2005) reported the above-ground biomass of eastern Himalayan alpine pastures between 3800 m and 4200 m range to be between approximately 130 g m$^{-2}$ and 310 g m$^{-2}$, with the peak in August; corresponding values for below-ground parts were 930 and 1100 g m$^{-2}$, with the higher value at the beginning and end of the vegetation period; litter was between 10 and 40 g m$^{-2}$, probably indicative of high levels of grazing-induced decomposition rates. Rikhari *et al.* (1992) reported a range of communities in the central Himalayas between 3100 and 3750 m—most of which are likely to be largely of anthropogenic origin. The range of values for the *Kobresia–Danthonia* sedge-grass heath were: above-ground biomass (30–115 g m$^{-2}$), dead standing and litter (70–120 g m$^{-2}$) and below-ground (870–950 g m$^{-2}$) and total annual net primary production of (112 g m$^{-2}$ year$^{-1}$ above-ground and 95 g m$^{-2}$ year$^{-1}$ below-ground). The values obtained by Ram *et al.* (1989) for a *Danthonia*-dominated grassland phytomass in the central Himalayas at 3600 m between April and October for 2 years, in an area that had been excluded from grazing for about 20 years more resemble those from lightly grazed ecosystems. The above-ground peak biomass in August over a 2-year period to be a in the range of 382–409 g m$^{-2}$, standing dead ranged from 286 to 508 g m$^{-2}$ over the growing season, and litter 188–376 g m$^{-2}$; below-ground parts amounted to between approximately 900 and 1100 g m$^{-2}$. At a similar altitude Rikhari *et al.* (1992) recorded much lower above-ground biomass values (approximately 245 g m$^{-2}$), and especially dead standing (approximately 40–110 g m$^{-2}$) and litter ( approximately 40–60 g m$^{-2}$), indicating, again the likely impact of grazing.

An unquantified impact on snowbed vegetation and surrounding high altitude *Kobresia* heaths is that from herbal medicine collectors immediately after snowmelt. The target of such collection—that recently has entailed camping by several hundred people simultaneously on relatively small stretches of alpine summer grazing areas—is *Cordyceps sinensis*.

*Cordyceps* is a genus of fungus of the Clavicipitaceae (ergot) family and is notable for its exclusively parasitising insects. *C. sinensis* grows mostly on the larvae of the Himalayan bat moth (*Hepialis armoricanus*). The fungus has been used in traditional Tibetan and Chinese medicine as a tonic. It has seen a recent increase in popularity and become a lucrative cash source. The ecology of this fungus—insect interaction and its impacts on regulating the population cycles / density of the insect, and the resulting impacts on vegetation are poorly understood.

## 6.5.1 The aeolian environment

The largest expanses of aeolian environments with their ice and rock habitats above the limits of plant life are found in the Himalayas (Fig. 6.20) and numerous other high ranges in Asia and Central and South America; they are also present on other continents to varying degrees. Primary producers are rare in these environments (except for some algae) and saprophytes live off wind-blown dust and organic debris. Predators are the highest living resident organism in the aeolian regions of high mountains.

As has been emphasized earlier, wind is an important biophysical and environmental factor. One important aspect relating to wind is its role as an agent in redistributing animals, in a broader sense, nutrients. One such example concerns the Salticid spider *Euphrys omnisuperstes* from the Himalayas that feeds on wind-distributed anthomyid flies and springtails or collembola up to elevations of 6100 m (Swan 1981). In the absence of *in situ* plants, collembolas live on wind-blown debris, made up of fragments

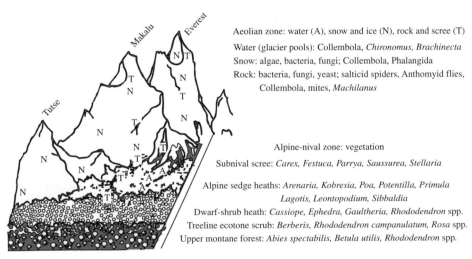

Aeolian zone: water (A), snow and ice (N), rock and scree (T)

Water (glacier pools): Collembola, *Chironomus, Brachinecta*
Snow: algae, bacteria, fungi; Collembola, Phalangida
Rock: bacteria, fungi, yeast; salticid spiders, Anthomyid flies,
　　　Collembola, mites, *Machilanus*

Alpine-nival zone: vegetation

Subnival scree: *Carex, Festuca, Parrya, Saussurea, Stellaria*

Alpine sedge heaths: *Arenaria, Kobresia, Poa, Potentilla, Primula*
*Lagotis, Leontopodium, Sibbaldia*
Dwarf-shrub heath: *Cassiope, Ephedra, Gaultheria, Rhododendron* spp.
Treeline ecotone scrub: *Berberis, Rhododendron campanulatum, Rosa* spp.
Upper montane forest: *Abies spectabilis, Betula utilis, Rhododendron* spp.

**Fig. 6.20**　　The zonation of the alpine habitats and vegetation, and a scheme of the abutting nival/aeolian zone in the Himalayas of the Nepal-Tibet border area. (Redrawn after Swan 1981.) Note the absence of vascular plants in the aeolian zone.

of plants, insects, and other organic material. Snow-dwelling organisms include bacteria, algae, ice worms (Annelida, Enchytraeidae), scavenging insects, arachnids, and birds. They feed on accumulated wind-blown organic debris, and birds on insects.

Other examples outside the Himalayas include the lizard *Sceloporus micro-lepidotus* about 200 m above the snowline at 4800 m on Citlaltépetl (5636 m), Mexico (feeding on wind-blown insects) and rattlesnakes at 4600 m that feed on the lizards. The endemic salamander *Pseudoeurycea gadovi*, 4500 m, lives under rocks and feed on predatory insects and arachnids that shelter under the rocks (Swan 1992). The best-studied examples are from North American mountain ranges, in relation to permanent and semi-permanent snowfields (Papp 1978; Spalding 1979; Mann *et al.* 1980; Swan 1981; Edwards 1987; Halfpenny and Heffernan 1992; Coulson *et al.* 2003). The total growing season nutrient input from wind-blown insects has been estimated to be approximately 21 mg carbon $m^{-2}$, 43 mg nitrogen $m^{-2}$, and 0.25 mg phosphorus $m^{-2}$ (Edwards 1972), an important amount for local functioning and processes, but a negligible figure in comparison with input by precipitation and atmospheric transport.

# 6.6 Subtropical sclerophyllous forest zone

This group is what is commonly known in the Northern Hemisphere as Mediterranean mountains on either side of the Mediterranean Sea in southern Europe and northern Africa (Sierra Nevada, High Atlas, Hellenids), the Taurus in Asia Minor, and the Zagros, Elburz and the Hindukush, and the Sierra Nevada of California; in the Southern Hemisphere this zone consists of the Maloti–Drakensberg in South Africa, the Australian Alps, and the central Chilean Andes (Fig. 2.1). The characteristic feature of these mountains is that many of them are subject to summer drought, which is reflected in their vegetation of hemispherical thorny cushion in the treeline ecotone in the European and North African mountains, and above that the parched and usually scant vegetation, except where late snowmelt provides water in the summer.

## 6.6.1 Treelines in subtropical mountains

The treeline in the European Mediterranean is sometimes difficult to delineate as trees may be lacking and their place is taken by thorny shrubs, such as in the Sierra Nevada of Spain (*Juniperus communis* ssp. *hemisphaerica*, *J. sabina*, *Genista baetica*), or Mount Etna (*Berberis aetnensis*) (Fig. 6.5). The treeline, in the High Atlas at approximately 2400 m is formed by *Juniperus excelsa* and *J. foetidissima*; in the Hellenids by *Pinus heldreichii*, and *Juniperus thurifera* in the Taurus; in the Sierra Nevada of California by *Pinus albicaulis*, and in the White Mountains

**Fig. 6.21**    *Eucalyptus niphophila* the Australian snowgum forming the treeline at about 1800 m in the Kosciuszko Mts, Australia.

of California by *P. longaeva*. The treeline in south-east Australia is at about 1800 m, formed by *Eucalyptus niphophila* (Fig. 6.21), a species that grows from about 1500 m (the lower limit of winter snow lie) upwards in the upper montane forest zone. In southern Africa, above *Podocarpus lat-ifolius* (1800 m), fynbos or scrub, formed by *Philippia*, *Erica*, and *Protea* grows up to approximately 2850 m.

## 6.6.2 The vegetation above the treeline in subtropical mountains

The High Atlas typifies the habitat types on Mediterranean mountains. Above the highest growing trees at approximately 2400 m (annual mean temperature 9°C; July 20°C; January 2.5°C), there is an extensive *garrigue* of hemispherical spiny cushion that can be divided into three distinct habitat types: (1) steep dry scree slopes with angular stones (dry); (2) gentler slopes with snow over 5–6 months (mix of dry and humid conditions); (3) deposits of fine soil in depressions where the snow lies for 6–8 months (humid) (Quézel 1957). Above 3500 m (annual mean temperature, 5°C; July, 15°C; January, –1.5°C) hemispherical cushions are no longer found. The vegetation is scant as there is very little soil. The main habitat creating factors are related to weathering rates, soil development, and hydrology. The lack of soil exposes large areas of rock and scree, which are colonized by chasmophytic and scree species. Characteristically, the zonal habitat is vegetated by a sequence of hemispherical thorny cushion communities. Importantly, small habitats occur that rely on all-year round irrigation from snow melt (Box 6.1). For some information on the flora and

---

**Box 6.1** The habitat types and their vegetation in the High Atlas (after Maire 1924, Quézel 1957)

**Zonal habitats and their vegetation**

The zonal vegetation is xerophytic spiny hemispheric cushion *gararrigue* and summer-dry grasslands. The hemispheric cushion vegetation (2400–3100 m) is scattered *Juniperus thurifera* with *Alyssum spinosum, Bupleurum spinosum, Cytisus balansae*. On rocky slopes and fixed screes, there are summer-dry rupicolous grasslands (two siliceous communities with *Arenaria pungens, Helyzaldia violacea, Spergularia flaccida; Vicia glauca* var. *rerayensis*, and three calcareous one: *Agropyron festucoides* var. *pseudofestucoides, Avena montana, Scrophularia ramosissima* var. *macrorhynca, Teucrium musimonum, Vella mairei*), and *Alyssum spinosum* and *Astragalus ibrahimianus* hemispherical cushion communities between 3000 and 3200 m. Very open high-altitude vegetation above 3500 m on scree (one type on siliceous rock with *Viola dyris* and *Linaria lurida* and four on calcareous bedrock (*Asperula litardieri, Cirsium dyris, Silene ayachica, Raffenaldia primulina, Veronica chantoni*).

**Azonal habitats**

Rupicolous chasmophytic vegetation (four communities on siliceous and 11 on calcareous rocks) with *Draba oreadum* and *Saxifraga pedemontana* var. *demnatensis* (3400–4165 m) in dry rock crevasses, up to the highest summits of the central High Atlas. Wet rock crevasses, irrigated by snowmelt have *Carex leporina; Sedum atlanticum*, in shaded, humid fissures there is an abundant species-rich bryophyte growth.

On mobile screes, there is a sparse growth of *Linaria lurida* and *Platycapnos saxicola*. In places with deep soil in depressions with temporal water supply from snow melt summer-dry meadows are found (similar to those described by Quézel (1967) from the Pindos and Olympus in Greece): *Festuca rubra* var. *yvesiana, Plantago coronopus, Trifolium humile*. Depressions with deep soil and all-year round irrigation bear closed alpine grasslands of *Botrychium lunaria, Carex capillaris, Cerastium cerastioides, Luzula spicata, Myosotis alpestris, Nardus stricta, Poa alpina*, and *Sagina nevadensis* (Alps–Sierra Nevada species). Waterlogged *pozzines* or mires are found locally on siliceous bedrock with *Carex fusca* and *Veronica repens*, and runnels and seeps have *Cirsium chrysacanthum*.

---

vegetation of the Elburz and Zagros see Noroozi *et al.* (2008) and of the Hindukush Breckle (2006).

The South African alpine vegetation in the Maloti–Drakensberg is a mixture of abundant mire types, rock outcrop and cliff vegetation, grassland, and dwarf-shrub heath (Killick 1978, 1997). The zonal vegetation is *Erica dominans–Helichrysum trilineatum* dwarf-shrub heath. The origin of the heavily grazed grasslands (*Merxmuellera disticha, Festuca caprina, Pentaschistis oerodoxa, Euryops evansii*) may be at historic burning, rare nowadays because of the overgrazed nature of the vegetation. There are numerous bogs (seepages and stream head flushes) and streams on undulating summit ground. On horizontal outcrops along the edge of the escarpment and summit of high peaks, there are cryptogamic communities, xeromorphic herb communities (*Merxmuellera stereophylla, Helichrysum retortoides*) and chasmophytic communities. The ecology of the hummocky bogs has been investigated with respect to potential freeze–thaw action and animal burrowing. Indeed, it seems that it is burrowing rodents (the common mole rat *Cryptomis hottentotus* and ice rat *Otomys slogetti*) that work in tandem to make and enlarge mounds that then become vegetated (Lynch and Watson 1992).

In Australia, the main alpine areas are found in the south-east of the mainland (Snowy Mountains–Kosciuszko and the Victorian Alps–Bogong High Plains) and in Tasmania (McVean 1969). Of these, the Kosciuszko is the largest (Costin *et al.* 2000). Snow cover is uneven and the steep western slopes, which are wind-exposed and insolated, starkly contrast with the leeward more gentle easterly slopes, where long-lying snow patches occur in sheltered locations (Edmonds *et al.* 2006). Topography and hydrology define habitats and their vegetation types. The zonal vegetation is *Celmisia costiana–Poa* spp. tall alpine herb field and *Chionochloa frigida* tussock grassland. Open vegetation of dwarf cushion and mat-forming species lie at both extremes of the snow gradient with wind-exposed sites bearing *Epacris microphylla–Chionohebe densifolia* formation and *Coprosoma niphophila–Colobanthus nivicola* being typical of snowbeds. Irrigation from snow-melt water is characteristic of the short alpine herb fields of *Plantago glacialis–Neopaxia australasiaca*. The main hydrological gradient, arising from drainage characteristics of soils along a topography gradient is associated with the spatial distribution of a number of habitats and their vegetation. Well-drained rocky sites support an *Oxylobium ellipticum–Podocarpus lawrencei* dry heath and an *Epacris glacialis* wet heath is typical near topogenous mires. The mesophilous tall alpine herb fields give way to *Poa costiana–Rhytidosperma nudiflora* tussock grasslands in wet valleys with restricted drainage and permanently wet habitats support soligenous (*Carex gaudichaudiana*) and topogeneous mires (*Carex gaudichaudiana–Sphagnum cristatum*, valley bog; *Epacris paludosa–Sphagnum cristatum*, raised bog).

The dynamic nature of exposed *Epacris* heath habitat, resulting in a contrast in soil–vegetation–invertebrate systems between exposed and local sheltered micro-habitats has been demonstrated by Costin *et al.* (1969). The underlying differences in the abiotic environment result from wind erosion and re-deposition of soil. The differences are observed along solifluction terrace surfaces and lobe end micro-slopes, and other contrasting small-scale exposed and sheltered micro-habitats: exposed stony and drier soils under *Epacris* are lower in soil nitrogen and macro- and micro-arthropods than finer more moist soils under grasses. The build-up of fine soil and replacement of dwarf-shrub by *Celmisia–Poa* vegetation may be accomplished in approximately 50 years. Wilson (1993) has also found that available nitrogen was higher in the more humid *Celmisia–Poa* grassland than in a drier treeline ecotone heath on well-drained rocky ground. The work by Wilson (1993) that used transplants to investigate competition effects suggested that strong below-ground competition in the *Celmisia–Poa* grassland (mean root mass 1380 g m$^{-2}$) was an effective control of potential establishment by *Eucalyptus niphophila*.

The alpine vertebrate fauna in Australia is poor, alpine specialist mammals are best represented by the mountain pigmy-possum (*Burramys parvus*), which occurs in the alpine zone and down to the winter snow-lie in the eucalypt montane forest; there are no alpine/mountain specialist birds (Green and Osborne 1994). Equally, anurans and reptiles are rarely confined to alpine elevations; however, there are some high mountain specialists, such as the snow skinks of Tasmania (*Niveoscincus greenii, N. microlepidotum*), or various skink species of south-eastern Australia (*Eulamprus kosciuskoi, Egernia* sp., *Pseudemoia cryodroma, Cyclodomorphus praelatus*), and the frogs *Pseudophryne corroboree* and *Philoria frosti*.

The different alpine habitats cater for widely differing faunas (Green and Osborne 1994). Exposed fell-fields, and scree have no vertebrates and their invertebrate fauna is rather depauperate, with mostly resident collembollas and mites, with noticeable absence of spiders, an indication of lack of prey species. Rocky habitats have a rich vertebrate fauna, including small mammals (e.g. the marsupial antechinuses *Antechinus swainsonii* and *A. stuartii*, and pigmy possum and rodents, such as broad-toothed rat *Mastacomys fuscus* and bush rat *Rattus fuscipes*), numerous birds that inhabit or feed in alpine habitats in the summer; most reptile species live in rocky habitats and boulder fields. The rich and abundant invertebrate fauna provides prey for the vertebrates. Snowbeds, with their short vegetation are little used by animals; however, in the past, domestic grazers have much degraded them. Alpine mires provide breeding places for anurans. The treeline ecotone dry and wet heaths are very rich in animal life.

The zonal *Poa–Celmisia* vegetation on mountain slopes and tussock grassland on undulating or flat areas offer relatively little to small mammals because of the lack of cover; however, many birds feed on insects, such as grasshoppers and moths in this habitat. The invertebrate soil fauna of the *Poa–Celmisia* vegetation is abundant, too (macrofauna approximately 200 individuals m$^{-2}$; mesofauna 125,000 m$^{-2}$ and microfauna (nematodes) 2.5 million m$^{-2}$ (Wood 1971), with about two-thirds of the biomass being contributed by the macrofauna—earthworms (41.2 g m$^{-2}$) (Wood 1974) and arthropods (15.3 g m$^{-2}$).

# 6.7 Aseasonal tropical alpine habitats

The aseasonal tropical alpine includes the northern Andes, the New Guinean and East African mountains, together with Mount Cameroon and Hawaii. The East African mountains, together with the northern Andean páramos have served to shape perception about typical tropical zonal vegetation, with their giant rosettes. In the alpine zone, described from Mount Kenya (Coe 1967), the presence of lobelias (*Lobelia keniensis, L. telekii* from 3300–3500 m to 4600–5000 m) is characteristic, together with *Senecio* (*Dendrosenecio*) species, appearing approximately at 3800 m (Young and Peacock 1992). In the Andean páramos a sequence of *Espeletia* species occurs (Monasterio 1980; Rangel-Ch 2000). On Mount Cameroon, there is a restricted fragmentary alpine zone above approximately at 3800 m with sparse grasses, such as *Pentaschistis mannii* and dwarfed sparse shrubs (*Blaeria mannii, Philippia mannii*) (Letouzey 1968). In Malesia, or New Guinea, there are no giant rosettes. The neotropical alpine areas are very species rich (in the region of 3000 vascular plants), while the Old World ones are rather poor (<200 in both Africa and New Guinea). Hawaii is one of the rare examples where the alpine zone has been affected by introduced species.

## 6.7.1 Tropical treelines

The treeline in the East African mountains is at about 4000 m, formed by trees of *Erica* and *Phillippia* species (Hedberg 1964; Miehe and Miehe 1994). On Mount Cameroon, the altitude that trees reach is much suppressed by fire and volcanic activity (Proctor *et al.* 2007). In the northern Andes the closed forest line is at between approximately 3200 and 3500 m; however, various species of *Polylepis* with *Escallonia* or *Hesperomeles* in Colombia (Rangel-Ch 2000), or *Gynoxis* in Venezuela (Monasterio and Reyes 1980) reach higher altitudes (up to approximately 4100 m). Both in Africa and the northern Andes some species of giant rosettes occur from the upper montane forest. The treeline in New Guinea (Mount Trikora) is formed by *Schefflera altigea*, *Saurauia alpicola*, and *Vaccinium dominans* in a scrub matrix at approximately 3400–3800 m (Mangen 1993). Tree ferns (*Cyathea* spp.) may or may not form natural vegetation types; however, they do not reach the alpine zone or the treeline. On Mount Kinabalu, Borneo scattered prostrate specimens of one of the upper montane tree species *Leptospermum recurvum* are found all the way to the summit. The highest growing trees of *Metrosideros polymorpha* on Hawaii reach approximately 2600 m and *Sophora chrysoplylla* 2900 m, where they are thought to be limited by night frost; water deficits and substratum age are also thought to be contributing factors (Mueller-Dombois and Frosberg 1998).

## 6.7.2  Habitats and vegetation in the tropical alpine zone proper

Hedberg (1964) classified the East African alpine vegetation into five broad classes: (1) *Dendrosenecio* woodland on deep soil with non-stagnant soil water; (2) *Helichrysum* scrub; (3) *Alchemilla* scrub; (4) tussock grassland; and (5) *Carex* bogs and related communities. The habitats and vegetation of the alpine zone in a glacial landscape in the East African mountains is exemplified by that described by Coe (1967). There is a sharp contrast between the lower alpine zone and the zones above, where glacial landforms dominate, on the one hand, and within the upper zones, between glacial valley walls (*Senecio keniodendron*) and bottoms (*S. keniensis*), on the other hand (Young and Peacock 1992). The subnival frost-heaved habitats and retreating glacier edges have open vegetation (e.g. Mizuno 1998; Mizuno 2005).

The páramo, or humid tropical alpine zone in Colombia (3600–4100 m) is characterized by the zonal vegetation of the flagship giant rosettes (*Espeletia s. l.*), tussock grasslands with *Calamagrostis effusa, C. recta, Agrostis tolucensis*, and bamboo thickets with *Chusquea tessellata*. In Venezuela, the zonal vegetation is similar, with the grasses *Calamagrostis* and *Cortaderia*, and *Espeletia* species. The so-called superpáramo (4100 m to the limit of nival zone), or subnival zone has patchy grasslands and dwarf-shrub vegetation in Colombia (Rangel-Ch 2000). In Venezuela, various *Espeletia*-dominated formations and *Draba chionophylla* dwarf-rosette communities characterize the range of habitats (Monasterio 1980). The azonal vegetation types in Colombia include the humid bamboo thickets with *Chusquea tessellata*, bog vegetation with *Distichia muscoides*, or *Oreobolus* spp.; other cushion vegetation with *Azorella* or *Arenaria* spp. In Venezuela, rock habitats and screes are inhabited by *Draba* spp. and *Espeletia* spp. (see the detailed account by Monasterio 1980).

In New Guinea, there are contrasting habitat conditions not only with regard to physical properties of the substratum but also in relation to successional status following glacier retreat (Wade and McVean 1969; Hope 1976; Smith 1980; Mangen 1993). Tall and closed shrublands inter-digitate with secondary grasslands, rich in shrubs from between approximately 3200–3800 m up to the treeline. They are followed by a sequence of habitats with different ages and drainage status. Dwarf-shrub heath (*Styphelia suaveolens, Tetramolopium* spp. with grasses and sedges, and moss cushions) on shallow rocky soils up to 4500 m on margins of recent moraines, and open *Racomitrium* moss heath (open *Styphelia suaveolens, Tetramolopium macrum* with closed mat of mosses, e.g. *Racomitrium lanuginosum* (Mount Wilhelm), *Tetramolopium klossii–Racomitrium crispulum* (Mount Jaya/ Carstensz)) occur on areas >30 years exposed after glacier retreat. A variety of grassland types (tussock, *Deschampsi klossii*, smooth, *Danthonia oreoboloides*, herb fields with cushions, *Astelia alpina*) are formed in topographies and soil ranging from free draining to shallow peats. There is an alpine moss-dominated open vegetation with *Poa callosa*, *Potentilla forsteriana*, and *Unciana* sp.; its wet variety is with *Ranunculus saruwagedicus*, and the dry variant with *Parahebe ciliata*. It is thought that the moss domination that forms approximately 50–100 years after ice retreat, later gives way to dwarf-shrub heath or tussock grassland; and on wet ground to forming mire. The mires include bogs with predominantly upper montane distribution, and hard cushion bogs on inundated flats (*Plantago polita*, *Oreobolus ambiguous*, *Astelia alpina*) and fens at lake and pond edges (e.g. *Carpha alpina*) up to 4300 m. Rock faces and scree are colonized by lichens rapidly after deglaciation.

The Hawaiian alpine peaks of Manua Loa, Manua Kea, and Haleakala lie above the trade wind inversion that causes a cloud base at approximately 2000 m and receive much decreased precipitation above 2000 m; the annual precipitation decreases from 800 mm on Haleakala (3056 m) to 146 mm on Manua Kea (4207 m). Manua Loa is an active volcano that has a low heath shrub of *Styphelia* and *Vaccinium reticulatum* at 2600–3000 m, followed by *Racomitrium lanuginosum* dominated landscape at 3000–3300 m, and an alpine stone desert from 3300 m to the summit (4172 m). Manua Kea has a Pleistocene glaciation history. The *Styphelia* scrub turns into open scattered landscape and finally peters out at 3400 m, followed by a bryophyte-dominated zone (*Grimmia haleakalae*), before giving way to the lichen (*Lecanora melaena*) crust from 3900 m upwards. On Haleakala, there is an *Argyroxiphium* alpine desert (2700–3000 m), topped by sparse *Styphelia* heath (Mueller-Dombois and Frosberg 1998).

### 6.7.3 The functioning of aseasonal tropical alpine ecosystems

Páramos, in areas that are not pastured, have a mean above-ground phytomass that is in the range of 3000 g m$^{-2}$, of which no more than one-third is live plant tissue (Table 6.9); however, there is very little information regarding other tropical alpine areas, particularly African alpine ecosystems. Little is known about nutrient cycling in tropical alpine ecosystems, especially in comparison with that in temperate and further northern areas. An interesting, but not unique aspect of the mineral nutrition of the pachycaul *Senecio keniodendron* is the partial decomposition of retained dead leaves (thought to be having an insulation function) and the appearance of adventitious roots in the decomposing mat (Beck 1994). A similar study failed to establish similar findings for *Espeletia lutescens* and *E. timotensis* the northern Andean páramos (Garay *et al.* 1983). However, the pattern of presence of decomposers (micro-arthropods) and nitrogen concentration in different parts in the dead leaves on the stem and in the litter under the plants suggested that soil nutrients were recycled in the vicinity of *Espeletia* plants. Soil nitrogen mineralization rates have been estimated by Rehder (1994) on hill slopes and valley bottoms in the Teleki

**Table 6.9** Above-ground phytomass in north Andean páramos and a New Guinean tussock grassland

| Vegetation type | Site | Geographic coordinates (°) | Altitude (m) | Above-ground biomass (g m$^{-2}$) | Dead standing and litter (g m$^{-2}$) | Above-ground phytomass (g m$^{-2}$) | *Espeletia* (g m$^{-2}$) | Source |
|---|---|---|---|---|---|---|---|---|
| *Calamagrostis effusa* bunch-grass páramo | Cordillera Oriental, Colombia | 4°33'N; 74°02'W | 3300–3400 | 314–1854 | 421–806 | 735–2660 | | Cardozo and Schnetter (1976) |
| *Chusquea tessellata* bamboo páramo | Cordillera Oriental, Colombia | 4°40'N; 73°48'W | 3650 | 959 | 1666 | 2625 | | Tol and Cleef (1994) |
| *Calamagrostis effusa* bunch-grass páramo | Cordillera Oriental, Colombia | 4°35'N; 74°04'W | 3620 | 603 | 2802 | 3405 | 1250 | Lutz and Vader (1987) |
| *Calamagrostis effusa* bunch-grass páramo | Cordillera Oriental, Colombia | 4°35'N; 74 °04'W | 3620 | 1374 | 2012 | 3386 | 2315 | Beekman and Verweij (1987) |
| Grazed bunch-grass páramo | Cordillera de Mérida, Venezuela | 8°47'N; 70°48'W | 3530–3560 | 149–427 | 56–278 | 206–651 | | Smith and Klinger (1985) |
| *Calagrostis intermedia* bunch-grass páramo | Cordillera Oriental, Ecuador | 1°48'S; 78°32'W | 3750 (3950) | | | 837 (794) | | Ramsay and Oxley (2001) |
| *Calamagrostis effusa* bunch-grass páramo | Cordillera Central, Colombia | 4°40'N; 75°10'W | 4100 | 420 | 2399 | 2820 | 666 | Hofstede et al. (1995) |
| *Deschampsia klossii* tussock grassland | Mount Wilhelm, New Guinea | 5°47'S; 145°01'E | 3400 (4350) | 490–722 | 2106–3063 | 2829–3553 | | Hnautiuk (1978) |

Valley of Mount Kenya at 4.5 g nitrogen m$^{-2}$ and 6.5 g nitrogen m$^{-2}$ in an incubation study over a period of 35 to 49 days. Rehder (1994) has used these estimates to predict primary productivity of 682 g m$^{-2}$ (289 g m$^{-2}$ above-ground), on slopes and 970 g m$^{-2}$ (372 g m$^{-2}$ above-ground) in valley bottoms. There are no measured data from Mount Kenya to compare with the predicted values; however, a word of caution is justified by way of using incubation studies on homogenized soil, which can increase the rates of mineralization in comparison with intact soil cores. Aside from soil nutrients, hydrological factors alone, or in combination with herbivory by rock hyrax (*Procavia johnstonii*) can cause dramatic fluctuations in the populations of giant lobelias (*Lobelia telekii* on well-drained slopes, *L. keniensis* in valley bottoms) as has been shown on Mount Kenya (Young 1994; Young and Smith 1994). Another giant rosette, *Senecio keniodendron*, a scree specialist and mast seeding species appears to have its own occasional lethal herbivore visitor in the form of elephants (*Loxodonta africana*) (Smith and Young 1994), or fire. Fire, either natural or manmade, is an important factor and it has not only affected the alpine zone itself, but contributed to its downward extension, for example, as documented on Mount Kilimanjaro (Hemp 2005, 2006). Its impact on the drier slopes of Mount Kenya, in combination with herbivores (e.g. zebra *Equus quagga* ssp. *burchelli*; eland *Taurotragus* sp.; buffalo *Syncerus caffer*; bush duiker *Sylvicapra grimmia*), attracted to tender regrowth after fire, results in an anomaly in that the grazed drier slopes may appear greener than the humid slopes where fire does not impact *Festuca pilgeri* that maintains large old tussocks which are mainly of standing dead culms and leaves (Young and Smith 1994). Herbivore impacts are also highlighted by the short carpet-like vegetation near rock hyrax colonies, and around rodent burrows, such as the mounds of mole rats *Tachyoryctes rex* and the runs of the common groove-toothed rats *Otomys* spp.

In the Andean páramo, the primary drivers are land use—the cultivation of tubers and cereals in the lower reaches of the páramo and livestock grazing in the upper reaches of glacial valleys, with, or without the use of fire (Mena *et al.* 2001; Hofstede *et al.* 2003). It affects vegetation, and it has also been reported to cause a decline in earth worms (Lumbricidae). It also appears that some ecological processes, such as decomposition might be affected by drought periods. Observations on soil arthropods—mostly mites Acari (64–76% of total soil arthropod fauna) and springtails Collembola (15–24%) have shown large fluctuations in their numbers with soil moisture. The populations appear to crash during wet periods in an otherwise dry high alpine to subnival páramo site, while a wetter site registers large fluctuations in the wet season, and lesser ones during dry spells (Diaz *et al.* 1997). No information is available how these fluctuations affect litter decomposition rates. The native vertebrate fauna of the páramos is rich (Diaz *et al.* 1997; Rangel-Ch 2000; Mena and Medina 2001): 24 anuran species in Venezuela and up to 90 species in Colombia—one,

*Gastrotheca riobambae* is of particular interest for being marsupial; five species of reptiles (*Stenocercus* lizards); 79 resident birds in Venezuela, and 49 mammals of which 25 are resident. Notable species are the Andean bear (*Tremarctos ornatus*) and mountain tapir (*Tapirus pinchaque*), both forest species that often visit the páramos for feeding. There are three species of Cervidae (white-tailed deer *Odocoileus virginianus*, little red brocket *Mazama rufina*, northern pudu *Pudu mephistophiles*), and carnivores include the puma (*Felis concolor*), pampas cat (*Oncifelis colocolo*), the Andean fox (*Pseudalopex culpaeus*), and the long-tailed weasel (*Mustela frenata*). Many species of rodents, including guinea pigs (e.g. *Cavia aperea*) are preyed on by the carnivores. Numerous flying insects and humming bird species are important pollinators of *Espeletia s. l.* species.

Without doubt, the introduction of ungulates and rodents on the Hawaii Islands that have no native terrestrial mammals has caused much impact, mostly in the lowlands but also in the treeline ecotone and alpine zones (Loope and Medeiros 1994). There are species that have become extinct or are threatened by extinction (e.g. *Argyroxiphium sandwicense* ssp. *sandwicense*) on Manoa Kea, and many introduced grasses cover extensive areas at the expense of native species.

# 6.8 Seasonal tropical alpine environments

In the Northern Hemisphere, seasonal tropical alpine environments are found in the Sierra Madre Oriental and Occidental in Mexico, above a *Pinus hartwegii* treeline at approximately 4300 m, largely represented by the zonal *Festuca tolucensis* and *Calamagrostis tolucensis* bunch grasslands (Lauer 1978; Islebe and Velázquez 1994). In the Andes, between approximately Cotopaxi in Ecuador and Lake Titicaca in Bolivia at approximately 17°S extends the humid puna. South of this line, to approximately 30°S in north-west Argentina extends the arid, or thorn and desert puna (Navarro and Maldonado 2002). The humid puna has an upper montane forest/scrub zone of *Polylepis besseri* and *P. racemosa*, largely replaced by various grasslands and scrub. The alpine zone is found between approximately 4000 and 4700 m with zonal grasslands, and azonal rock and scree habitats, and mires. In Bolivia, the zonal vegetation is of *Azorella diapensiodes–Festuca dolichophylla*. This vegetation, where excessively burned and grazed is turned to a *Pycnophyllum molle–Festuca rigescens* type; eroded areas can be colonized by *Aciachne acicularis* and *Stipa pungens*. In the subnival zone (4600–5000 m) shortgrass vegetation with *Calamagrostis minima* dominates, above which, up to about 5300 m, scattered plants inhabit the cold and cryoturbation affected habitats. Of the azonal habitats, block scree has a *Senecio rufescens* type dwarf-shrubby vegetation. The fissures of rocky outcrops have chasmophytic species, such as *Saxifraga magellanica*. The permanently wet valley bottoms in depressions have *Eleocharis–Festuca* fens, and the

**Table 6.10** Some vertebrate taxa that are confined to, or have their optimum distribution in the puna. Some typical species with wide Andean distribution are also included

| Class/family | Species |
| --- | --- |
| Mammalia | |
| Canidae | Andean fox (*Pseudalopex culpaeus andina*) |
| Felidae | Puma (*Felis concolor*), Andean mountain cat (*Felis jacobita*) |
| Camelidae | Lama (*Lama glama*), alpaca (*Lama pacos*), vicuna (*Vicugna vicugna*) |
| Cervidae | Taruka (*Hippocamelus antisensis*) |
| Muridae | White-bellied grass mouse (*Akodon albiventer berlepschii*), Bolivian grass mouse (*Akodon boliviensis*), altiplano grass mouse (*Akodon lutescens lutescens*), Puno grass mouse (*A. subfuscus subfuscus*), Andean mouse (*Andinomys edax edax*), painted big-eared mouse (*Auliscomys pictus*), Andean big-eared mouse (*Auliscomys sublimis*), Andean vesper mouse (*Calomys lepidus carillus*), pleasant bolo mouse (*Bolomys amoenus*), altiplano chinchilla mouse (*Chinchillula sahamae*), Jelski's Altiplano mouse (*Chroeomys jelskii ochrotis*), *C. jelskii pulcherinnus*, Andean gerbil mouse (*Eligmodontia puerulus*), Andean swamp rat (*Neotomys ebriosus ebriosus*), leaf-eared bunchgrass mouse (*Phyllotis osilae phaeus*) yellow-rumped leaf-eared mouse (*P. xanthopygus rupestris*) |
| Chinchillidae | Viscachas (*Lagidium viscacia cuscus*, *L. viscacia cuvieri*) |
| Cavidae | Montane guinea pig (*Cavia tschudii nana*), common yellow-toothed cavy (*Galea musteloides auceps*), Andean mountain cavy (*Microcavia niata*) |
| Octodontidae | White-toothed tuco-tuco (*Ctenomys leucodon*), highland tuco-tuco (*Ctenomys opimus*) |
| Dasypodidae | Andean hairy armadillo (*Chaetophractus nationi*) |
| Didelphidae | Pallid fat-tailed opossum (*Thylamus pallidior*) |
| Mustelidae | Lesser grison (*Galictis cuja*) |
| Reptilia | |
| Colubridae | Snake (*Tachymenis peruviana*) |
| Iguanidae | Lizards (*Ctenoblepharis schmidti*, *Liolaemus alticolor*, *L. forsteri*, *L. ornatus*, *L. pulcher*, *L. signifer*, *L. simonsi*, *L. variegatus*, *Stenocercus marmoratus*, *S. variabilis*) |
| Teiidae | Lizards (*Proctoporus bolivianus*, *P. guentheri*) |
| Amphibia | |
| Leptodactylidae | Frogs (*Pleurodema marmorata*, *Telmatobius albiventris*, *T. atacamensis*, *T. coleus*, *T. gigas*, *T. huyra*, *T. marmoratus*) |
| Bufonidae | toad (*Bufo spinolusos*) |

Data after compilation for Bolivia in Navarro and Maldonado (2002) (pp. 246 and 460) and for Peru after Ramirez *et al.* (2007).

peat bogs are either of the flat (*Plantago rigida* or *P. tubulosa*) or hard cushion type (*Calamagrostis jamesonii–Distichia muscoides*, and in the subnival zone *Werneria marcida–Distichia filamentosa*).

The arid puna is a rather heterogeneous collection of habitats, which, in addition to an alpine climate and topography, are determined by the large expanses of salt flats on the Altiplano. It is also peculiar in that trees can reach near to the upper limit of vascular plant distribution (Fig. 3.9). The zonal vegetation is therefore rather complex: (1) high Andean tussock grassland (*Stipa, Deyeuxia*); (2) fragments of *Polylepis tarapacana* open woodland and scrub; (3) xerophytic thorny scrub (*Trichoceres, Mutisia*); (4) xerophytic dwarf-shub (*Fabiana, Parastrephia*); (5) dwarf-shrub (*Parastrephia*); and (6) high Andean mesic grasslands (*Deyeuxia*).

The azonal habitats include frost-heaved open scree (Andean frost desert with, for example, *Werneria, Nototriche*) at the upper end. The salt-affected habitats can have herb meadows with *Anthobryum*, halophytic scrub (*Atriplex, Suaeda*), and open vegetation with prostrate cushions and succulents (*Sarcocornia*). The wetlands or hard cushion bogs not affected by salinity contain *Calamagrostis jamesonii* and *Distichia muscoides*, and other, salt affected wetlands have *Scirpus* and *Puccinellia*. It is clear from the above the variety of habitat factors that affect the functioning of the arid puna ecosystems, rarely encountered in juxtaposition in an alpine setting. Typical faunal examples, confined to, or finding their optimum in the alpine zone in Bolivia are listed in Table 6.10.

The single most important perturbation in the humid puna region of the Andes has been heavy grazing pressure (Adler and Morales 1999; Preston *et al.* 2003) since at least Incaic times. The colonization of the central Andes further increased grazing pressure after the introduction, from Europe, of large herbivores, such as sheep, cattle, and horses. Grazing is also an important factor in the arid puna (Braun Wilke 2001; Buttolph and Coppock 2004), where, in addition, the extraction of different types of salt has been affecting the salt flats and their vegetation.

# 6.9 Conclusions

This chapter has given an overview of the variety of alpine ecosystems in the world and some aspects of their functioning. A common theme running through it has been evoking the factors that create characteristic habitats. In general, in any climate, there are topographic and hydrological gradients that largely determine soil physical and chemical properties and thereby growth conditions for living organism and below- and above-ground processes. Along these gradients one may discern extremes, either in the form of a shortage or excess of resources. Specialist organisms usually populate such habitats. The habitats that reflect the average climate (growing conditions) usually best reflect the regional character-giving vegetation, such as the alpine grasslands (sedge heaths) in the European Alps, or tropical alpine heaths and grasslands with pachycaul rosettes. On a world-wide scale, most alpine habitat types are common to all mountains regions; however, the organisms that inhabit them are very diverse, as they reflect the history and evolution of the organisms (and that of the Earth, see Chapter 7). In addition to common features, there are many particularities that belong to certain climates and geographies, such as the extreme aeolian environments in the highest reaches of mountains, above the alpine-nival zone, the cyclic population peaks of small rodents and their impacts on vegetation in boreal mountains, or the extensive salt flats in the arid part of the Andean Altiplano. Our understanding of alpine habitats and the functioning of their ecosystems is largely based on research in few well-studied places in temperate alpine ecosystems. It appears that the functioning of alpine ecosystems is tightly regulated between the below- and above-ground components—microbes and plants. Such seasonal regulation is well established for northern mesic alpine areas, but its details need elaborating in seasonally dry and arid, and aseasonal tropical and subtropical alpine regions. A further challenge is to establish the elasticity of alpine ecosystem functioning in the face of external interventions, such as those grouped under the term 'global change'.

# 7 Biogeography, adaptation, and evolution of alpine organisms

## 7.1 Introduction

The discussion of alpine habitats and their plant communities and animal assemblages so far has represented a snapshot of their present spatial arrangement, our treatment having been deterministic, based on environmental factors and on assumed 'preferences' of living organism found in those habitats. However, these patterns are not static and have undergone large changes at various scales of geography during the history of life on Earth. The processes that result in dynamic patterns of distribution are speciation, ecological diversification, and extinction, together with various dispersal processes. At the level of the individuals and communities, adaptations of various kinds may be used to interpret alpine ecosystem patterns and processes. This chapter summarizes the consensus that is commonly thought to be behind today's patterns of distribution of alpine species that populate alpine habitats world-wide. This is largely made in anticipation of using the insight gained from the past to better predict the future spatial arrangements of alpine habitats, communities, and species (e.g. Crisci 2001). A geographic characterization of high mountain ranges and their climate, hydrology, and soils was given in Chapters 2–5, while Chapter 6 reviewed the plant communities and some of the animal assemblages from a geo-ecological point of view. Here we attempt to give an outline of historical biogeographic explanations, including the formation of alpine floras in relation to historic events. In addition, tolerances and adaptations by organisms that inhabit alpine environments are scrutinized in search of explanations for observed patterns and processes. Chapter 9 on global change impacts adopts an approach of environmental or ecological (biogeographic) considerations.

## 7.2 The backdrop: orogenesis and past climatic changes

To illustrate to what extent geographical barriers have operated in shaping the alpine floras one can compare their composition across high mountain ranges world-wide. For example, about half of the approximately 1500 species strong arctic flora (Abbott and Bochmann 2002) occurs in the Holarctic (Northern Hemisphere extra-tropical) alpine ranges; one readily observed pattern is its decrease in the proportion it contributes to alpine floras from arctic–alpine mountains to Mediterranean latitudes (Fig. 7.1). Such a relationship is the result of the increasing distance from the arctic on the one hand, and is much influenced also by past climatic changes that have affected the floristic richness and composition of mountain regions (Table 7.1).

Species distributions in cases where alpine areas are linked by dry land, or were at some point in the past (e.g. between the British Isles and continental Europe; between Far Eastern Russia and Alaska) may be explained by tectonics, or climate history. Inter-mountain areas are likely to have acted as bridges between alpine 'islands' during glaciations when the alpine zone descended toward lower elevations. Even some of today's shallow sea floor—exposed during glacial events—is likely to have been used by migrating plants and animals.

The distribution and origin of alpine floras has an important, but at a first glance perhaps hidden relevance for the biology of alpine habitats. For example, is there a link between orogenic age and the age and composition of the alpine floras and faunas? What is the relative importance of the above in relation to the evolution of alpine floras in response to successive climatic changes, especially to the one we are witnessing? To answer these questions, one must go back in time and briefly recount the events of the past approximately 65 million years, the epochs of the Tertiary and Quaternary (Willis and McElwain 2002). Most importantly, in those epochs plate tectonics resulted in the last major event of mountain formation, the emergence of the so-called alpine system, including many prominent mountain chains. The same period saw a major peak in global temperatures, followed by a strong cooling. In the early Palaeocene to middle Eocene (approximately 65–55 Ma), while the general configuration of continents had already been established, it is thought that there still existed a land link between North America and Europe, by way of Greenland, between North America and Asia via Beringia, and Australia was attached to Antarctica. The latter was also linked to southern South America via an archipelago. These assumptions have played an important part in explaining contemporary floristic distributions in general, as well as interpreting the origins of today's alpine floras and faunas. The configuration of the continental plates finalized in the middle Eocene to the end of the Miocene (approximately 45–5 Ma). The mountain building events between approximately 55 and 40 Ma included the uplift of

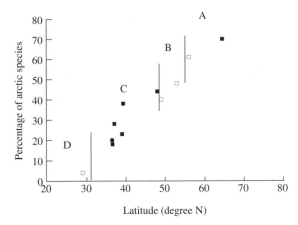

**Fig. 7.1**    The change in the proportion of arctic species in the Holarctic alpine floras of North American (filled symbols) and Asian (empty symbols) mountain ranges. Vertical lines indicate approximate delineations of floral subregions: A, arctic and subarctic floral subregion; B, euro-siberian/western and central asiatic; C, north American; and D, western and central Asiatic/Sino–Japanese/Indian (Holarctic/Palaeotropical). Data from compilation by Ozenda (2002).

the Himalayas and the Tibetan Plateau, and the Cordilleras (including the Rocky Mountains); a second period, approximately 35–25 Ma formed the Alps, Pyrenees, Carpathians, Caucasus, and the Zagros. The last uplift of the alpine system of the European mountain ranges occurred between 10 and 2 Ma ago. A number of other alpine ranges are also of relatively young age (Pliocene to Pleistocene, 4–2 Ma), such as the northern Andes, the New Guinea, and the New Zealand ranges. Some of the events along the above chronology are shown in Fig. 7.2. The Quaternary (1.8 Ma to present) has seen recurrent cycles of glaciation and thaw, with some areas being covered (repeatedly) by ice for long periods. Compared with the climate of the Earth before the alpine mountain formation events, today it is much colder and the carbon dioxide concentration is by far lower (but about twice as high as it was during extensive glaciations).

The arctic and alpine type vegetation is thought to date back as far as approximately 40 Ma in the Southern Hemisphere, where the first glaciations occurred in Antarctica and approximately 10 Ma in the Northern Hemisphere (Arctic); however, the only fossil evidence is from between 2 and 3 Ma. Overall, it is assumed that the basis for the current broad geographic configuration of biomes, including alpine floras and vegetation formations became established approximately 10 Ma ago. The basic distribution of alpine areas has not changed; however, the Quaternary, with its glacial–interglacial cycles caused recurring temporal and spatial rearrangement, making the history of the alpine a cyclic succession of extinction and re-colonization over large expanses, while some refugial

**Table 7.1** Some examples of floras (total and alpine) of mountain regions. The definition of 'alpine' species much varies (see e.g. Ozenda and Borel 2003). Some authors consider only those that are exclusive to the alpine zone, others include those that, although also occur below the alpine zone, have their optimum distribution in the alpine (and treeline ecotone or upper montane zone); yet others base their figures on enumerations of species above the treeline. To illustrate how widely such figure will vary, in Corsica there are 11 species that exclusively occur in the alpine zone, a further 23 have their optimum there (11 + 23 = 34), and the total number of species that has been enumerated in the alpine zone is 113 (Gamisans 2003). Consequently, the figures below are for information and the original sources need to be consulted for definitions

| Mountain range | Estimated area (alpine) (km$^2$) | Total vascular flora | Alpine vascular flora (exclusively alpine) | Mean species richness (*, 4 m$^2$;[†] 4–100 m$^2$;[‡] 25 m$^2$;[§] community richness in samples) |
|---|---|---|---|---|
| Arrigetch, Brooks Range, Alaska | (200) | | 234 (Cooper 1989) | |
| Scandes | | 900 | 463 (Virtanen 2003) | 6–33 |
| Scottish Highlands | (1600) | 1117 | 224 (Nagy 2003) (58) | 3–44;* (9–78[§]) |
| Western Sayan | | | 600 | |
| Altai | 170,000 | c. 3000 | 332 | |
| Signal Mountain, Canadian Rockies | (13) | | 157 (Hrapko and La Roi 1978) | |
| Pyrenees | 40,000 (1760) | 3500 | 420 (Ozenda 2003), 803 (Gómez et al. 2003) (11) | 8–30[†] |
| Alps | 200,000 (15,000) | 4530 | 850 (Ozenda 2002) | 10–40[§] |
| South-east Carpathians | 66,000 (7000) | 1650 | 663 (Coldea 2003; for the south-eastern Carpathians) 300 (Ozenda 1994) | 26–92[§] (total number per plant community; Coldea and Cristea 1998) |
| Corsica | 8748 (100) | 2090 | 131 (total number for (11) habitat/vegetation types; Gamisans 1999) | 19–50§ |
| Apennines | | 3091 | 728 (Pedrotti and Gafta 2003), (Ozenda 2003) 190–220 | 13–31[†] |
| Rila | (740) | 400 | 211 (Roussakova 2000) | |
| Caucasus | 540,000 (80,000) | 6300 | 760 (Nakhutsrishvili 2003) | 2–20; 6–39[‡] |
| Olympic Range | | | 120 | |
| Rocky Mountains | | | 510 | |
| Tushar Mountains, south-west Utah | (19.3) | | 171 (Taye 1995) (13) | |
| Greece | | 5700 | 246 (Strid et al. 2003) | |
| Himalayas | 500,000 | 12,000 | 3000 (Ozenda 2002) | |
| Hawaii | 16,644 | 2180 | 119 (Price 2004) (13) | |
| Kosciuszko, Australia | (100) | 8846 | 212 (Costin et al. 2000) | |
| New Zealand | 58,000 (8000) | 2450 | 613 (Halloy and Mark 2003) | |

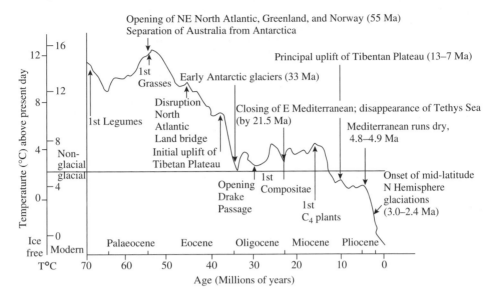

**Fig. 7.2**   Reconstructed temperature trends in the Tertiary and corresponding geological and botanical milestone events (redrawn from Willis & McElwain 2002)

areas remained vegetated. The retreat of the last continental ice sheet and the subsequent plant colonization, expansion, and vegetation succession dynamics have important lessons for making predictions about future changes in alpine flora and vegetation (see Chapter 9 on climate change impacts).

Species richness of the alpine zone in each mountain is a result of a number of factors, such as the extent of the alpine zone, its proximity to other similar environments, its degree of isolation by way of distance or other means, glaciation history, and the length of an environmental gradient within the alpine zone. In whole mountain ranges, species richness and diversity are determined by the extent of the ranges and the degree to which they are restricted to a particular climate zone, or if they straddle climatic zones such as the Himalayas, Andes, Alps, or the Pyrenees; for a classification of ranges see Grabherr (1995). The composition of the floras in each of the alpine mountain ranges is of various origins and distributions. An often-used method to classify floras, including those of alpine areas, is by geographic area (*chorology*, *khoros* gr. place) of known species distribution. This can be made at a range of scales and can result in a confusing diversity of systems that is difficult to use in global comparisons. For example, the system applied to Scotland, based on latitude (arctic-montane, boreo-arctic montane, wide-boreal, boreal-montane, boreo-temperate, wide-temperate, temperate, southern-temperate) and longitude (oceanic, suboceanic, European, Euro-Siberian, Eurasian, circumpolar) by

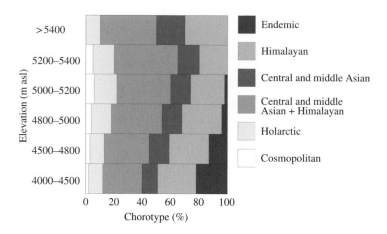

**Fig. 7.3**     Chorotypes of the flora in the Hindukush Mts. Data from Agakhanjanz & Breckle (1995).

(Preston *et al.* 1997) provides a useful summary of the areal distribution (and origin) of the species that make up the Scottish flora. An example at a larger scale is that by Agakhanjanz and Breckle (1995) for the Hindukush (Fig. 7.3). However, for global comparisons less detailed chorologies are in use. A widely used hierarchical system divides the world into floral regions (kingdoms) and subregions. The main subregions that lie over alpine areas and therefore have largely contributed to the formation of alpine floras are listed in Table 7.2; for details on province level information for extratropical Eurasia and for the chorology of the central European flora see Meusel *et al.* (1965). The Holarctic region (North America and Eurasia) with its numerous subregions forms a very large unit, which geographically contains the majority of the high mountain areas of the world (see Fig. 2.1). The separation of Alaska from eastern Siberia is relatively recent and it is likely that the sea floor of the Bering Strait served as a land bridge several times during Quaternary glaciations. The Holarctic mountains have had a long shared history and they are likely to have closer biogeographic and evolutionary connections with each other than with mountains in other flora regions. An example of shared geographic origin of vascular plant, bryophyte and lichen genera and species is that by Qian *et al.* (1999) for two widely separated Holarctic mountains, the Indian Peaks, Rocky Mountains, and the Changbaishan, China (Fig. 7.4). A large body of literature has been synthesized on the biogeographical connections of the alpine zone of Holarctic mountains by Ozenda (2002). The Palaeotropical alpine floras and vegetation have been reviewed by Hedberg (1969, 1970, 1986) for Africa, and by Smith (1982) for Malesia and New Guinea, Price (2004) has provided an account for Hawaii, and Smith and Cleef (1988) have made a pan-tropical review. A comparison along the Andes from

**Table 7.2** The classification of alpine mountain regions by flora region and subregion

| Flora region | Subregion with alpine relevance | Mountain ranges |
|---|---|---|
| Holarctic | Arctic and subarctic | Iceland, Svalbard, northern Scandes, Chibiny Mountains, Novaya Zemlya, Severnaya Zemlya, Polar Urals, Gory Byrranga (Taimyr), Chukhotskoye Nagorye, Koryakskoye Nagorye, Brooks Range, the mountains of Ellesmere, Axel Heiberg, Baffin, Greenland; |
| | Euro-Siberian | Scottish Highlands, south Scandes, Urals, Cantabrian Mountains, Pyrenees, Alps, Carpathians; Caucasus, Putorana, Verkhoyansk, Chersk, Kolmsky, Sredinnyy Khrebet (Kamchatka), Dzhungdzur, mountains of the Baikal region; |
| | Sino-Japanese | Changbaishan, Sikhote-Alin, Japanese Alps; |
| | West and central Asiatic | Sayan, Altai, Tianshan, Pamirs, Hindukush, Kunlunshan and the Plateau of Tibet, Karakoram, Zagros, Elburz; |
| | Mediterranean | Sierra Nevada of Spain, High Atlas, Corsica, Apennines, Dinarids, Hellenids, Taurus, Lebanon; |
| | Atlantic north American | Alaska Range, Canadian Rockies, Labrador Mountains, Newfoundland, Laurentian Mountains; |
| | Pacific north American | Wrangell–St Elias Range, Mackenzie Mountains, Cascades, Rocky Mountains, Sierra Nevada, Sierra Madre Oriental |
| Palaeotropical | North-east African highland | Ruwenzori, Ethiopian Highlands, Kilimanjaro, and other volcanoes; |
| | South African | Maloti–Drakensberg; |
| | Indian | Himalayas; |
| | Malesian | Mount Kinabalu, New Guinean Highlands; |
| | Hawaiian | Volcanoes of Hawaii |
| Neotropical | Caribbean | Sierra Madre del Sur, Sierra Nevada de Santa Marta; |
| | Andean | Cordilleras Occidental, Central and Oriental of Colombia, Cordillera de Merida, Venezuela, Andes of Ecuador and N Peru, central Andes, Cordillera de la Atacama, south Chilean and Argentine Andes |
| Australian | North and east Australian | Australian Alps |
| Antarctic region | New Zealand | New Zealand Alps; |
| | Patagonian | Cordillera Patagonica |

aseasonal wet Colombian páramo, via dry puna in the central Andes, to the Tierra Fuego alpine in the Antarctic flora region is available (Simpson and Todzia 1990); however, an up-to-date in-depth analysis across the Neotropical region is yet to be made (but see Hughes and Eastwood 2006, Box 7.1). The New Zealand and Patagonian alpine within the Antarctic region have been discussed by Mark *et al.* (2001) and the Australian region has been treated by Kirkpatrick (2002).

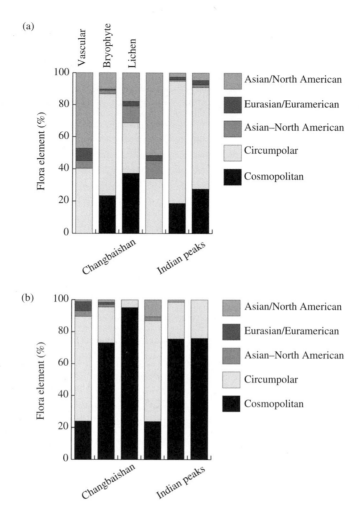

**Fig. 7.4**    The comparison of the biogeography of the vascular, bryophyte, and lichen floras of the alpine summits of Changbaishan, China–N Korea border and the Indian Peaks, Colorado Rockies. The total number of species (top panel, a) recorded on the two summits were: 129 vs. 244 vascular plants, 128 vs. 118 bryophytes, and 67 vs. 109 lichens. The corresponding numbers for genera (bottom panel, b) were 87 vs. 122 vascular plants, 67 vs. 74 bryophytes and 21 vs. 50 for lichens. (Data from Qian *et al.* 1999)

# 7.3  The biogeography of regions

## 7.3.1  **The Holarctic region**

At the level of vascular plant families the alpine zone does not differ from those at lower altitudes. European, central Asian, or Japanese (Hokkaido) alpine floras share Asteraceae, Brassicaceae, Caryophyllaceae, Cyperaceae,

**Box 7.1** Phylogeography

Phylogeographers seek to establish the factors, historical and contemporary, controlling the current distribution of different genealogical lineages within and between plant and animal species. The tools employed fall into two main categories: historic biogeography and genetics. The latter has a focus on geographically discernible patterns within and among lineages or species. Apart from identifying past population expansions/migrations, phylogeography may be connected to the estimation of the rate of speciation (e.g. Kadereit *et al.* 2004; Hughes & Eastwood 2006).

A number of phylogeographical studies have been carried out in alpine systems recently. For example, Comes & Kadereit (2003) have studied the origin of the alpine flora of European high mountains. They have established that a collection of species of Mediterranean and Asian origin (or affinities) that today geographically co-occur in European high mountains colonized the alpine zone relatively recently (Pleistocene). Interestingly, the species appear to have originated from lowland taxa and their colonizing the alpine zone did not follow a common route, or was uniformly accomplished in a single event. Another group of plants, those that survived in refugia in the Alps and recolonized after the retreat of ice was studied by Tribsch & Schönswetter (2003). In a similar vein, Ikeda & Setoguchi (2007) have studied the endemic *Phyllodoce nipponica* in the alpine zone of the Japanese mountains. In the Rocky Mountains alpine refugia, DeChaine & Martin (2005) have concluded that populations of *Sedum lanceolatum* had persisted across the latitudinal range throughout the Quaternary glacial and interglacial cycles. The observed genetic differentiation was most likely to have been caused by putative short-distance migration along altitude in response to climate change, and by species traits such as low dispersal rates and the capacity to reproduce vegetatively. For another species with limited dispersal, *Arabis alpina*, interesting, if predictable, patterns were confirmed by Ehrich *et al.* (2007) in three different alpine systems (north Atlantic with full glaciation during the last Ice Age—resulting in a single lineage, central European and east African with successive glaciations and, both with high genetic diversity). The east African *Lobelia gibberoa*, a giant rosette plant of the upper montane zone appears to have maintained three different strongholds (Ethiopia, and mountains on either side of the east African Rift) for a very long time (Kebede *et al.* 2007).

Fewer estimates are available on rates of species diversification, e.g. Hughes & Eastwood (2006) estimated in the high Andes within the genus *Lupinus*—with 81 species endemic to the Andes—rates in the range of 2.49–3.72 species per Ma. An intriguing finding by Kadereit *et al.* (2004) was the different rates of speciation from that of extinction for European alpine plants, in a temperature-dependent manner. Without a generalizable pattern as yet, temperature appears to be correlated positively with extinction and negatively with diversification. If that holds

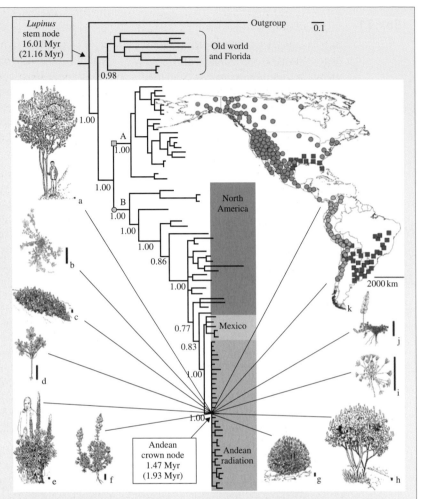

**Fig. 1**     The variety of *Lupinus* (81 species endemic to the Andes) is illustrated in the growth forms the species have assumed: (a) treelet, *Lupinus semperflorens*; (b) prostrate herb, *Lupinus* sp. nov; (c) perennial woody shrublet, *Lupinus smithianus*; (d) ephemeral annual herb, *L. mollendoensis*; (e) giant stem rosette, *Lupinus weberbaueri*; (f) woody perennial shrub, *Lupinus sp.* nov; (g) acaulescent rosette, *Lupinus nubigenus*; (h) perennial woody shrublet, *Lupinus* sp. nov; (i) dwarf acaulescent rosette, Lupinus *pulvinaris*; (j) prostrate herb, *Lupinus prostratus*. (Scale bars for drawings: 5 cm). Phylogeny of *Lupinus*—posterior probabilities of Bayesian analysis for major clades are shown below nodes (for details see Hughes & Eastwood 2006, where the figure is redrawn from). (A) Eastern New World; (B) western North America, Mexico, and the Andes.

true, speciation (diversification) is more likely to have taken place in low altitude glacial refugia, as opposed to high-altitude interglacial refugia. Little glaciation impact was reported on the phylogeographical patterns of the European Alpine salamander, *Salamandra atra* (Riberon *et al.* 2001).

Poaceae, Ranunculaceae, and Rosaceae as the most important families in terms of contributing species (Ozenda 2002). Below the similarity at the family level lies more diversity when comparisons are based on the genus or species levels. About one-third of the alpine floras in Europe are made up of species that belong to 15 non-graminoid genera (Table 7.3); a similar proportion is found for the Himalayas. The same genera contribute about one-sixth of the flora in central Asia (Altai), or in North America. Species-level similarity in comparison with a reference area (e.g. Swiss alpine flora) appears to be inversely correlated with geographical distance (Fig. 7.5), but can be relatively high even between such far apart locations as the Indian Peaks in the Rocky Mountains (244 vascular plant species) and the Changbaishan summit (129 species) that share 72 species; 43% of the Changbaishan species are present on the Indian Peaks and 22% of the Indian Peak species are found on Changbaishan (Qian *et al.* 1999; Fig. 7.4). The differences between floras are, in part, due to different local stock, and also to their local evolution. One facet of the latter is the relatively high occurrence of endemics (taxa not found elsewhere) in most alpine floras. However, not all mountain ranges are rich in endemics. The Scandes or the Scottish Highlands have very few, presumably because of the relatively recent origin of their floras (post last glacial maximum). In contrast, there are nearly 400 endemics in the whole of the European Alps, some of them restricted to the very east, the south, and the south-west, others occurring across the whole of the range; the figure for the Carpathians and the Pyrenees is 116 and 114, or about 20% of their alpine floras. Open habitats, such as rock and scree are by far the richest in endemics, in the European Alps 35–40% of all endemics are found in these habitats. In the Caucasus, there are 1600 endemics in total, with 203 alpine (>1900 m); nine of the 15 Caucasian endemic genera are exclusively alpine (Nakhutsrishvili 2003). The European Mediterranean mountains, which were relatively little impacted by glaciations, have an even higher proportion of endemic species in their alpine floras (e.g. Sierra Nevada, 36%; Corsica, 38%; Hellenids, 40%). Endemism reaches its peak (approximately 50%) in the Himalayas

**Table 7.3** The most species-rich non-graminoid genera in the alpine zone of European and Himalayan mountains

| Region | Genus |
| --- | --- |
| Europe | *Androsace, Artemisia, Astragalus, Cerastium, Campanula, Draba, Gentiana, Pedicularis, Phyteuma, Potentilla, Primula, Ranunculus, Salix, Saxifraga, Viola* |
| Himalayas | *Anaphalis, Gentiana, Leontopodium, Pedicularis, Polygonum, Primula, Rhododendron, Saussurea, Saxifraga, Swertia* |

*Source:* compilation by Ozenda (2002).

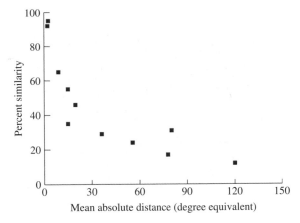

**Fig. 7.5**   The percentage similarity of the floras with reference to the Swiss alpine flora. The percentage values are those of the Swiss flora elements found in each of the local floras. Note the relationship between geographic distance and species similarity. Data from Jerosch (1903).

that lie at the intersection of climatic and flora regions. A similarly high value is found among the highest growing species in the central Caucasus; however, the overall rate of endemism there is about 25%. In central Asia, there seems to be a decline in the proportion of endemism with aridity (22% in the western Tianshan versus 1.7–4.6% in the Pamirs).

## 7.3.2 North American ranges

It is thought that North American alpine floras descended from those evolved in the central Asian Highlands and the Rocky Mountains chain during the period of orogenesis in the Miocene and Early Pliocene (Bliss 1985, and references therein). These centres provided the alpine stock for the North American alpine regions through migration during the Late Tertiary and Pleistocene. The accession of local elements from lower altitudes has provided a further complement. Many arctic species are present, most prominently in the northern Appalachians and in the Rocky Mountains (Figs 7.1 and 7.6).

Analyses of similarity of floras along the Rocky Mountains chain have shown an increasing dissimilarity with latitude distance; other factors related to dissimilarity were various measures of isolation and the presence of barriers to migration (Hadley 1987). A distance-related (but not latitude) dissimilarity was also found by Taylor (1977) in the Cascades–Sierran ranges, and by Cooper (1989), who has found that the flora of the Arrigetch Peaks Region, Brooks Range shared 21–35% species in common with other arctic and Alaskan and other North American alpine ranges

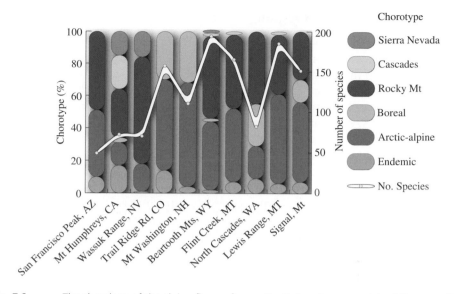

**Fig. 7.6** The chorology of the alpine floras of some North American mountains (after compiled data from Bliss 1985).

north and south, and as far as the Rocky Mountains, and the Altai in central Asia. Most importantly, this study has demonstrated the close links between the North American and the Asian alpine floras, as 78% of the 234 vascular plant species recorded in Arrigetch also occur in Asia. Cooper took his work further and analysed the geographical and habitat affinities of dominant and characteristic taxa, according to their distribution. He found that late-lying snowbeds, ridges, screes, south-facing bluffs, and meadows were dominated or characterized by Beringian or Alaska–Yukon endemic taxa. In contrast, circumpolar arctic-montane taxa characteristically occurred on mobile valley slopes, fens, marshes, and sometimes on rock outcrops. The author suggested that the alpine plant communities of the continental and semi-arid portions of interior and northern Alaska provide a reasonably good analogue of the vegetation that is likely to have existed in Beringia at the end of the last glacial maximum and in the early Holocene. Palynological records from the same period tallied well with the herb taxa that are present in today's arctic–alpine vegetation.

The level of endemism in North America appears to be lower at approximately 5–10% than that found in Eurasian temperate mountain ranges (compare above cited and Fig. 7.6). One exception is in the Arrigetch, Brooks Range where Cooper (1989) recorded 19% Alaska–Yukon endemics. This latter figure is similar to the degree of endemism reported for most Eurasian ranges and is far higher than the rate of endemism found in northern European ranges, presumably owing to contrasting glaciation histories.

## 7.3.3 The Palaeotropical region

Two main groups of mountains belong to the Palaeotropical region: the high mountains of Africa, and Malesia with New Guinea. The alpine areas in Africa are found largely in the East African mountains and the Ethiopian Highlands, and a small ill-defined area on the West African Mount Cameroon. The alpine zone in the Malesian subregion comprises the New Guinean Highlands and Mount Kinabalu. The altitude of both Mount Cameroon and Kinabalu makes them borderline cases for being considered as alpine. In addition, Mount Cameroon is an active volcano and its eruptions have limited the development of an alpine vegetation, although there is a distinctly high mountain suite of species on the high ground. Mount Kinabalu represents a treeline ecotone type vegetation. The high elevation zones on two of the Hawaiian Islands make a special case for island biogeography (Price 2004).

Most palaeotropical mountains are relatively young (Table 7.4). An analysis of the composition of the genera that make up their alpine floras has shown that most have distributions to either to the northern or southern temperate zones (and between one-third and a half of them if fact grows in both hemispheres). The northern temperate genera are more frequent in Africa in general (21% in Africa versus 15% in Malesia), but not on Mount Cameroon, and the southern temperate ones in Malesia (28% versus 15% in Africa); genera that are found in both hemispheres make up 48% in Africa versus 35% in Malesia (Smith and Cleef 1988). This leaves a total of 16–22% for regional endemic and tropical genera to contribute to the alpine floras (although they make up 42% of the taxa combined for endemic afromontane and south African elements for East Africa (Hedberg 1986; Table 7.5). Given such composition of the genera, it has been suggested that the alpine floras of the Palaeotropics developed from immigrants from temperate regions. Smith (1982) has argued that Quaternary glaciations reduced distances for dispersal within New Guinea; in addition, despite large distances from potential sources in Malesia, Australia, or New Zealand, the very occasional (Smith quotes one in every 8000 years) success in large distance dispersal could account for today's alpine flora in New Guinea.

Such 'homogeneity' of the origin of the alpine flora at the generic level should not convey the impression of similarities at the species level either within the Palaeotropics, or with alpine areas outside the tropics. During the glacial periods most land in New Guinea over 3600 m had ice caps, some with valley glaciers (Hope and Peterson 1975) and the area of the alpine zone during and following the last glacial maximum was manyfold larger than today (Hope 1980). The alpine flora of Malesia can be thought of as a single flora with local variants (Smith 1980), found on the high mountain grasslands of New Guinea, and to a lesser extent on other high mountains of the region. Most species above the forest zone are non-specialists, with a broad ecological tolerance, i.e. they occur in a

**Table 7.4** The age of some palaeotropical alpine mountain ranges

| Mountain range | Geology | Tectonic age (age of flora) | Glaciation history |
|---|---|---|---|
| *Hawaii* | | | |
| Hawaii | Volcanic | 0.6 (0.23)* | Pleistocene |
| Maui | Volcanic | 2.0 (1.2)* | Pleistocene |
| *East Africa* | | | |
| Ruwenzori | Precambrian | Mid Tertiary | Extant |
| Mount Elgon | Volcanic | 15 Ma | Pleistocene |
| Mount Kenya | Volcanic | 2 Ma | Extant |
| Mount Meru | Volcanic | 0.2 Ma | – |
| Virunga group | Volcanic | Active, some ongoing cone building | – |
| Kilimanjaro | Volcanic | 4–2 Ma | Extant |
| Ethiopian Highlands | Volcanic | Mid Tertiary | Pleistocene (Bale) |
| Mount Cameroon | Volcanic | Active, some ongoing cone building | – |
| *South-east Asia* | | | |
| New Guinea | Sedimentary intrusive | Approx. 4 Ma | Pleistocene/Extant |
| Mount Kinabalu | Intrusive | 1.5 Ma | – |

*The age of flora is different from the Orogenic age where it is known that a mountain range has been entirely glaciated. In such cases the age of the flora is taken as post Last Glacial maximum (Price 2004).

**Table 7.5** Flora elements in the African alpine flora (Hedberg 1986)

| Flora element | Number of taxa (%) |
|---|---|
| Endemic afro-alpine/montane | 82 (32) |
| South African and Cape | 25 (10) |
| Southern temperate | 6 (2) |
| Northern temperate | 34 (13) |
| Mediterranean | 18 (7) |
| Himalayan | 8 (3) |
| Pan-temperate | 87 (33) |
| Total | 260 (100) |

number of different habitats. Many species and most genera are common to the alpine floras in more than one island, but few occur in lowland habitats below about 1000 m. No comparable megaphytes with those of Africa and the Andes are found in New Guinea. The flora is largely composed of Malesian endemic species (74%); however, most of the species belong to genera that are cosmopolitan or distributed on both the northern and Southern Hemispheres (Smith 1982). Interestingly, single mountain endemism is low at 2% on Mount Wilhelm, and 30% of the species grow outside New Guinea. This is in stark contrast with that found on East African mountains where 69% of the species are single mountain endemics and only 3% grow outside East Africa (Hedberg 1969). Smith (1980) has attributed the low level of single mountain endemism to the short time elapsed since isolation of established plant populations in the alpine zone of the New Guinean mountains. In addition, it has been argued that rapid evolution of endemics was restricted to a single or few mountains and migration during glacial periods had a major role in the composition of today's alpine flora (Smith 1980). The high level of polyploidy in the New Guinea alpine flora (67% versus 49% on Mount Cameroon, and 45% in East Africa; Morton 1972) suggests relatively recent history.

Hedberg (1970) has suggested that long distance dispersal (from outside Africa) was behind the initial composition of the alpine flora on the East African mountains. The establishment of such a flora was then followed by *in situ* speciation, hence the large proportion of single mountain endemics. The high degree of endemism indicates that the isolation of the mountains from each other and from temperate areas has not been broken for a very long time. Some of the main characteristics of the African alpine flora include its paucity in terms of species numbers (a total of 260; Hedberg 1997), the occurrence of the character-giving pachycaul species of *Lobelia* (Campanulaceae) and *Senecio* (Asteraceae), and the wide altitude distribution of many of the constituent species (Hedberg 1969). There is an interesting contrast between *Senecio* and *Lobelia* on Mount Ruwenzori, where species of *Lobelia* form an altitude sequence: *L. gibberoa* (upper montane), *L. lanuriensis* (subalpine), *L. bequaertii* (upper subalpine, mires), *L. wollastonii* (alpine); meanwhile, a single *Senecio, S. adnivalis* occurs from the treeline ecotone to the subnival zone.

The Hawaiian alpine flora, totalling 119 species, of which 84% are Hawaiian endemics, is distributed on two islands. Interestingly, the treeline ecotone–alpine area on Maui is about 10-fold smaller than on Hawaii, yet it has more species (102 versus 90), a higher percentage of single island endemics and a higher rate of speciation. These differences are attributable to the approximately sixfold older substrate age on Maui (Price 2004).

It is difficult to evaluate the immediate adaptive significance of the origin and evolution of the flora for withstanding the vagaries of long-term climate change. Morton (1972) has argued that the lower hills of West Africa can harbour plant assemblages that are normally found at higher altitudes (see also Wood 1971) and they can act as stepping stones during species migrations in response to climate change. Pleistocene climatic changes have caused the forest line to ascend or retreat, and the alpine areas to decrease or increase. A decrease in area might have forced the existing species into close proximity, which enhanced cross-breeding or hybridization (such as in the case of the *Ranunculus montanus* group in Europe; Landolt 1954, 1956). On forest retreat and expansion of the alpine zone these hybrids might have evolved into distinct taxa. Fundamental to predicting species responses to ongoing climate change would be to understand what adaptive traits make successful colonizers/survivors with regard to habitat change and dispersal ability and available dispersal agents. Relatively little is known about the dispersal ability of species that occur in the alpine zone. Hedberg (1970) has attempted to classify East African species by their mode and agent of dispersal and emphasized the importance of presumed zoochory (by mammals—propagules dispersed through attachment to animals' feet in mud) and anemochory, totalling over 80% (see also the section on aeolian environments in Chapter 6 for the importance of wind dispersal).

## 7.3.4 Neotropical region: the Andean subregion

The Andes in South America span aseasonal humid tropical, semi-seasonal and arid tropical, subtropical, temperate and boreal lowland climates, which provide a varied biogeographical framework for species origins and distributions. The floras of the separate biogeographic regions of the high Andes are the products of these geographical, ecological, and geological factors: distances from source areas of propagules, climates, and differential effects of Pleistocene glaciations. The Andes fall in two major biogeographic subregions: the Andean and the Patagonian, the former in the Neotropics and the latter in the Antarctic region. Within the Andean subregion there are two main subdivisions (*páramo* and *puna*). The páramo extends from 11°N at Nevado de Santa Marta, Colombia to approximately 9°S in Peru; the puna ('the high grounds of the cordilleras' in the Quechua language) of the central Andes, not only is a biogeographic region, but it has been the centre of civilizations, with a high past population density and associated human activities (Monasterio 1980). The geographic distribution of the tropical alpine flora of the humid páramos at the genus level include 27% endemics, 24% Southern Hemisphere temperate, 36% with both Northern and Southern Hemisphere distribution, 4% Northern Hemisphere, and about 9% pan-tropical (Smith and Cleef 1988). Simpson and Todzia (1990), who have compared potential origins for the Colombian páramo, north-west Argentine puna, and the Tierra de Fuego 'southern temperate/boreal alpine' (Table 7.6) suggested that the differences observed were caused by (1) the source of the origin of the floras, and (2) the relative uniformity of the habitats of the puna and the southern boreal alpine. In the semi-arid puna and the southern boreal it appears that the main source of taxonomic diversity was immigration by way of the random arrival of propagules. In contrast, in the humid tropical Andes there had been a high level of autochthonous speciation

**Table 7.6** Percent spectrum of the potential sources of the genera in the high Andean flora (after Simpson and Todzia 1990)

| | Origin of flora element | Páramo–Colombia (n = 185 genera) | Puna–north-west Argentina (n = 171 genera) | Southern boreal alpine–Tierra del Fuego (n = 110 genera) |
|---|---|---|---|---|
| 1 | Austral | 9.7 | 3.5 | 20.0 |
| 2 | South American temperate | – | 6.4 | 16.3 |
| 3 | Andean temperate | – | 4.1 | 4.5 |
| 4 | Andean neotropical | 18.6 | 14.6 | 5.5 |
| 5 | Neotropical | 14.6 | 11.1 | – |
| 6 | Subtotal: south American | 33.0 | 36.2 | 26.4 |
| 7 | Cosmopolitan tropical | 8.1 | 9.4 | 0.9 |
| 8 | Subtotal: tropical | 41.0 | 35.1 | 6.4 |
| 9 | Cosmopolitan general | 9.2 | 9.4 | 3.6 |
| 10 | Cosmopolitan temperate | 23.2 | 10.5 | 21.8 |
| 11 | Subtotal: cosmopolitan | 40.4 | 29.2 | 26.3 |
| 12 | American | 7.0 | 17.5 | 3.6 |
| 13 | Holarctic | 9.7 | 13.4 | 23.6 |
| 14 | Subtotal: northern temperate | 33.0 | 24.0 | 45.4 |

6 = 2 + 3 + 4 + 5; 8 = 4 + 5 + 7; 11 = 7 + 9 + 10; 14 = 10 + 1

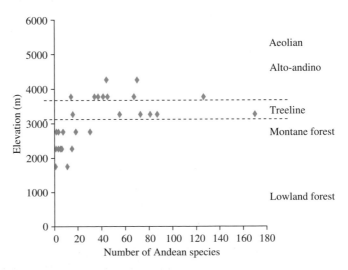

**Fig. 7.7**     Speciation in genera of putative Northern Hemisphere origin (each symbol) in the Andes (redrawn from Hughes & Eastwood 2006)

in the páramos. It is clearly apparent that Northern Hemisphere genera have undergone high levels of speciation in the treeline ecotone–alpine zone of the Neotropical Andes (Fig. 7.7). Simpson and Todzia (1990) have also demonstrated that similarities in climate are at least as important as geographical distance in determining the composition of high Andean floras. Regional studies in the Ecuadorian páramo have agreed that distance among mountains alone did not explain floristic similarities. Altitudinal range of the species was found indicative of their geographic distribution: the wider the altitude distribution of a species is the more likely that it is geographically widely distributed.

## 7.3.5 Australian region

The geological age of the Australian alpine zone is old, dates back to Ordovician deposits and Ordovician to Devonian uplift, followed by a second uplift in the Tertiary (Ollier 1986). Today's alpine flora has strong endemic (Australian), cosmopolitan, and southern temperate elements (Fig. 7.8). In the Mount Kosciuszko area, which has a former heavy grazing history following European settlement, there are 14 introduced species that have established, making up approximately 7% of the alpine flora.

## 7.3.6 Antarctic region

This region is very varied as it encompasses the Andes from the subtropical Mediterranean middle Chilean domain to temperate and boreal latitudes all the way to Tierra de Fuego in South America and the oceanic temperate New Zealand Alps. While the Neotropical Andean region has closer links with North America, the southern South American biota were shown to have closer connections with Australia–Tasmania, New Guinea, and New Zealand (Crisci *et al.* 1991). The Andean (highland) genera of the Asteraceae of Chile show strong Neotropical and amphi-hemispherical connections, along with high endemism (Moreira-Muñoz and Muñoz-Schick 2007; Table 7.7).

The New Zealand alpine habitats are relatively young (having uplifted between approximately 5 and 2 Ma) and have been shaped by repeated Pleistocene glaciations. It is interesting that at the species level, 90% of the alpine flora are endemic, but only 10% are at the genus level (Mark and Adams 1995); in addition to the endemics, there are Australian, southern temperate, and cosmopolitan elements. This indicates a high degree of speciation (Fig. 7.9), after being colonized mainly by plants of Northern Hemisphere origin, but derived from Australia, as a result of long-distance dispersal (e.g. Raven 1973; McGlone *et al.* 2001; Winkworth *et al.* 2005). It has also been suggested that part of the endemic New Zealand alpine flora might have been derived from an

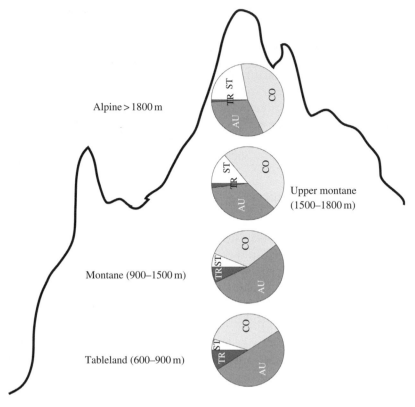

**Fig. 7.8**    Geographic flora elements in the Australian montane and alpine flora. Based on data from the Monaro Region, New South Wales (Costin 1954). AU, Australian; CO, cosmopolitan; ST, southern temperate (Antarctic); TR, tropical (Malesian).

Antarctic mountain flora, dating back to the time before New Zealand and Australia separated from Antarctica about 60 million years ago (Mark and Adams 1995).

## 7.4  Species richness and the applicability of the theory of island biogeography to the alpine zone mountains

The theory of island biogeography is often alluded to in relation to species diversity on alpine mountain peaks (e.g. Körner 2000). This is based on the superficial similarity to an island-like existence of alpine ecosystems in many geographical localities. Typically, questions are asked in relation to 'island size' and distance to contributing gene sources. For example, Hadley (1987) reported that the area above the treeline, distance to other alpine areas and latitude were moderately correlated with species richness

**Table 7.7** The distribution and origin of Asteraceae found in Chile (Moreira-Muñoz and Muñoz-Schick 2007)

| Element | Subelement | Definition of area | Total genera | List of genera |
|---|---|---|---|---|
| 1. Pantropical | | Tropics | 8 | *Achyrocline, Centipeda, Conyza, Cotula, Mikania, Sigesbeckia, Spilanthes, Wedelia* |
| 2. Australasiatic | 2.1 Australasiatic | Southern Hemisphere: America, Australasia | 4 | *Abrotanella, Lagenophora, Leptinella, Trichocline* |
| 3. Neotropical | 3.1 Wide Neotropical | North-west USA, Mexico to Chile | 19 | *Ambrosia, Baccharis, Galinsoga, Gamochaeta, Heterosperma, Schkuhria, Stevia, Tagetes, Coreopsis, Grindelia, Trixis, Viguiera, Villanova, Haplopappus, Helenium, Verbesina, Erechtites, Ageratina, Chaptalia* |
| | 3.2 Andean | Costa Rica, Colombia to Chile | 8 | *Aristeguietia, Chuquiraga, Cuatrecasasiella, Diplostephium, Mutisia, Perezia, Xenophyllum, Werneria* |
| | 3.3 Altiplanic | Altiplano Peru, Chile, Bolivia, Argentina | 17 | *Aphyllocladus, Chaetanthera, Chersodoma, Helogyne, Leucheria, Lophopappus, Lucilia, Luciliocline, Mniodes, Nardophyllum, Nassauvia, Pachylaena, Parastrephia, Plazia, Polyachyrus, Proustia, Urmenetea* |
| | 3.4 South Amazonian | Andes and southern Amazonia | 8 | *Chevreulia, Dasyphyllum, Facelis, Micropsis, Noticastrum, Ophryosporus, Picrosia, Tessaria* |
| 4. Antitropical | 4.1 Wide antitropical | Cool regions, both hemispheres | 10 | *Adenocaulon, Antennaria, Artemisia, Aster, Erigeron, Hieracium, Hypochaeris, Pluchea, Solidago, Taraxacum* |
| | 4.2 Circum-Pacific | Temperate regions: North America, South America and Australasia | 4 | *Flaveria, Gochnatia, Microseris, Soliva* |
| | 4.3 Pacific-antitropical | Chile–western USA | 12 | *Agoseris, Amblyopappus, Bahia, Blennosperma, Encelia, Flourensia, Gutierrezia, Lasthenia, Madia, Malacothrix, Perityle, Psilocarphus* |
| 5. South-temperate | | Temperate: Chile/ Argentina | 10 | *Belloa, Brachyclados, Doniophyton, Chiliophyllum, Chiliotrichum, Eriachaenium, Gamochaetopsis, Lepidophyllum, Macrachaenium, Triptilion* |
| 6. Endemic | | Continental: Chile/ Chilean islands | 17 | *Acrisione, Calopappus, Centaurodendron, Dendroseris, Guynesomia, Gypothamnium, Leptocarpha, Leunisia, Lycapsus, Marticorenia, Moscharia, Oxyphyllum, Pleocarphus, Podanthus, Robinsonia, Thamnoseris, Yunquea* |
| 7. Cosmopolitan | | Worldwide, most continents | 4 | *Bidens, Centaurea, Gnaphalium, Senecio* |

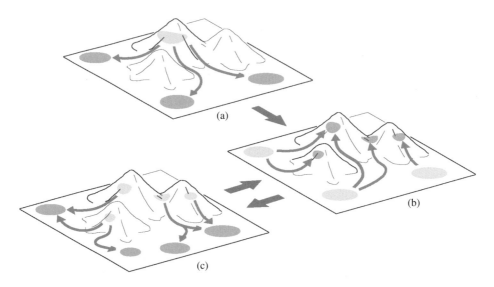

**Fig. 7.9**     A conceptual model offered by Winkworth *et al.* (2005) to hypothesise the mechanism behind the diversification of alpine plant lineages during the Pliocene and Pleistocene (2–4 M years ago) in New Zealand. Such a model has a general relevance for alpine speciation: a–range expansion and establishment of isolated daughter colonies; b–range contraction and establishment of population in new habitats (different from those in a), chance of hybridisation; c–repeated alternation of a and b leading to further chances of speciation (from Winkworth *et al.* 2005).

among 13 alpine areas in the Rocky Mountains chain; the area of alpine best explained species richness, followed by proximity and latitude. To go further and explore immigration and extinction equilibria, however, is near impossible. Analyses of apparent floristic dissimilarities and their relation to environmental variables usually do not provide an explanation. The explanation is substituted by references to glaciation history and to the contraction and displacement towards higher elevations of the alpine zone after post-glacial maximum extension (e.g. Fig. 7.10). Extinction is immeasurable in the absence of one or other form of palaeo evidence, or at best patchy because of the problems related to identifying palaeo taxa. An alternative, or complementary to using palaeo evidence is the not entirely risk-free approach of using an analogous contemporary climate and its corresponding flora, applied as a baseline, against which changes in the flora of an area of interest (for estimating rates of immigration and loss over a period of time) is compared. The alternative estimation of rates of immigration and speciation of extant taxa is being targeted by way of applied phylogeographic methods (e.g. Kadereit *et al.* 2004; Hughes and Eastwood 2006; see Box 7.1).

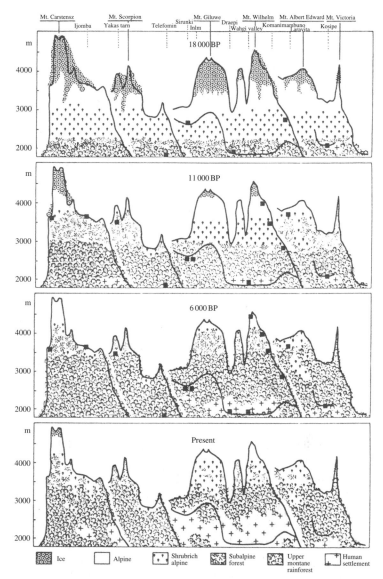

**Fig. 7.10**   A schematic reconstruction of major vegetation changes in the New Guinea Highlands over the last 18,000 years (from Hope 1980). Notice the contraction of glaciers and the alpine zone, and the concurrent expansion of forest. Black squares indicate sites sampled for the reconstruction.

What is usually left of the island biography approach in an alpine context is to relate species richness to the extent of alpine areas. The size of the alpine area is linked to the length of the gradient within the alpine zone, where a species–area power function is applied. However, as the number and proportion of azonal habitats is not necessarily related to area *per se*, and as azonal habitats occur below the alpine zone, azonal habitats can contribute disproportionately more species on an area basis than zonal habitats (see e.g. McGlone *et al.* 2001 for New Zealand). The total vascular plant species richness of the Andean páramo ecosystem (which occurs above the upper montane zone, above altitudes of approximately 3500–3800 m) is estimated to be between 3000 and 4000, with approximately 60% of them being endemic (Luteyn *et al.* 1992). The estimated number of species in Europe's alpine areas, including the treeline ecotone, and excluding the Caucasus, is 2500 species and subspecies (Väre *et al.* 2003); the exclusively alpine zone species for the Alps are estimated to be 800–1000 species (Ozenda and Borel 2003). This is in stark contrast with the alpine zone, above the ericaceous zone on the isolated mountains of East Africa, where vascular species richness varies between 77 and 182 on individual mountains and totals approximately 240 altogether (Hedberg 1992).

Using an island biogeography approach to describe the relationship between the extent of the alpine zone (within a mountain, within a range, or among ranges at the continent-scale) and species richness is not straightforward. A species–area relationship determined empirically for a particular set of mountains (islands) by using the power function ($S = cA^z$, where $S$ is number of species; $A$ is area; $c$ and $z$ are the fitted intercept and slope constants) may not be a good predictor for other areas as the relationship changes over scales (see chapter 5 in Hubbell 2001); this was demonstrated by Halloy and Mark (2003), who sought a suitable predictor for the future size of the New Zealand alpine flora under forecast warming scenarios. Even with the best fitting function at the relevant scale, the attached confidence intervals are broad (i.e. the accuracy of the predictions is low), as a result of a variety of reasons connected to heterogeneity, such as latitude and associated differences in environmental history (e.g. see Fig. 7.11 for the Arrigetch, Brooks Range, Alaska from Cooper 1989). A further point to consider is the foundation of the island biogeography theory, which is a neutral theory in that it does not differentiate among species in their ability to colonize, survive, or become extinct. In other words, its use in predicting future species presences in the alpine is restricted, as different species do carry different adaptations (see Section 7.5 below) by which they can live, for example in a broad temperature range (most alpine species, while they have their optimum in the alpine zone, are not restricted to it and occur in the upper montane forest zone or in avalanche tracks penetrate the montane zone). For example, McGlone *et al.* (2001) proposed a split of 228 versus 343 alpine specialist versus non-specialist for the New Zealand alpine flora and showed that non-specialist were found, on average, in about three times as many regions as specialists. This may be related to the existence of the

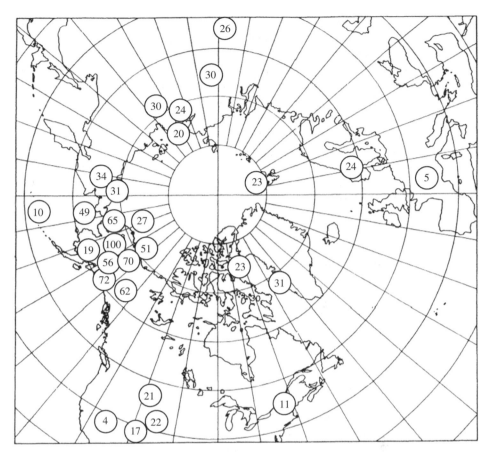

**Fig. 7.11**    Flora similarities between the Arrigetch, Brooks Range, Alaska and of selected Holarctic arctic and boreal mountains (reproduced after Cooper 1989). Values are percent of Arrigetch flora in other floras. The high similarity with other north Alaskan and Yukon mountains rapidly falls off with distance, indicating the high incidence of Beringian-Yukon endemics at Arrigetch.

vast open gravel beds along rivers that have contributed to the evolution of the young alpine New Zealand flora. Another example to demonstrate the limits of predictions based on a species–area relationship is illustrated by that of the contemporaneous alpine flora of Scotland. The equilibrium assumptions about immigration and extinction of the island biogeography theory clearly do not apply if one considers the extent of the alpine zone and its species richness alone that has remained since the last full re-colonization after melting of the ice shield. If one reconstructed the flora and vegetation after the last glacial maximum and modelled the successive diminution of area in the last 18,000–15,000 years (Birks 1986; compare with Fig. 7.10 from New Guinea) one could predict the complete disappearance of the flora in Scotland by today (Nagy *et al.* 2008). In fact, there are about 50 vascular plant species that are still surviving, some in very small numbers and restricted to particular local habitats. This is probably a result of local cool oceanic climate that depresses the treeline and has allowed the maintenance of open alpine habitats on the one hand (few species covering relatively 'large' areas), and the topography-related habitat diversity that has allowed a number of rare species to survive ('many' species with a restricted niche). The consequence that arises from this example is that the application of a species–area relationship to extant alpine areas across latitudes needs to be made with care. Areas that have been undergoing unidirectional change (e.g. north-western Europe) as a result of a major long-term environmental driving force are different from areas that have been in a quasi-balance state with regard to immigration (and speciation) and loss (extinction), such as the Alps, for example.

The degree of isolation can vary and can be manifest temporally and spatially. Spatial segregation of lone volcanic cones in East Africa is a textbook example of the impact of isolation on speciation. While overall species richness is low, the composition of the individual mountain floras is rather different because of speciation that followed initial colonization (see various references by O. Hedberg).

At the continent scale, glaciation history can much influence species-richness patterns in that relatively young floras (post-glaciation establishment, e.g. Scandes in Europe) are smaller than those that partially survived glaciations in local refugia, such as that in the Alps. Nevertheless, such comparisons are complicated by the size of the alpine areas. While the Scandes may be compared with the Alps, there is no European Mediterranean massif comparable in size whose flora could be meaningfully compared with those of the Scandes or the Alps. Sample-based comparisons (in the absence of taking into account the size of the total flora) for Europe demonstrated that the history of mountains (both in terms of geology and climate) could have overriding impacts on species richness (Virtanen *et al.* 2003). The authors have also found that different geologies had different local species-richness–regional species-richness relationships, and community types (habitats) also differed in this respect. Only snow-protected communities (mesic sedge heaths) on calcareous substratum had a positive correlation between local and regional species richness; most communities on acid soils/bedrock had no positive relationship. Few instances are available on the impact of sustained long-term grazing on taxonomic and growth form composition. A comparison of the Rocky Mountains and Caucasus tentatively suggested that plant family patterns and morphology might reflect the long presence of domestic livestock in the Caucasus (Bock *et al.* 1995).

# 7.5  Adaptation, survival, and evolution of alpine organisms

Plants are adaptable in a general sense because of their modular organization. Alpine biological diversity is underpinned by the multitude of contrasting mosaics of habitat conditions such as microclimate, parent rock, hydrology, and land use. Adaptation to high mountain conditions can be perceived in manifesting in a variety of 'strategies', reflecting the varied habitat conditions, rather than in adaptation to a 'general alpine' environment. Several species that grow in the alpine zone also occur at lower elevations and some species occur across wide latitude ranges, and consequently experience drastically different environments, such as *Arabis alpina*, whose range extends from the Arctic (with a short growing season and 24-hour light) to tropical East Africa (with a growing season that is all year round, but has a high diurnal temperature variation). The variety of the origins and history of alpine species, plants and animals, the varied environments they grow in under the umbrella term of 'alpine' have resulted in a diverse array of adaptations (Tables 7.8 and 7.9; Fig. 7.12). Adaptations are required at the individual level to survive and optimize growth and development. This is achieved by physiological means in the first instance, and in many instances building on pre-adaptation via morphological/anatomical specialization (e.g. Table 7.9). Genetic means of adaptation underlie adaptation from generation to generation. During this evolutionary process 'strategies' can vary again; species at the infraspecific level can develop adaptive traits with regard to their physiology, or, conversely, they may maintain no genetic variation that can be correlated with environmental variation. As will be apparent, generalizations about adaptations by plants to alpine life conditions are difficult to make, as when one discovers a trait that is clearly perceived as an adaptation in one group (species) the majority of its congenerics or confamilials, growing side-by-side with it, may appear to grow perfectly well without the same kind of adaptation (e.g. Fig. 7.13).

In general terms, plant adaptation to low temperature and low available nutrients in temperate alpine conditions include the incidence of growth before, and rapid growth after snowmelt, early burst of growth, sometimes using stored carbohydrates—mostly in below-ground organs (see high root to shoot ratio), selection of life form (chamaephyte and hemicryptophyte), and morphological features such as flower shape (e.g. parabolic shape is perceived as a way of concentrating heat in the reproductive parts of flowers), pubescence (protection against temperature and water stress), and low stature (cushion, prostrate shrubs) (e.g. Wielgolaski 1997c). Two excellent volumes by Körner (2003) and Larcher (2003) provide an in-depth discussion of the ecophysiological basis for plant functioning in alpine environments; plant adaptations to cold, drought and fire, plant

**Table 7.8** Some broad general features of alpine areas and their plants by climate zones

| | General feature | Growing season | Typical growth forms/anatomy/physiology | Root:shoot/reproduction |
|---|---|---|---|---|
| Arctic | Strongly seasonal temperature and light; low precipitation; uneven snow distribution, long snow-lie | 30–70 days (depending on altitude and topography); continuous light; little loss of heat during the night | Evergreen dwarf-shrub | High, up to approx. 10; Predominantly vegetative; flower primordia are produced in the year(s) before flowering |
| Boreal | Strongly seasonal temperature; uneven snow distribution, long snow-lie | 80–150 days | Evergreen dwarf-shrub | High, approx. up to 10; Largely vegetative; flower primordia are produced in the year(s) before flowering |
| Temperate | Strongly seasonal temperature; uneven snow distribution, long snow-lie | 50–220 days (depending on altitude); cold nights | Graminoids, dwarf-shrubs; abundant clonal growth | High approx. up to 10; Largely vegetative, flower primordia are produced in the year(s) before flowering; good years can produce large numbers of seeds/seedlings |
| Subtropical | Seasonal, with or without summer drought; uneven snow distribution, long snow-lie | 150–250 days, with or without summer drought | Hemispherical spiny cushion-like shrubs; water conservation, anti-grazing defence | Approx. 0.5 |
| Arid subtropical | Seasonal temperature; very little precipitation throughout, shallow snow cover (cold desert) | Short, arid | Bunch grass (insulation by sheaths of the inner), sclerophyllous shrub (water conservation, deep rooting, e.g. *Fabiana* spp. of the tolares) | ? |
| Seasonal tropical | Slight but increasing seasonality in precipitation and temperature with distance from Equator; little occasional snow-lie | Little restriction on year round growing season length | Bunch grass; sclerophyllous shrub | Bunch grasses: <1; some other species >1 |
| Aseasonal tropical | Aseasonal temperature and precipitation (mostly rain), and light; high amplitude diurnal temperature variation—no snow-lie; little wind (and snow and ice abrasion) | 365 days—all year round | Giant rosette; bunch grass; acaulescent rosette (night closure of leaves around meristem, supercooling), longevity | Bunch grasses: <1; some other species >1 |

**Table 7.9** Adaptation in insects to arctic conditions

| Adaptation | Attributed advantage | Examples |
|---|---|---|
| *Colour and structure* | | |
| Melanism | Solar heat gain in cool environments | Many insects |
| Hairiness | Conserves heat gained, in larger insects | Many insects |
| Robustness | Moisture conservation | Some ichneumonids |
| Reduced wings | Energy saved (flight limited by cool conditions) | Several flies, moths |
| Reduced eyes | Normally nocturnal forms are active in light in Arctic | Some noctuid moths |
| Reduced antennae of males | Detection of females during swarming flights is not necessary when flight is limited | Some chironomid midges |
| Small size | Species can use limited resources | Many species |
| *Activity and behaviour* | | |
| Selection of warm habitats for activity or progeny production | Activity allowed in cool environments in locally favourable sites | Most species |
| Selection of protected sites for overwintering (e.g. plant clumps) | Protection against cold or wind-driven snow | Many species (but some species overwinter fully exposed) |
| Low temperature thresholds for activity | Activity allowed in cool environments | Several species |
| Opportunistic activity | Activity whenever temperatures are permissive | Bumble bees, midges, etc. |
| Basking | Heat gain in sunshine | Many butterflies, etc. |
| *Physiology and metabolism* | | |
| Adjustment of metabolic rate | Life processes continue in cool environments | Some species of various groups |
| Resistance to cold | Survival during cold winters | All species |
| Resistance to starvation | Survival when food supplies are unpredictable | Some moth larvae, some springtails |
| *Life cycles and phenology* | | |
| Multi-year life cycles | Life cycle completed although resources are limited | Moth larvae, midge larvae, etc. |
| Rapid development | Life cycle completed before winter supervenes | Mosquitoes, etc. |
| Abbreviation of normally complex life cycles | Life cycle completed before winter supervenes | Aphids, etc. |
| Earliest possible emergence in spring | Life cycle completed before winter supervenes | Midges, etc. |
| Brief and synchronized reproductive activity | Life cycle completed before winter supervenes | Many moths, etc. |
| Dormancy | Activity stops before damaging winter cold; individuals held dormant through summer to emerge as early as possible the next spring | Chironomid midges, etc. |
| Prolonged dormancy for more than one season | 'Insurance' against a summer unsuitable for reproduction | Various species of flies, moths, sawflies, etc. |

**Table 7.9** (continued)

| Adaptation | Attributed advantage | Examples |
|---|---|---|
| *Food range* | | |
| Autogeny (development of eggs without a blood-meal in biting flies) | Offsets shortage of hosts; or short season | Mosquitoes |
| Different food plants | Survive although a particular food plant is absent | Few species |
| Polyphagy | Survival despite low density of foods | Many species, especially saprophages/detritivores and predators |
| *Genetic adaptation* | | |
| Parthenogenesis | Reduced mating activity in harsh conditions; well-adapted genotypes buffered against change | Midges, black flies, caddisflies, mayflies, scale insects, etc. |

*Source*: Huntington 2000.

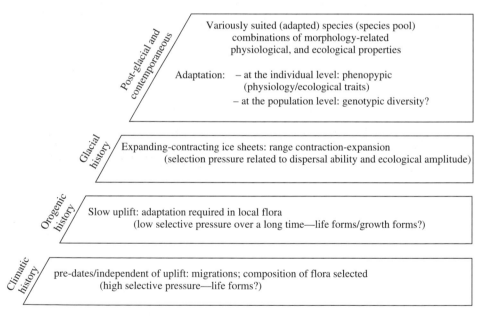

**Fig. 7.12**   Adaptations to alpine existence at large time scales. Large fluctuation in ambient carbon-dioxide concentration between glacial and inter-glacial periods has clearly prompted an adaptive response in the number of stomata on leaves. For example, *Salix herbacea* has been shown to have decreased the density of stomata in its epidermis since the penulti-mate full glacial, about c. 0.14 Ma (a mean of 180 stomata mm$^{-2}$; $CO_2$ cc was about 170 ppm) to today (a mean of <100 stomata mm$^{-2}$; $CO_2$ c. 350 ppm) (Beerling *et al.*, 1993). Adaptations are often a compromise to optimise as in the case of stomatal density. High density of stomata in a $CO_2$-limited environment offers an advantage in acquiring carbon, however, it also risks losing more water than at low stomatal density.

(a1)  (a2)

(b)

**Fig. 7.13**     It has been argued that nutant capitula in some *Senecio* (a1, a2) and some species in
other genera of the Asteraceae in the high Andes have evolved as an adaptation to
protection against snow and ice that can occur frequently overnight. Sklenár (1999) has
suggested that such protection might be an adaptive trait that maintains a more stable
and higher temperature in the floral parts than that of the ambient. It is also plausible
that a nutant habit prevents icing over and meltwater entering the inner floral parts.
Whilst this certainly is a logical explanation, the question arises why most species do
not have this kind of habit (e.g. *Xenophyllum* sp., also in the Senecioneae section of
Astereraceae, b). Such apparent morphological differences are insufficient explanations
as adaptations. As Sklenár (1999) has suggested such hypothetic differences should
be validated by comparing the biology of reproduction across a range of species with
different floral insertion. (Photos: L. Nagy)

form and functioning in tropical alpine environments are discussed by
Rundel *et al.* (1994).

Insects and other arthropods have adopted a variety of ways to live in
seasonal alpine conditions (Bale *et al.* 1997; Sømme 1997). These characteristics may include cold hardiness, supercooling, or rarely, tolerance

of freezing, anaerobiosis in response to ice crusting, increased rates of metabolism, and resistance to desiccation (Table 7.9). There are specialists that make use of winter habitat differences associated with snow cover and resulting temperature. Morphological adaptations include reduction in size, wing reductions, melanism and thermoregulation, nocturnal activity (to avoid overheating in melanistic species). Life cycles may be prolonged (e.g. Carabids, Acari, *Pardosa* of Arachnidae), or favourable microhabitat occupancy may be taken advantage of to complete life cycles in a single year (see contrasting examples within the genus *Pardosa,* or arctic aphids (e.g. Bale *et al.* 1997). Alpine aquatic invertebrates, especially those living in glacier-fed waters represent another example of specialization (Füreder 1999). In tropical mountains, shelter seeking is characteristic, e.g. in scree or in senescent leaves of giant rosette plants such as *Espeletia* spp. (Andes) or *Senecio* spp. (Africa); there is a limited number of freezing tolerant species and different degrees of desiccation tolerance exist. Specialist predators and scavengers with well-developed thermal and moisture regulation use aeolian environments (e.g. Loope and Medeiros 1994). As with plants, in animals, too, features perceived as adaptive in an alpine animal species may well exist in its lowland conspecifics, such as freezing tolerance in both the alpine (*Hemidenia maori*) and lowland wetas in New Zealand (Brent *et al.* 1999), questioning the applied value of some of the perceived adaptations.

## 7.5.1 Levels of adaptation

Adaptations to high mountain/alpine environments encompass a broad spectrum, as these environments, apart from sharing being cold, vary in terms of seasonality, atmospheric and soil hydrology and geochemistry. Accordingly, adaptations must provide for optimizing growth and fitness in a variety of combinations in a generally low temperature environment (which, however, can incidentally turn lethally hot during the daytime), along a gradient from drought to waterlogging. These patterns can be observed at global or continental scales, along latitude (e.g. seasonality from Poles to Equator), or along oceanity–continentality gradients from coast to inland. Equally, such gradients are present within most climate zones, and are the causes of habitats, their vegetation and animal life. Evidently, there is no single jack-of-all-trade alpine adaptation to fit all these criteria, not for all plants or animals, not in terms of life form, growth form, physiology, or ecology. For a general discussion on trade-offs of traits and biodiversity see Wilkinson (2006) p. 41–56.

Conventionally, adaptation is demonstrated by focusing on the presence of certain morphological or functional traits in alpine organisms, or by comparing morphological/anatomical features and physiological functioning of individuals between contrasting environments, or along environmental gradients (for plant examples, see Körner (2003); for animals, see Table 7.9). However, it is equally important that beyond the ability of individual plants or animals to function in alpine environments, they are able to maintain their populations, at a time-scale beyond the lifetime of an individual. It may be taken by one step further to the evolutionary time-scale where adaptations need to be considered. Accordingly, when discussing adaptations, three levels need to be distinguished: individual, population, and taxon.

## 7.5.2 Individuals: plant life forms versus growth forms

Life-form classification is largely made by using the system that Raunkiaer (1934) proposed and is based on the position of the perennating organ and the protection of the growing points or meristems. Raunkiaer's system has often been used to make comparisons between biomes, or lesser bioclimatic zones in terms of their flora and vegetation structure. In an alpine context, phanerophytes (trees and shrubs with their buds in the air) are important components of the treeline ecotone. In the alpine zone proper, most chamaephytes (bearing their buds just above the ground) are prostrate dwarf-shrubs or cushion plants. Locally, they can be important components of the vegetation: in wind-exposed conditions (e.g. *Loiseleuria procumbens* in the Alps, *Dracophyllum muscoides* in New Zealand, *Baccharis* spp. in the Andes, *Diapensia* in the Taisetsusan, Japan), or as cushions, typically at the upper end of the alpine zone in open scattered vegetation. These cushions can be rather long-lived, in excess of 300 years, as has been estimated for *Silene acaulis* from their size distribution in the Rocky Mountains by Morris and Doak (1998). *Silene acaulis* is an interesting species in that it produces two types of individuals: one that bears hermaphrodite flowers and another that bears female flowers only—the latter type being over four times more prolific in seed production, without apparent differences in growth form morphology. Hemicryptophytes form a rather mixed group of plants whose shoot apices perennate in the soil surface: graminoids, forbs, and rosette plants. They make up the highest proportion in alpine floras, typically about 70% in Europe (Table 7.10). Rosette plants are peculiar in that they appear to need to reach a minimum rosette size before they flower. For example, in the Scottish Highlands the rosette size of the flowering specimens of *Lychnis alpina*, an arctic–alpine species, was reported to be about double that of those that did not flower (Nagy and Proctor 1996), and the same applies to giant lobelias in East African alpine habitats (Young 1994). Size is dependent on growth conditions, such as nutrient availability, which, in turn, is related to temperature-driven nitrogen and phosphorus mineralization rates. Size and nutrient status has also been demonstrated from other growth forms (see e.g. Karlsson and Jacobson 2001 for reproduction in *Rhododendron lapponicum*, a dwarf-shrub). Cryptophytes perennate under the soil surface by way of rhizomes, bulbs, stem tubers, or root tubers (the latter especially in the Mediterranean alpine zone: e.g. *Arum*, *Chionodoxa*, *Crocus*, *Fritillaria*, *Iris*, *Scilla*, *Tulipa*). One has to remember that the characterization of floras by their life-form spectrum (most often reported in the literature) would be different if one included the abundance of species in the vegetation. For example, the dominance of the hemicryptophytes in the flora (e.g. Klimeš 2003) would likely to be matched by dominance of various rhizomatous plants (mostly graminoids—sedges, grasses, and rushes) that make up most of the closed alpine sedge heaths. Therophytes, or annual plants are mostly associated with perturbation and the few that are found in alpine environments are hemiparasites, such as those in the genera *Castilleja* or *Euphrasia*, or grow in disturbed areas such as *Koenigia islandica* in boreal and some middle latitude ranges, or *Polygonum minimum* (Sierra Nevada, California).

The alpine flora of tropical mountains includes growth forms that appear peculiar and not to readily fit an account on temperate alpine adaptations. East Africa is renowned for its giant senecios and lobelias, the humid tropical Andes have its giant rosettes of *Espeletia* species, replaced in the central Andes by *Puya*; Hawaii has its *Argyroxiphium sandwicense*. Hedberg and Hedberg (1979) classified humid tropical alpine plants into five growth forms: giant rosette, tussock grass, acaulescent rosette, cushion, and sclerophyllous shrub (Fig. 7.14). These were based on assumed morphological specializations by the plants in response to low night-time temperatures on the one hand, and on their ability to tolerate high insolation during the day. Hedberg and Hedberg (1979) interpreted their results as evidence for parallel adaptive evolution in response to environmental triggers. Despite the visually striking nature of tropical alpine vegetation, the above growth forms all fit into the Raunkiaer system: phanerophyte (sclerophyllous shrubs), chamaephyte (cushions), and hemicryptophyte (acaulescent rosette, tussock grass), again emphasizing the importance of the hemicryptophytic life form in alpine environments. Giant caulescent rosettes are somewhat peculiar and perhaps represent a transition towards the phanerophyte form. Given the history of human activities in the upper montane zone in, for example, the humid tropical Andes, it might be tempting to attribute the abundance of giant rosettes to human influence; however, while some giant rosettes occur in the upper montane forest (e.g. *Lobelia gibberoa*, *L. lanuriensis*, *Senecio erici rosenii* on Mount Ruwenzori), they are much more prominent in the alpine zone. It is worth noting that curiously, the Malesian (New Guinea, Mount Kinabalu) alpine zone, or the humid West African Mount Cameroon do not harbour giant rosette plants.

**Table 7.10** Raunkiaer life form spectra in European alpine mountains

|  | Scandes | Scottish Highlands | Alps* | South-east Carpathians | Apennines | Pyrenees | Caucasus[†] | Corsica |
|---|---|---|---|---|---|---|---|---|
| Phanerophyte | 5 | 1 | 0 (2.8) | 2.5 | 3.7 | 1.9 |  | 1.5 |
| Chamaephyte | 18 | 16 | 24.5 (14.7) | 13.6 | 14.3 | 21.5 | 6.8–13.1 | 18.3 |
| Hemicryptophyte | 59 | 67 | 68 (68) | 70 | 66.8 | 64.6 | 83.3–87.7 | 71.8 |
| Cryptophyte | 16 | 12 | 4 (10.9) | 7.5 | 11.1 | 9 | 3.6–5.5 | 7.6 |
| Therophyte | 2 | 4 | 3.5 (3.6) | 6.5 | 4.1 | 3 |  | 0.8 |

*>2500 m, Switzerland (excludes treeline ecotone), second figure for the Italian part of the Alps (includes treeline ecotone).

[†]Upper alpine (subnival).

## 7.5.3 Morphological adaptations

In addition to the generally applied life forms, other morphology-based classifications have been proposed to characterize vegetation types within biomes, such as those by Whitmore (1998), or (Grubb 1977) for tropical rainforests. Whitmore's system uses leaf size and shape as a guide to classify forests into types from lowland evergreen rainforest (dominated by mesophylls with drip tips) to upper montane (microphylls with small, often coriaceous leaves). Classifications that are related to ecological functionality are gaining strength, and by being incorporated into databases of taxonomic collections, they are opening up traditional taxonomic collections to ecologists (e.g. Pendry *et al.* 2007). For alpine plants Halloy and Mark (1996) undertook a classification, based on leaf morphology characters of species where they compared New Zealand species with those from the Andes, from tropical to southern boreal latitudes. Their main finding was that overall, with increasing elevation, leaves were smaller, rounder, softer, and hairier and were more often folded. One prominent lesson from this study is that there is no uniform alpine type of morphology, but a distribution of character spectra that typifies alpine plant communities (Fig. 7.15). This seems to apply whether these spectra represent species across all habitats, as was in the case of the study by Halloy and Mark (1996), or they are grouped by habitat type, e.g. in the study by Pyankov *et al.* (1999). In addition, similar spectra may arise in alpine environments that differ in their presumed constraining environmental factors. For example, the spectra in the oceanic-humid alpine zone of the New Zealand do not seem to differ from those in the arid cold mountains of the eastern Pamirs–Karakoram (Fig. 7.16; Pyankov *et al.* 1999). It seems probable that such species spectra represent a collection of species, some of which are present in the alpine zone because of

| Growth form | Structure | Stem height | Diameter | Frost protection | Photo example |
|---|---|---|---|---|---|
| Giant rosette | Core: soft stem with a wide pith<br>Outer: dry leaves remain attached | 0.8–1.8 (up to 5) m | 0.5–1.0 m | Insulation of stem by dry leaves; of apex by closing of leaves at night |  |
| Tussock grass | More or less densely packed stems | c. 0.5–0.7 m | Individual bunches of 0.1–0.5 m | Insulation of central portion inside tussock; 5–8 °C temperature difference |  |
| Acaulescent rosette | Ground-hugging rosette leaves; more or less tightly packed | Short term; deep rootstock | Very variable (few cm to few dm) | Reducing extremes (night cold, day heat) |  |
| Cushion | Deep, extendible root system. Tighly packed foliage | Prostrate stem | Variable (few cm to several m) | Decoupling from temperature in free air |  |
| Sclerophyll-ous shrub | Small, hard leaves, hairy, or folded; thin stem; extensive deep root system | Variable (0.2–0.7 m) | Variable (c. 0.5 m) | No frost protecion by structural means: reduced transpiration |  |

**Fig. 7.14** Classification of tropical alpine plant growth forms after Hedberg & Hedberg (1979). (Photo: pachycaul giant rosette–*Lobelia rhynchopetala*, H. Pauli; tussock grass–*Festuca orthophylla*, rosette–*Nototriche* sp., cushion–*Azorella compacta*, sclerophyllous shrub–unidentified, L. Nagy)

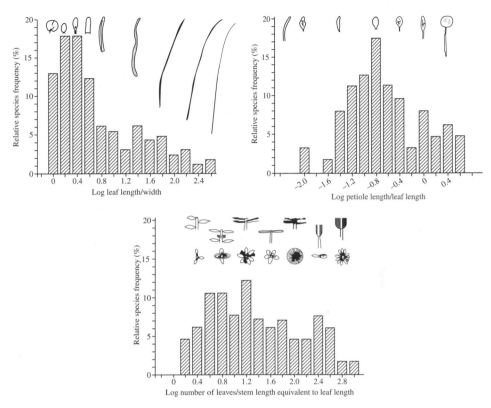

**Fig. 7.15**    The frequency distribution of leaf morphological parameters, Mt Burns, New Zealand (Halloy & Mark 1996)

their success related to their growth form, others combine morphology, physiology, or life history to be ecologically successful, i.e. be able to maintain a viable population.

Morphological spectra reflect adaptations to a range of constraining factors and are related to anatomy and physiological functioning. An often-used measure of anatomical specialization is specific leaf mass (leaf mass per unit leaf area, SLM), which generally decreases with altitude. The typical values of SLM in alpine and arctic environments are about (400)–600 mg dm$^{-2}$ (e.g. Barinov 1988; Diemer *et al.* 1992) and slightly higher in arid alpine with a mean of 700 mg dm$^{-2}$ (ranging between approximately 500 and 950 mg dm$^{-2}$; Pyankov *et al.* 1999), values about 10-fold higher than in tropical forest tree leaves. A high SLM may be caused by small cell size at the stage of leaf initiation because of growth constraints, such as drought, or nutrient deficiency, the former bearing a geographic determination and the latter being a frequently cited feature of cold-limited environments. Nutrient addition experiments have demonstrated the existence of such

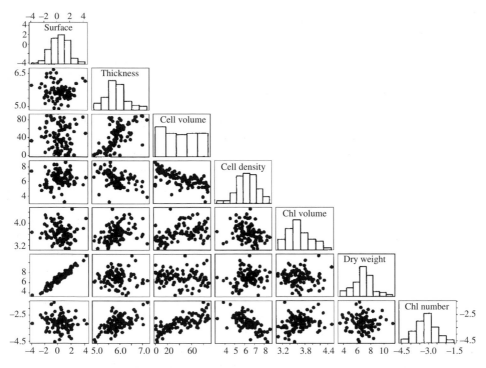

**Fig. 7.16**    Leaf attribute frequency distributions (and their pair-wise relationships as scatter plots) based on the means of samples of 94 plant species in an arid alpine environment between 3800–4750 m asl in the eastern Pamirs, Karakoram, Tadjikistan (Pyankov *et al.* 1999). Values are log$_e$-transformed.

limitations; in most cases larger plant size, higher palatability to herbivores, increase in shoot:root ratio, and an increase in flowering resulted (e.g. Nagy and Proctor 1997; Bowman and Seastedt 2001; Jonasson *et al.* 2001; Heer and Körner 2002). Whatever the cause of a high SLM is, it means a low area per unit mass that correlates with low relative growth rates, often reported for alpine plants. It also puts anatomical changes into a different perspective in regard to adaptation: Are they the cause or rather the consequence? In practice, it appears that harmonization in morphology, anatomy, and physiology occurs in successfully functioning individuals. For example, in *Espeletia schultzii* leaf cell size and intercellular space decrease with altitude (both leading to decrease the probability of nucleation and ice crystal formation in the leaves), and supercooling occurs at increasingly lower temperatures, tracking the lapse rate in air temperature with altitude (Rada *et al.* 1987).

## 7.5.4 Morphology, phyllotaxis, and growth habit in some tropical life forms

In general, in cold environments, plants by growing smaller in stature compensate for decreasing temperature and thereby experience little reduction in bud/meristem temperature along an altitude range, as a result of increasing aerodynamic resistance (e.g. Wilson *et al.* 1987). All the more peculiar is then to find relatively large plant structures (some approaching the size of treeline trees) in most tropical alpine areas. Such large stature plants employ morphology, leaf arrangement, and diurnal movements of their leaves to optimize their growth. Espeletias of the Andes and senecios in East Africa obtain heat insulation to their stem by dead leaves, and live leaves provide protection to the meristems (e.g. Beck 1994). Dead leaves protect the upper part of the stem and live leaves form a so-called night bud around the meristem by closing from their open state during the day, resulting in a temperature difference between stem or meristem and outer leaf surface of 5–8°C. A similar temperature differential is found in other typical tropical growth forms, most notably in large tussock grasses (e.g. *Calamagrostis, Festuca, Stipa*), where the outside of the tussock, often with dead culms and leaf sheaths, provides insulation to the middle (Beck 1994). Insulation works both night and day and just as it is useful in reducing radiative heat loss it can counter the impacts of high insolation as well. For example, the heavy hairiness of *Argyroxiphium sandwicense* on Hawaii has indeed been shown to keep daytime leaf temperature at or below that of ambient air, while the parabolic arrangement of leaves reflects heat towards the shoot apex, the temperature of which can be 25°C above ambient temperature (Melcher *et al.* 1994). However, such a leaf arrangement may have a cost in that lower altitude distribution of the species may be restricted as the same parabolic effect that can stimulate growth at low temperatures can lead to heat stress at high temperatures. Sensitivity to heat stress in boreal alpine plants has also been suggested as the main factor restricting their distribution (Dahl 1951, 1998). In subtropical environments water shortage, coupled with high daytime and low night-time temperatures favours xeromorphic dwarf-shrubs, extreme forms of which are found in the desert regions of the high Andes (Fig. 7.17).

## 7.5.5 Physiological adaptations

The physiological functioning of plants in cold environments is restricted to short growing seasons in seasonal climates (Billings 1974; Wielgolaski 1997c; Körner 2003; Larcher 2003). Adaptations to ensure survival and optimum functioning require mechanisms that operate from molecular to organ and whole plant levels, not independently of morphological growth forms (Fig. 7.18). The complexities of internal and environmental control of plant physiology are described in great detail by Körner (2003,

**Fig. 7.17**     *Acantholippia seriphioides* (Verbenaceae) 'desert trees' at c. 4200 m near Tolar Grande, Salta Province, north-west Argentina, just south-east of the Atacama desert. These tree shaped bushes display features of adaptation to one of the most extreme environments on Earth where they are continually challenged by water stress during the day and cold stress at night. In physiological terms, however, both these stresses require similar adjustments, i.e. coping with symplastic water stress. Notice the sculpting of the stem by wind-blow sand abrasion. (Photo: L. Nagy)

pp. 101–220). The main, often-cited features of physiological characteristics (differences between plants in alpine zone and lower elevations), such as the production of antifreeze agents, the capacity to photosynthesize at low temperatures (sometimes even under snow), low respiration rates, and the specifities of allocation of assimilates are summarized in Table 7.11. Of all physiology related adaptations, perhaps low temperature tolerance is the most characteristic, and apparently regulated by day length and not other, more temporally variable factors, such as snow cover (Bannister *et al.* 2005). The same appear to control the onset of hardening and dormancy (Bliss 1985). Frost resistance varies across plant species and their spatial distribution is correlated in seasonal climates with snow cover. Species with the highest frost resistance occur in ridge positions with little snow protection (e.g. *Dracophyllym muscoides* in New Zealand, *Loiseleuria procumbens* in Europe, or *Dryas octopetala* in North America and northern Europe), while those that require deep snow have a lower level of frost tolerance (e.g. *Celmisia prorepens* in New Zealand, *Soldanella* in Europe,

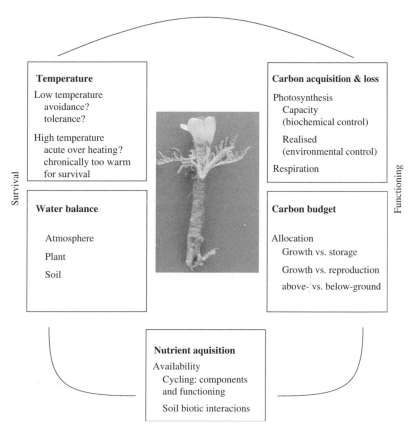

**Fig. 7.18** Components of putative physiological adaptations in alpine plants (compiled largely after Körner 2003) (unidentified cf. *Nototriche* sp., Photo: S. Halloy)

or *Sibbaldia procumbens* in the Northern Hemisphere). Contrasting cold tolerance mechanisms—supercooling (frost avoidance mechanism) in, for example, espeletias (Rada *et al.* 1987) versus frost tolerance in, for example, bunch grasses (Márquez *et al.* 2006)—characterize different life forms in the tropical Andes, where there are no snow-protected habitats.

## 7.5.6 Longevity and reproduction

No one is surprised to learn about the old ages of grand trees; however, the same, often are incredulous about the life span of some alpine plants. Although different life and growth forms dominate the alpine zone from those in wooded vegetation types, it is not unlikely that a survey of the age distribution of species, communities, and ecosystems in comparable natural states and locations would produce surprisingly similar results in terms

**Table 7.11** General features of alpine plants (A) related to their environmental physiology in comparison with lowland plants (L)

| Temperature | Mechanism | Comment | A versus L |
|---|---|---|---|
| *Low temperature* | | | |
| Avoidance | | | |
| Osmotic adjustment | Accumulation of solutes (e.g. TNC) | Low efficiency | ND |
| Supercooling | Avoidance of nucleation allows tissues to cool below freezing | Efficient, short-term avoidance of relatively light frost (e.g. tropical alpine) | > |
| Tolerance | | | |
| True tolerance | Cellular dehydration by passing water into intercellular space to freeze | Efficient, for longer lasting freezing (temperate, tropical) | > |
| Resilience after frost damage | Various | | ?, <, > |
| *High temperature* | | | |
| Acute overheating | Related to frost tolerance (membrane stability) | ? | ND |
| Chronic warmth | Acclimation of respiration | No acclimation leads to excessive respiratory loss in sensitive taxa—explaining distributions limits | |
| Water availability | | | |
| Plant level control | Stomatal control, succulence, desiccation tolerance | Strongly connected with morphological and anatomical features | ND? |
| Nutrient acquisition | Various | | ND? |
| Carbon acquisition | | | |
| Photosynthetic capacity | Higher stomatal density; mesophyll anatomy | | > (ND) |
| Realized photosynthesis | Thermal acclimation, low minimum temperature for net photosynthesis, high light saturation of photosynthesis | Limitation by lack of light | ND |
| Respiration | Thermal adjustment and increased number of mitochondria | Requires respiration acclimation in warm periods/environments | > (?) < |
| Allocation | | | |
| Non-structural versus structural carbohydrates | High TNC and lipids | Cold related: defence/excess to growth? | > |
| Storage organs | | | ND |
| Growth versus reproduction | | Comparison based on total mass to combined stem and flower mass | > |
| Leaf versus total | Various (temperate species level 8–48%; community approx. 20%) | | ND |
| Root versus total | Fine root mass ratio | On total mass basis | > |

Largely after Körner (2003, pp. 110–240).

of longevity of the characteristic species. Most information is available about cushion plants as they are amenable to repeat size measurements from which their growth rate can be calculated, or size–age relationships can be modelled. In addition to the *Silene acaulis* (Caryophyllaceae), referred to above, cushions of *Azorella compacta* (Apiaceae; Fig. 7.19) have been aged between 400 and 1000 years from the subtropical and 850–3000 years from seasonal tropical Andes (Halloy 2002). Other cushion species, such as *Pycnophyllum convexum* (Caryophyllaceae) were also estimated to reach between 25 and 190 years in the Andes. Many rosette-forming species are long lived, too, and clonal species can maintain their genet indefinitely (Fig. 7.20). Similarly to the grass species in Fig. 7.20, the sedge *Carex curvula* follows a growth pattern that changes from an early tuft to a 'a fairy ring' form (Grabherr *et al.* 1988; Grabherr 1997). The long-lived alpine vascular plants, as captured in the conceptual model by Molau (2003) are predominantly graminoids with clonal growth, complemented by the long-lived cushions (Fig. 7.21). An interesting question in regard to clonal species is how the evolutionary age of the genet may affect growth and success of a species in the long term.

**Fig. 7.19**    Cushions of *Azorella compacta* Apiaceae (large humped cushion in front of large boulder) and *Pycnophyllum* sp. (small, light coloured flat cushion to the right of *Azorella*), Cerro Yasasuni, Nevado Sajama, Bolivia at c. 4800 m. These species grow radially about a mean of 4 mm a year (Halloy 2002) and they can reach ages up to c. 200 years (*Pycnophyllum*) and >1000 years (*Azorella*). (Photo: L Nagy)

**Fig. 7.20**     Many clonal plants grow in expanding concentric circles in alpine environments, such *Festuca orthophylla* on Nevado Sajama at c. 4800 m. These clones may break up and the individual tussocks continue the radial expansion; in principle clonal species may never reach a definite age. Ageing, in evolutionary terms, however, may be factor that disadvantages individual clones in the long-term.

## 7.5.7  What kind of adaptation after all?

These different types of 'adaptation' provide an array of combinations that allow individuals to exist, and species to maintain their populations and coexist with other species. It is a matter of conjecture to what extent each type of adaptation, or combinations of adaptation is deterministic, and to what extent the organization of alpine ecosystems is stochastic. It would seem to be of interest to analyse species spectra to characterize the contributing morphs for the mechanisms that allows them to (co)-exist in their particular type of alpine ecosystem. Most of all, as it has been emphasized throughout, a large number of species in today's alpine environments have experienced large climatic fluctuations and underwent local extinctions and re-colonization events. A rigid alpine-type 'hang-on-for-dear-life-and-survive' specialization probably would have been disadvantageous. Adaptation, being after all about plasticity (phenotypic or genotypic), needs to be looked at today in terms of the capacity of species, growth forms, and life forms to respond to environmental cues by altering their functioning/anatomy, reproduction, or selecting for the genotypes that are best adaptable. This is a key interest for population geneticists,

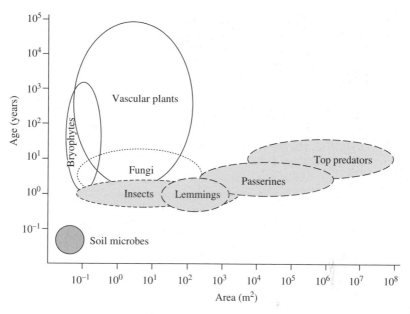

**Fig. 7.21**  A conceprual model of space and time occupancy by organisms in the alpine zone (reproduced from Molau 2003)

ecologist, and conservation scientists. There is a plethora catalogued infra-specific genetic diversity in alpine plants (e.g. INTRABIODIV for the Alps and Carpathians in Europe, http://intrabiodiv.vitamib.com/); however, it is not known what such diversity means in adaptive terms. It remains to be established even in cases where a clear ecological segregation is discernible among cytotypes, as has been shown for *Senecio carniolicus* in the European Alps by Schönswetter *et al.* (2007) and for *Lotus alpinus/corniculatus* by Gauthier *et al.* (1998). Holderegger (2006) and Holderegger *et al.* (2008) have offered a way on how to further making inferences about the adaptive value of genetic differentiation at the landscape scale. One may witness soon the collaboration of geneticists, physiologists, and ecologists in understanding better genetic adaptation in the field.

What is the relative importance of phenotypic adaptation to genotypic in alpine plant populations? Many plants that grow in the alpine zone are long-lived perennials, some of them are extremely so, and can live in the range of hundreds to thousand(s) of years. For such long-lived individuals, adaptation *sensu stricto* appears rather limited at the phenotypic level and it is available, mostly via physiological (and also by morphological means; e.g. see *Loiseleuria procumbens* in different habitats; Grabherr 1979). It seems that, at least in lowland ecosystems (genotypically different) clones have different growth rates and traits and they respond to different environmental cues in an individualistic fashion (e.g. Fridley *et al.* 2007). In the absence of knowledge about the tolerances of such long-lived species one cannot make inferences about the history of the environment during the lifetime of the plants, i.e. disentangling environmental stability from physiological tolerance is necessary first. Analyses of species with a wide altitude distribution range may appear to be a good proxy for estimating climatic tolerances in species. *Espeletia* species in the Venezuelan Andes occurs from the montane forest zone at 2400 m (*E. neerifolia*) up to 4600 m (*E. moritziana*) close to the permanent snowline (Monasterio 1980). Some individual species have very wide distribution ranges such as *Espeletia schultzii*, which occurs between 2600 and 4200 m (Rada *et al.* 1987). This is equivalent to a mean temperature range of about 10°C. In this instance, adaptation or tolerance is through physiological means; the increasing capacity of *E. schultzii* to supercool and avoid frost damage with decreasing minimum temperature. The ability of this species in having individuals distributed along such a broad temperature gradient is not equivalent to an individual of the same species being able to tolerate an identical temperature range. For very long-lived individuals it is difficult to reconstruct lifetime climate variation, except in few cases. None the less, the existence of such long-lived individuals, in the presence of firm environmental proxy data provides opportunities for understanding the nature and scale of physiological adaptability (tolerance) of these individuals at temporal scales rarely considered in ecology (see e.g. Halloy 2002 for *Azorella compacta*).

At the population and species levels, altitude ranges (narrow versus broad) do have implications for adaptation to climate change impacts, past, present, or future. In fact, Wood (1971) has suggested that the wide altitude range of the afro-alpine species on Mount Kilimanjaro and Mount Kenya reflects the historic adaptation of the species to long-term climatic change (but see the case of vicariant *Lobelia* species on Mount Ruwenzori). For example, approximately 75% of the flora on Mount Kilimanjaro could persist within (at least part of) their existing altitude range if the limiting factors were raised or lowered by 650 m, an equivalent of change in temperature of about 4°C. The significance of a wide distribution range is also important for changes in the genetic structure of a species in its elevation range by affording local adaptation through high differentiation (see e.g. *Polemonium viscosum* in the Rocky Mountains; Galen *et al.* 1991) and through gene flow in the direction of environmental change. The latter may manifest in a genetic type replacement at the edges of a species' range under a long-term increase in temperature. Wide altitude distribution is not restricted to tropical alpine environments; it is prominent in temperate alpine floras as well. In New Zealand about 60% of the alpine species occur outwith the alpine zone and similarly, few species are exclusive to the alpine zone in the European ranges, one extreme being the Pyrenees, where a mere 22 of approximately 800 species that are found in the alpine zone are alpine exclusives (Gómez *et al.* 2003). It must be borne in mind, however, that such figures are based on enumerations of species, without attaching weights to their optimum versus marginal distributions or their occasional outlying outpost. This has caused inconsistency in defining 'alpine' species (see e.g. Ozenda and Borel 2003). A wide elevation range may reflect the true ability of a species to grow at a range of temperatures (e.g. *Hypochaeris robertia* in Corsica occurs from approximately 700 m up to approximately 2600m in a range of habitats from open forest to grassland, scree, and rocks), or it may reflect the distribution of equivalent habitat conditions, such as the so-called 'freezing cellars' where e.g. alpine *Rhododendron* species can grow in an otherwise submediterranean prevailing climate, or *Poa laxa* that in the Ötztal Alps occurs down to 1500 m on small block fields with cold air drainage. Temperature is not the only factor that determines range sizes. Azonal habitats with their shortage or excess of resources (see Chapter 6) can extend climatic ranges locally on mountains, and geographic distributions of species. Habitat characteristics, including hydrology and fragmentation have been found, for example, to be correlated with species distribution ranges in the eastern Sierra Nevada, California (Kimball *et al.* 2004). The authors have found that the regional distribution ranges of species were related to the distribution pattern of habitats: mesophilic species had larger distribution ranges and were centred on the western side of the mountain range, with large expanses of continuous mesic habitats; xerophilic species with small ranges had their distribution centred in the eastern Sierras with discontinuous dry habitats.

Less extreme plant ages than those that characterize very long-lived cushion plants or clonal plants bring phenotypic and potential genotypic adaptation closer. Apart from age, environmental cues that prompt sexual reproduction—not very often for many alpine species—can facilitate genotypic adaptation at the species level. It is possible to foresee a scenario where a favourable period that allows seed setting, dispersal, and establishment, may each time reshape the potential of a species for adaptation. We do not know the extent of adaptive potential in existing populations of alpine plants. For example, some species occur in a range of habitats—are their populations in all habitats the same, or do they have a genetic structure that was selected in an adaptive way to the local environment? There seems to be some evidence to support both, for at least some species. For example, Crawford *et al.* (1995) have reported high levels of genetic variation among populations of *Saxifraga oppositifolia* that grew in contrasting microthermic conditions on Svalbard. On the other hand, the contrary is true for some other species, such as *Sesleria albicans* that can grow in different habitats and ecosystems without much apparent genetic differentiation (Reisch *et al.* 2003).

The mechanisms that characterize plants of long-term persistence during historic epochs are also relevant for historic adaptations of plants to alpine conditions, with their climatic instabilities. These mechanisms include hybridization, polyploidy, asexual reproduction, and longevity and dormancy of propagules. Hybridity, in fact may be viewed as encompassing everything that arises from sexual reproduction from closely related individuals of rather similar genetic constitution at one end of the range to cases where an offspring is the result of mating between genetically distant parents. This latter crossing the genetic barrier produces new forms, which if successful may stabilize and outperform their parents. However, as hybrids are often sterile their supremacy in relation to their parents may be transient, unless they are able to reproduce asexually, or stabilize. The cases where hybrids stabilize and become fertile are linked to polyploidy, whereby a doubling of the chromosomes can restore fertility to hybrids. Polyploidy can itself be a first step in a new species forming within the existing range of the species and without geographic isolation. An interesting case is the prevalence of apomixis in genera such as *Alchemilla*, *Hieracium*, and *Taraxacum*, whose micro-species count in the hundreds and are frequent in alpine environments (see Hörandl *et al.* 2008). In its simplest form, during apomixis, an embryo develops from an unfertilized diploid egg cell, which had not gone under meiotic cell division during macrogametogenesis. The case of the apomictic polyploid *Antennaria rosea* that has derived from eight distinct diploid progenitors in the Rocky Mountains is particularly interesting. The diploid species occur in distinctive habitat sequences from lowland to alpine (Fig. 7.22), while *A. rosea* spans the entire elevation range. Bayer (1997) has proposed that this wide ecological niche breadth might be the result of former hybridization between the polyploid and several diploid species. Quite why apomixis happens, and in these particular genera, is not well understood. None the less in a way—only in terms of adaptive value, not in biological terms—it might be thought of as a transition between sexual and asexual reproduction.

Asexual reproduction is prevalent in alpine plants, whether by clones of linked underground structures, or by other means that permit the dispersion of the vegetative propagules, for example, bulbils (Fig. 7.23). The main advantages of asexual reproduction are twofold: the potential to maintain sterile hybrids in the long-term (i.e. prolonging the chance for sterile plants to regain fertility), and to be able to reproduce and persist in adverse environmental conditions in the absence of cues for flowering (Willis and McElwain 2002). In addition to persistence by asexual reproduction, the seed bank *in situ*, or extrazonally may provide a source of propagules for accession. All these mechanisms together form an important character, another source of adaptation (persistence, resistance, and resilience) in

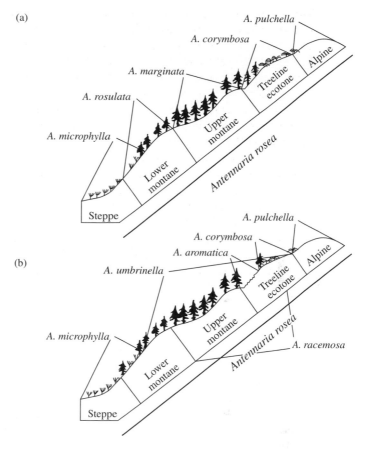

**Fig. 7.22**   The distribution of *Antennaria rosea* extends from steppe to the alpine, along the full gradient shown; those of its progenitors in the southern (a), and northern (b) Rocky Mts are redrawn from Levin (2002).

alpine plant communities. The major difference in this respect between plants and animals is the absence of the capacity of asexual reproduction in most animals, which may make them more vulnerable to change in their environment than plants.

## 7.5.8  Outlook: Do perceived adaptive characters work in practice?

To better understand what makes a species successful in an alpine habitat in relation to its physical environment and neighbouring species, a simultaneous analysis of species characters or combinations of characters is required. Species rank abundances in a particular habitat may serve as a starting point for quantifying success, followed by an evaluation of each

**Fig. 7.23**    *Saxifraga flagellaris*, or spider plant, Hockley Hills, Selawik Wildlife Refuge, Alaska. The bulbils that are borne at the end of long stolons, and separate from the mother plant to become free-living individuals, exemplify an instance of asexual reproduction, frequent in alpine plants. (Photo: L. Nagy.)

character perceived as potentially conferring an adaptive value (e.g. in terms of life and growth form, morphology, physiology, anatomy, life history). Such an analysis may conceivably be carried out by using a method, where site-by-species, site-by-environment, and species-by-trait are ordinated simultaneously (e.g. Dolédec *et al.* 1996). It would identify whether presumed adaptive characters indeed have a role in ecological success in alpine habitats, or these characters have a meaning when comparisons are made between alpine and non-alpine habitats. It would also help understand better the structuring and ecological functioning of alpine communities, the subject of the following chapter.

## 7.6  Conclusions

The current broad pattern of the distribution alpine climate zones has been in place for about 10 million years. Pleistocene glaciations have repeatedly caused local/regional extinction, or caused large spatial and temporal dynamics in distribution and composition of alpine floras and faunas in many high mountain ranges. The interpretation of present patterns in the distribution of the alpine climate zone with its habitats, communities, and species needs to be considered in view of this intense past dynamics. Predictions about temporal changes in the distribution of alpine organisms in the future could benefit from using a flexible combination of general biogeographic and ecological models.

Plant and animal species that permanently inhabit the alpine zone of high mountains have developed a number of traits that can be construed as adaptations. No apparent general alpine type plant or animal exists and adaptations at individual, population, and evolutionary scales show a great variety. The various physiological and morphological characteristics of adaptation to alpine environments show spectra that characterize habitats and their plant communities and animal assemblages. The role of genetic variation in adaptive traits (physiological and ecological) is yet to be unravelled.

# 8 Temporal and spatial dynamics

## 8.1 Introduction

The discussion of temporal and spatial aspects of the dynamics in alpine plant communities (and some examples of animal assemblages) begins with outlining processes during colonization. For expediency, the colonization of bare substrata is considered first, followed by plant community and animal assemblage formation (primary succession), and then by secondary succession. Subterranean and microbiological processes that pre-date and co-occur with the succession of higher life forms are considered. This phase (colonization and space filling) is highly dynamic in the absence of a climatic constraint, such as found on glacier forelands. In contrast, established alpine communities are characterized by generally slow dynamics (Fig. 7.21).

The second part is about change and its drivers in community composition in time and space: from inter-annual variability to long-term change at the patch and habitat scales. The main variables considered include recruitment, persistence, and sexual versus vegetative reproduction. The interactions that contribute to the dynamics of alpine plant communities are examined by looking at competition, facilitation, and mutualism. Finally, the impacts on species and vegetation dynamics of temperature, nutrients, and herbivores are discussed.

## 8.2 Alpine primary succession

The most often cited example of primary substratum available for colonization is till material in glacial valleys after glaciers melt. However, typical alpine primary succession is that found in the aeolian zone (see Chapter 6) and at the interface between the aeolian environment and the edge of permanent snow fields. Additional primary substratum is available for colonization after volcanic activity, or catastrophic mass movement events, such as landslides. Primary succession in these circumstances much depends on the extent of the exposed substratum and on the properties of the surrounding ecosystems, but usually is comparable with that on glacier forelands. The main characteristics of primary succession from colonization to establishment and growth are summarized in Fig. 8.1.

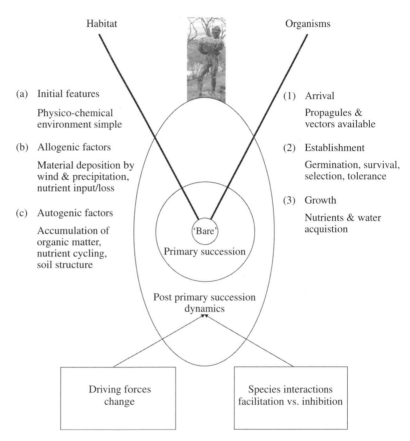

Habitat      Organisms

(a) Initial features

 Physico-chemical environment simple

(b) Allogenic factors

 Material deposition by wind & precipitation, nutrient input/loss

(c) Autogenic factors

 Accumulation of organic matter, nutrient cycling, soil structure

(1) Arrival

 Propagules & vectors available

(2) Establishment

 Germination, survival, selection, tolerance

(3) Growth

 Nutrients & water acquistion

'Bare'

Primary succession

Post primary succession dynamics

Driving forces change

Species interactions facilitation vs. inhibition

**Fig. 8.1** The main features of primary succession (Bradshaw 1993) and post-colonisation processes. If 'bare' or unoccupied and previously uncolonised 'habitats' are available, and propagules (microbial, fungal, plant, animal) can reach them, primary succession begins. Alpine examples include aeolian environments with their cryptoendolithic communities (e.g. Vestal 1993) or microbes in recently deglaciated till material (Nemergut *et al.* 2007; Schmidt *et al.* 2008). Vascular plants may arrive relatively early, however, their establishment and survival are rather precarious and readily measurable changes occur only when successful establishment and growth occur relatively frequently, such as during a sustained period of higher than normal temperature (e.g. Pauli et al. 2007). The picture insert is that of a sculpture 'Traðamaðurin / Crofter', by Hans Pauli Olsen in Torshaven, Faroe Isles (Photo: L. Nagy)

The types of alpine primary succession are rather varied in their scope. The first instance of primary succession is that in aeolian environments, which are a rather hard nut for colonizers to crack. Aeolian environments lie above the limit of vascular plant life and therefore colonization is restricted to a small number of organisms and the developing food-web is usually rather simple. However, aeolian environments are also known to break rules and have sometimes, at the macroscopic level at least, a reverse colonization order, whereby predators (e.g. salticid spiders in the high Himalayas) appear before the building up of the food-web. Such an anomaly owes its existence to the main driver of the aeolian (wind-derived) environment where suitable prey species are transported by katabatic winds from low elevations regularly (Halfpenny and Heffernan 1992). A type of fascinating succession-like process also occurs on snow and ice; however, by the very nature of these substrata, the end is always near and succession does not progress very far—none the less on melting they contribute propagules to succession on exposed mineral substrata.

A second instance of primary succession is associated with the permanent snowfield edges. Succession is not a simple unidirectional process in such environments as the line of permanent snow can fluctuate, i.e. fluctuations reminiscent of glacial–interglacial periods may occur locally at short time-scales. Landscape-scale trends since the end of the Little Ice Age (approximately 1860) have been an altitude increase in the snow line—e.g. from 3050 m in the early 1960s to 3200 m in 2000 in the Ortles-Cevedale of the European Alps (Pedrotti and Gafta 2003)—which has allowed the range extension and migration of species. As a result, in well-identifiable localities it is possible to follow changes in the biota, e.g. vascular plants in the European Alps recently (Pauli *et al.* 2007). We shall return this later when the ecological impacts of climate change are discussed.

The third case concerns succession after glacier ice melts. The specificity of glacier foreland colonization concerns the fact that glacial tongues often protrude into valleys whose climate would otherwise allow the development of ecosystems similar to those in the surrounding landscape (Fig. 8.2). In such cases, glacial retreat is followed by rapid simultaneous colonization by all kinds of organism (Erschbamer *et al.* 1999; Kaufmann 2001; Bardgett *et al.* 2005; Tscherko *et al.* 2005), somewhat contrary to the textbook examples of gradual build-up of the chemical environment and subsequent amelioration of conditions that finally allow vascular plants to colonize. None the less, contemporary glacial till colonization is useful in illustrating temporal and spatial dynamics in soil (structure, chemistry, microbes, and animals), and food-web development from plant community formation to the impact of herbivores and predators.

Alpine examples of detailed studies of primary succession on volcanic substrata are relatively few and include those in the upper montane zone

**Fig. 8.2**     Vegetation unaffected by Little Ice Age glaciation (foreground and left side of the photograph) and largely unvegetated lateral moraine (middle plan) of the retreating glacier (background) at c. 5100 m asl at the head of Sibinacocha, Cordillera Vilcanota, Peru. (Photo: L. Nagy)

on Mount Fuji (Nara *et al.* 2003), the Cotopaxi, Ecuadorian Andes (Stern and Guerrero 1997), and perhaps the Karymsky, Kamchatka (Dirksen and Dirksen 2007), the latter probably being more of a secondary succession type. The various substrata (lava, pumice, scoria, pyroclastic flows, lahars, tephra) provide diverse conditions for colonizers (for details see del Moral 1993); the importance of the substratum variability in terms of different physical characteristics has also been pointed out by Jochimsen (1970) for glacier moraines.

## 8.2.1 Early succession: cryptoendolithic microbionts of aeolian environments

To illustrate the classic sequence of successive turnover of organisms during colonization a set of examples is considered from the Antarctic. The Antarctic is excluded from most accounts of alpine biology, none the less, as a result of a long-term international research engagement there (Miles and Walton 1993); it offers a fascinating insight into succession, comparable with what is characteristic in high alpine aeolian environments. Here we consider the colonization of rock and skeletal soil by microbial organisms, and by cryptogams. Perhaps the most incredible of all colonizations is that by cryptoendolithic organisms (cyanophytes and lichens that live in the pores of certain rock types, such as sandstones, quartz, limestones, and shale). This kind of habitat and their biota have received little attention

**Table 8.1** Physical environment and physiological activity of cryptoendolithic communities

| Environmental variable | Characteristics | Habitat |
| --- | --- | --- |
| Moisture | Variable, from dry to moist | Rock matrix can hold water |
| Temperature | Fluctuating from low (< 0) to high | Direct solar irradiation heats rocks in cold; thermal inertia dampens temperature fluctuations |
| Radiation | High fluxes of visible and UV radiation | Rock attenuates all radiation |
| Wind | Variable abrasion | Protection inside rock |

Modified from Vestal (1993).

outside Antarctica; however, it widely occurs in rocky alpine (Bell *et al.* 1986) and arctic (Omelon *et al.* 2006) environments. The cryptoendolithic habitat provides special conditions for microbial communities (Table 8.1). The metabolically active period (temperature >0°C; moisture and light availability) can vary from approximately 50–300 h year$^{-1}$ in Antarctica (Vestal 1993) to about 2500 h year$^{-1}$ in the Canadian Arctic (Omelon *et al.* 2006). This type of succession involves colonization either by cyanophytes (alkaline pH), or lichens (acid pH) and by saprophytic bacteria and fungi. Succession, *sensu stricto* (i.e. turnover of organisms) does not occur as the colonizers remain in place until their activities result in rock exfoliation, part of the weathering process. It is not known how the rates of weathering are related to cryptoendolithic activity; however, it is unlikely that forecast changes in climate will have a major impact on the rates of cryptoendolith-mediated weathering in the future. It is probably safe to assume that today's rocks will not offer a fundamentally different habitat to plant and animal growth in the short-term.

## 8.2.2 Primary succession proper

On rock exfoliation cryptoendolithic organism are able to disperse or be dispersed, and may become the initial colonizers in the process of stabilizing skeletal stone-sorted soils. The first visible stage of microbial presence is the formation of crusts on the surface, made up of cyanophyte and algal filaments binding mineral soil particles (Wynn-Williams 1993; Belnap and Lange 2003; Türk and Gärtner 2003). Unlike the space-confined cryptoendoliths, crust-forming cyanobacteria can respond to amelioration of climatic conditions by enhanced growth. Even the simplest crusts, such as those studied in Antarctica, consisting of filamentous cyanophytes and globular green algae, build up a layered structure that offers a suitable environment for flagellate protozoan grazers. The presence of certain cyanophytes enables such ecosystems to fix nitrogen that can lead to the progression of the colonizing process, whereby further autotrophic organisms such as lichens and bryophytes, and heterotrophs such as bacteria, yeasts, and fungi, and finally invertebrates and decomposers can appear (Wynn-Williams 1993). The biological soil crusts described from the Alps by Türk and Gärtner (2003) reflect such a complex system (Fig. 8.3).

However, for such complex systems to build up there is a long way, and there are sometimes shortcuts. Cryptogams (some lichens and bryophytes) can establish on rock surfaces or in fissures, and do not require soil. In addition to their role in weathering, by establishing on fine mineral fragments, cryptogams contribute to binding particles, which is rather important in soil formation and stabilization in periglacial environments. An important feature of bryophytes is their ability to scavenge and accumulate nutrients from the air, or precipitation (e.g. Woolgrove and Woodin 1996). The main impacts of cryptogams by modifying their substratum affect microclimate

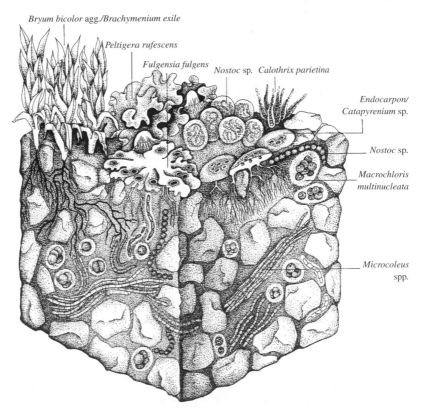

**Fig. 8.3**   A generalised scheme of a biological soil crust (3 mm of thickness). Biological soil crusts form in a variety of environments, including the alpine zone of mountains across the world. (From Belnap & Lange 2003)

(dampened amplitudes of temperature and moisture), albedo (increased radiative heating), chemical properties (solubilization and mineralization), retention of soil particles, and provision of habitat for invertebrate and epiphytic organisms (Walton 1993). There is a large number of spores in glacier ice and soil; however, their fate and that of fragments of shoots depends on a number of environmental factors (Fig. 8.4). One striking outcome of the simultaneous availability of higher temperature and lower relative air humidity (and no abrasion from wind-born particles) in an Antarctic environment was the development of a near complete cover of bryophytes from a soil-borne propagule bank in 4 years (Lewis Smith 1993). This is of interest for climate change responses in high alpine and nival (aeolian) environments and merits further field research and modelling studies. Cryoconite holes in ice contain alga and moss gametophytes, protozoans, tardigrades, rotifers, and nematodes, all of which may land on ice-free mineral deposits after ice melt, and thereby supply propagules for colonization. In addition, long-distance transport of spores, even as far as that between South America and Antarctica, or between the fumaroles of Volcán Socompa at

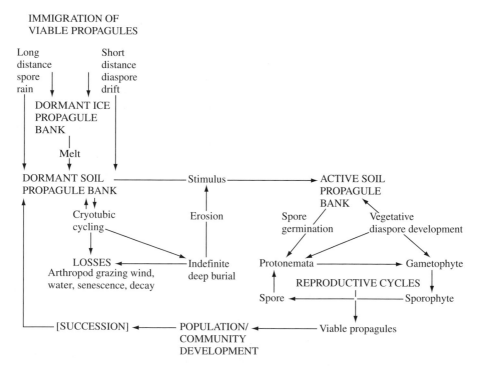

**Fig. 8.4**    A schema of bryophyte soil propagule banks in Antarctica by Lewis-Smith (1993). The dynamic processes in dispersal and contribution to different fractions of the propagule bank and populations are analogous to those by vascular plants.

5700–6000 m (surrounded, for hundreds of kilometres, by salt flats and deserts at approximately 4000 m altitude in the Andes) adds to the diversity of early colonizers during succession. By extension, what is of particular interest for succession in the today's aeolian zone in a warmer future is the availability of vascular plant propagules. It appears timely to evaluate the incidence and magnitude of dispersal events locally or regionally from the vegetated alpine zone, or from long-distance sources.

Studies on species range extension have become popular in the wake of concerns about the impacts of climate change on alpine ecosystems. A number of cases have shown an upward range extension of organisms in relation to previously known records. For example, Grabherr *et al.* (1994) have shown that vascular plant species richness increased on most high alpine summits surveyed above 2800 m altitude in the Austrian and Swiss Alps. The authors calculated a maximum altitudinal advance of approximately 4 m per 10 year on average by vascular plant species in the twentieth century; however, they emphasized the importance of landscape features for different rates of increase in species richness. In an independent survey in the Swiss Alps, Walther *et al.* (2005), reported an increase in the number of species on high alpine summits, and higher maximum altitudes after 90 years for a number of species off the summits. Such studies are useful for confirming changes in vascular plant distribution at the frontier of plant life, particularly those that have detailed spatial information about particular localities (e.g. Pauli *et al.* 2003). Additional information on vascular plant species from trait screening (e.g. Grime *et al.* 2007), information about other taxonomic groups and about ecosystem properties that could be used to piece together the jigsaw of characteristics that determine processes during primary succession in alpine habitats is largely missing at present. One programme that has the potential to fill these hiatuses in the long term is that run in the Tyrolean Alps where vegetation is periodically being recorded in the subnival zone, near the altitude limit to vascular plants (Pauli *et al.* 2007).

## 8.3 The case of glacier forelands

Some classic studies of succession on glacier forelands include those from Glacier Bay, Alaska (Burrows 1990)—a non-alpine example, the Rotmoosgletscher, Alps (Jochimsen 1970; Erschbamer *et al.* 1999; Kaufmann 2001; Tscherko *et al.* 2005); several sites in the southern Scandes (e.g. Matthews 1992; Crouch 1993; Vetaas 1994) and in North America, such as the Lyman Glacier in the North Cascades (see a list in Walker 1993), and the Tyndall glacier on Mount Kenya (Mizuno 1998; Mizuno 2005). These study sites span a range of bioclimatic zones, from boreal lowland forest (Glacier Bay), boreal upper montane to alpine (southern

Scandes), temperate alpine (Rotmoosgletscher), to tropical alpine (Tyndall), and altogether they show up some general features of succession after glacier recession.

### 8.3.1 The initial conditions and rapid succession after the retreat of glacier ice

The moraines left after the retreat of glaciers (Fig. 8.5) undergo a relatively rapid colonization. Colonization and succession begin in the raw mineral material in the vicinity of the ice face (Fig. 8.6) by microbial organisms. This mineral material is low in organic carbon, nitrogen, and phosphorus (Table 8.2), and hosts an increasing diversity of microbial organisms (including abundant nitrogen fixers) with distance from the glacier snout; e.g. Nemergut *et al.* (2007) has shown this along an estimated chronosequence of 0–20 years, at the Puca glacier, Cordillera Vilcanota, or Tscherko *et al.* (2004) in the Alps. From such initial conditions a marked change occurs within approximately 100 years during the development of soil (Egli *et al.* 2006) and vegetation (e.g. Erschbamer *et al.* 1999; Erschbamer 2007): a near closed vegetation canopy forms over a soil that has accumulated organic matter and undergone much acidification (Figs. 5.17; 8.5),

**Fig. 8.5**    The lateral moraines (near side of photograph with large boulders, and in afternoon shadow in the righth-hand side) of the debris covered glacier that is formed by the meeting of the Ngozumpa and Lungsampa Glaciers (light coloured morainic hillocks in the middle of the photograph) are covered by *Kobresia*-dominated vegetation at c. 4900 m asl, 86°40′ E, 28°75′ N, Khumbu Himal, Nepal. (Photo: L. Nagy)

**Fig. 8.6**    The melt of glacier ice leaves exposed mineral material available for colonisation, poor in nutrients, such as at the head of Sibinacocha, Cordillera Vilcanota, c. 4900 m asl (Photo: L. Nagy)

**Table 8.2** The composition of glacial mineral material after the retreat of ice in two glacier forefields in the high Andes and in the Austrian Alps

| Soil characteristics | Puca, Vilcanota, Andes | Rotmoos, Ötztal, Alps |
|---|---|---|
| Age (year); distance from snout (m) | 0–4 (0–100) | 4 (near; not specified) |
| Organic C (mg g⁻¹) | 1.14 (0.18)–1.29 (0.34) | 10.32 (1.00) |
| Nitrogen total (mg g⁻¹) | 0.08 (0.00)–0.09 (0.00) | 0.28 (0.03) |
| C/N | 12.9 (2.23)–15.3 (3.20) | 39.07 (6.80) |
| P (µg g⁻¹) | 1.71 (0.17)–2.52 (0.85) | 3.20 (1.86) |
| pH | 7.50 (0.20)–7.60 (0.10) | 7.45 (0.04) |
| Age (year); distance from snout (m) | 20 (500) | 20 (100) |
| Organic C (mg g⁻¹) | 1.59 (0.63) | 7.00 (0.53) |
| Nitrogen total (mg g⁻¹) | 0.11 (0.01) | 0.28 (0.02) |
| C/N | 15.3 (7.50) | 26.00 (3.09) |
| P (µg g⁻¹) | 1.53 (0.36)* | 0.00[†] |
| pH | 7.50 (0.10)[‡] | 7.51 (0.04)[§] |

*Resin extractable.

[†]Acetate-lactate extraction.

[‡]In water.

[§]In 0.01 mol l⁻¹ CaCl²; values in parentheses are standard deviation for the Andes and standard error for the Alps. Data from Nemergut *et al.* (2007) and Tscherko *et al.* (2004).

Organic matter content recorded at 20 years at the Lyman glacier, North Cascades was 2.3–3.3 (0.5) mg g⁻¹ Ohtonen *et al.* (1999).

has changed and increased in its microbial composition, and its inverte-brate diversity, both below- and above-ground (Erschbamer *et al.* 1999; Kaufmann 2001; Bardgett *et al.* 2005). However, it must be remembered that such a classical sequence might not always be apparent under cer-tain circumstances, such as heavy cryoturbation and grazing, reported for northern Iceland (Wookey *et al.* 2002).

Vegetation succession after glacier retreat begins with a relatively rapid colonization, where available, by wind-dispersed species (e.g. Stöcklin and Bäumler 1996), together with propagules available from transport by downwash, which then give way to slower colonizing species that arrive mostly from nearby vegetation (Svoboda and Henry 1987; Jones 2005; Caccianiga *et al.* 2006; Raffl *et al.* 2006). There is a rapid build-up of veg-etation cover that, depending on local conditions can reach approximately 70% after 20–40 years (e.g. Raffl *et al.* 2006). In this process, cryptogams have no classic role (Box 8.1) in pre-dating and ameliorating growth con-ditions for vascular plants. On moraines, a space-filling process occurs from neighbouring ecosystem(s), in contrast to succession on substrata in marginally 'sterile' environments such as polar or alpine aeolian envi-ronments; hence, the rapid development of the below- and above-ground ecosystem complex. Landslides and other newly formed primary substrata that appear in a well-defined non-limiting climatic environment are simi-lar to recently deglaciated moraines in that they can undergo very rapid succession (e.g. Grabherr 2003). Avalanche tracks are a particular case as they may result in total denudation of soil in their upper half (and lead to primary succession), and mixing and accumulation of soil-forming mate-rial in the lower half that gives rise to secondary succession.

There are few studies, where real-time recording of glacier recession and plant advance is available, as opposed to community succession, based on chronosequences and historical records on previous extensions of glaciers. The work by Mizuno (1998, 2005), building on earlier records from the Tyndall glacier, Mount Kenya, has followed the rates of ice recession and colonization by a set of species in real time over a period of nearly 50 years (Fig. 8.7). There was a trebling in the rate of glacier ice recession after 1997, a phenomenon also observed elsewhere. Interestingly, plants in the vicin-ity of the glacier appeared to have followed two different kinds of response to ice melt: one group has closely tracked the rate and at a close distance to the snout of the glacier (e.g. *Arabis alpina, Senecio keniophytum*); the other group, which includes giant rosette-forming species (*Lobelia telekii, Senecio keniodendron*) remained stationary between 1958 and 1997, but showed approximately 16 m year$^{-1}$ rates of advance afterwards. Another contemporaneous example of species range extension in response to tropi-cal glacier retreat concerns anurans in the seasonal tropical Andes, where breeding populations have recently reached altitudes of 5200 m (Seimon *et al.* 2007). This study is particularly notable as it highlights the potential

## Box 8.1   Cryptogams in succession on glacier forelands

Cryptogams, while having no necessary 'role' in the primary colonization and succession of glacier forelands, do colonize available ice-free substratum. For example, species of the saxicolous lichen *Umbilicaria*, were found to increase in thallus density from 0 to >1000 along a chronosequence of about 250 years on a glacier forefield in southern Norway (Hestmark *et al.* 2007). Interestingly, the six species of *Umbilicaria* have been found to have similar relative thallus abundances at all ages, and also on non-glaciated substratum; species ranks did not change and were related to life history (initial growth rate and size-related maturation).

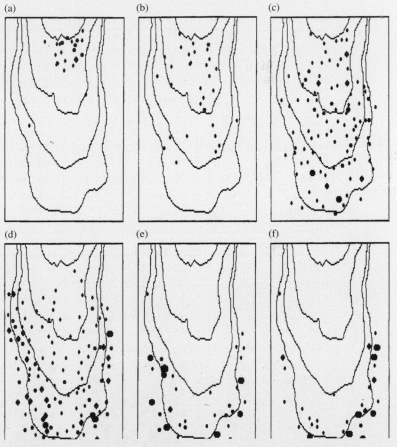

(a)   (b)   (c)

(d)   (e)   (f)

**Fig. 1**    Different distribution patterns shown by some cryptogamic species in relation to substratum age mapped on Storbreen, Norway (from Crouch 1993). (a) *Pohlia bulbifera*; (b) *Pannaria pezizoides*; (c) *Racomitrium canescens*; (d) *Cetraria islandica*; (e) *Hylocomium splendens*; (f) *Pleurozium schreberi*. Smallest symbol, 1% cover; largest symbol, 64–100%; seven categories altogether. Sample areas size: 2 m × 2 m.

**Box 8.1** *Continued*

Not all cryptogams respond to colonizable ground the same way. For example, lichen and bryophyte species have been shown to have a variety of distributions in a glacier forefield in southern Norway, in relation to age since deglaciation (Crouch 1993; Fig. 1, this box): the lichen *Pannaria pezizoides*, the moss *Pohlia bulbifera* colonized most recently deglaciated areas, contemporaneously with the vascular plants *Saxifraga cespitosa* and *Oxyria digyna* (Fig. 2, this box). Other cryptogams (*Hylocomium splendens*, *Pleurozia schreberi*) appear together with boreal heath species on the oldest parts of the moraine (100–200 years). Overall, there was an independent pattern of

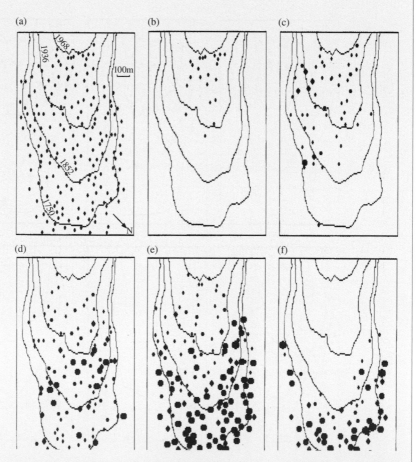

**Fig. 2**     Different distribution patterns shown by some vascular species in relation to substratum age (a) mapped on Storbreen, Norway (from Crouch 1993). (a) location of all sampling sites; (b) *Saxifraga cespitosa*; (c) *Oxyria digyna*; (d) *Salix lanata*; (e) *Empetrum nigrum* spp. *hermaphroditum*; (f) *Betula nana*. Smallest symbol, 1% cover; largest symbol, 64–100%; seven categories altogether. Sample areas size: 2 m × 2 m.

**Box 8.1** *Continued*

distribution of cryptogam and vascular plant species, and communities (compare Figs. 1 and 2). Crouch (1993) interpreted this as an indication of higher sensitivity of cryptogams than that of vascular plants to changes in environmental factors. Some cryptogams may indeed increase and others decrease with age since deglaciation and the start of vegetation development, as a function of biotic factors gaining importance, such as the establishment of canopy-forming species that may occlude cryptogam habitats.

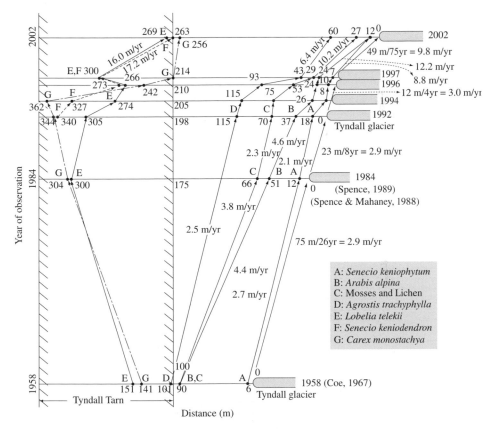

**Fig. 8.7**  Glacier ice recession and succession by a suite of species at the Tyndall Glacier, Mt Kenya (Mizuno 1998, 2005). There is a noticeable increase in the rate of glacier retreat after 1997 (about three times the rate of that recorded between 1958 and 1997).

range extension of and transgression of previous barriers (that have existed in the shape of glaciers) by pathogens, such as chitrids in the case of many Andean anurans.

Surprisingly little is known about the succession of invertebrates. From the few documented cases it appears that invertebrate assemblages build-up relatively rapidly. For example, in the European Alps it was found that most change in invertebrate assemblages occurred within 50 years after deglaciation (Fig. 8.8) (Kaufmann 2001). Interestingly, it appeared that colonization and succession largely followed predictable patterns, with random effects having a lesser importance for the outcomes. It is remarkable

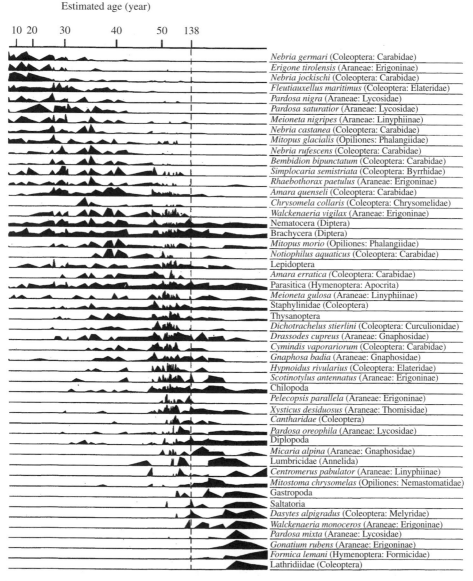

Estimated age (year)

10 20    30        40      50  138

Nebria germari (Coleoptera: Carabidae)
Erigone tirolensis (Araneae: Erigoninae)
Nebria jockischi (Coleoptera: Carabidae)
Fleutiauxellus maritimus (Coleoptera: Elateridae)
Pardosa nigra (Araneae: Lycosidae)
Pardosa saturatior (Araneae: Lycosidae)
Meioneta nigripes (Araneae: Linyphiinae)
Nebria castanea (Coleoptera: Carabidae)
Mitopus glacialis (Opiliones: Phalangiidae)
Nebria rufescens (Coleoptera: Carabidae)
Bembidion bipunctatum (Coleoptera: Carabidae)
Simplocaria semistriata (Coleoptera: Byrrhidae)
Rhaebothorax paetulus (Araneae: Erigoninae)
Amara quenseli (Coleoptera: Carabidae)
Chrysomela collaris (Coleoptera: Chrysomelidae)
Walckenaeria vigilax (Araneae: Erigoninae)
Nematocera (Diptera)
Brachycera (Diptera)
Mitopus morio (Opiliones: Phalangiidae)
Notiophilus aquaticus (Coleoptera: Carabidae)
Lepidoptera
Amara erratica (Coleoptera: Carabidae)
Parasitica (Hymenoptera: Apocrita)
Meioneta gulosa (Araneae: Linyphiinae)
Staphylinidae (Coleoptera)
Thysanoptera
Dichotrachelus stierlini (Coleoptera: Curculionidae)
Drassodes cupreus (Araneae: Gnaphosidae)
Cymindis vaporariorum (Coleoptera: Carabidae)
Gnaphosa badia (Araneae: Gnaphosidae)
Hypnoidus rivularius (Coleoptera: Elateridae)
Scotinotylus antennatus (Araneae: Erigoninae)
Chilopoda
Pelecopsis parallela (Araneae: Erigoninae)
Xysticus desiduosus (Araneae: Thomisidae)
Cantharidae (Coleoptera)
Pardosa oreophila (Araneae: Lycosidae)
Diplopoda
Micaria alpina (Araneae: Gnaphosidae)
Lumbricidae (Annelida)
Centromerus pabulator (Araneae: Linyphiinae)
Mitostoma chrysomelas (Opiliones: Nemastomatidae)
Gastropoda
Saltatoria
Dasytes alpigradus (Coleoptera: Melyridae)
Walckenaeria monoceros (Araneae: Erigoninae)
Pardosa mixta (Araneae: Lycosidae)
Gonatium rubens (Araneae: Erigoninae)
Formica lemani (Hymenoptera: Formicidae)
Lathridiidae (Coleoptera)

**Fig. 8.8**    The reconstructed succession of invertebrates along a chronosequence in the glacier foreland of Rotmoosegletscher, Tirol, Austria (from Kaufmann 2001)

that the first groups to appear are predatory beetles and spiders on the largely bare mineral substratum, while, understandably, herbivores and detrivores do not appear until suitable foraging conditions become available with the gradual build-up of soil and vegetation (Fig. 8.9).

## 8.4 Secondary succession

Secondary succession, in contrast to primary succession, occurs on a substratum that is not fundamentally different from what an ecosystem had functioned on prior to a major perturbation event that largely changed its structure (and composition). The soil, where previously present, is not destroyed and microbes, plants, and animals, even if in different proportions and qualities are present immediately after the disturbance. Most examples of secondary succession in alpine environments relate to post-fire processes or are associated with human land use, or both, and accordingly, vascular plant succession and associated animal assemblages may follow a number of possible routes (Fig. 8.10). Secondary succession after infrastructure building is well-documented (e.g. Brown *et al.* 2006). Plant cover (re-)establishment and species compositional changes appear to follow slightly different routes and dynamics in comparison with intact vegetation (e.g. see Fig. 8.11). A special case of secondary succession is that observed after agricultural land use, for example, the cultivation of potato in the páramo of the Venezuelan Andes. On abandonment, the fallows have a low species richness and the majority of the plant cover is formed by

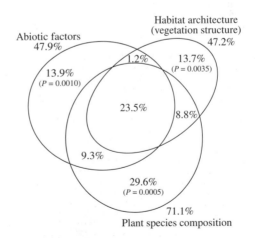

**Fig. 8.9**   The relative contribution of abiotic habitat conditions to plant species composition and vegetation structure, as explanatory variables for invertebrate composition in the glacier foreland of Rotmoosgletscher, Tirolean Alps, Austria (Kaufmann 2001)

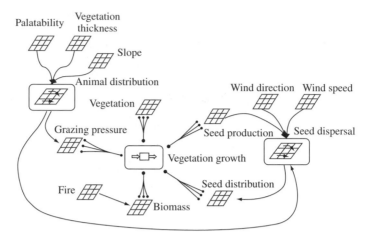

**Fig. 8.10**   The complexities of factors driving vegetation development after fire and in the presence of grazers. From a modelling study in non-alpine Mediterranean ecosystems by Mazzoleni *et al.* (2004)

introduced species (Sarmiento *et al.* 2003). Species richness doubles from the initial mean of 10 over about 3 years and then remains stable up to 12 years (usually the end of the fallow cycle); in the meantime, the cover contributed by native species builds up from approximately an initial 30% to 90%, a value just slightly lower than in páramo with no recorded cultivation history. It has also been observed that mycorrhizal infection increases with fallow age during secondary succession (Montilla *et al.* 1992). A frequent and not necessarily land use related phenomenon in the European Alps is the formation of small landslips (*Blaiken*). In such cases the alpine turf is uprooted, but the subsoil remains intact and is subject to secondary colonization. Such blemishes in the landscape form continually and 'heal' relatively rapidly, mainly through the action of stolonifeorus grasses, such as *Agrostis schraderiana*, that can grow up to 40 cm in length a year (Grabherr *et al.* 1988).

# 8.5 Post-succession vascular plant establishment, growth, and dynamics

In general, compositional dissimilarity, species turnover, and the rates in the change of plant cover decrease with age since first colonization (Foster and Tilman 2000), i.e. the above measures will indicate a highly dynamic system initially and a slow(er) changing one later on. How this works in the alpine zone is illustrated next.

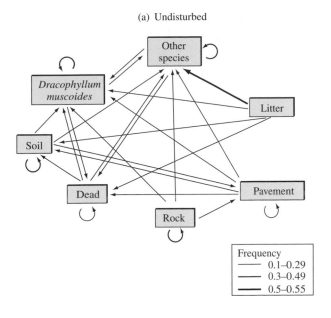

(a) Undisturbed

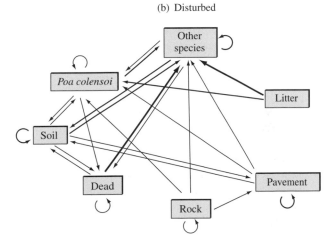

(b) Disturbed

**Fig. 8.11** Transition probabilities among several surface cover types in an undisturbed (a) and disturbed (b) cushion-dominated alpine area of New Zealand. Different thickness of the lines indicates different probabilities. From Brown *et al.* (2006)

A widespread view of succession that has shaped perceptions is based on the presumption that early stages of succession are nutrient-limited; therefore, nitrogen-fixing species (e.g. *Dryas* spp. in the arctic/alpine, *Alnus* spp. in the treeline ecotone) are likely to be at an early advantage in growing after colonization. This is usually taken one step further by claiming the facilitative nature of nitrogen fixers for later stages in vegetation development (e.g. see overview by Svoboda and Henry 1987). This seems logical; however, there are other factors that influence the arrival and the sequence of establishment of species, such as distance and dispersal ability. For example, if one examines the dispersal mode of *Dryas*, *Alnus*, and conifer trees (taxa that achieve successive dominance in cover in many mountain ranges on deglaciated ground that is below the treeline; e.g. Fastie 1995), it is clear that despite all three taxa being wind dispersed, some will be more efficient than others: *Dryas* seeds being *plumate* are most efficiently dispersed by wind; *Alnus* and conifers are *alate*, or winged, and are less efficient than plumate seeds. The time required for reaching maturity to produce copious seed is another important factor that affects the seral succession (*sensu* Clements 1928) of communities on moraines in terms of the apparent replacement of corresponding physiognomies, from open, cryptogam-rich communities to the local zonal vegetation, such as alpine sedge heath, dwarf-shrub heath, open upper montane scrub (e.g. Vetaas 1994), or closed forest (Box 8.2). For example, *Dryas drummondii* takes about 5 years (Lawrence *et al.* 1967), *Alnus* approximately 6–8 years, and conifers, e.g. *Picea sitchensis* approximately 20–40 years (Ruth 1965). Therefore, despite the early presence of conifers and other non-coniferous large tree species in the colonizing mix, their expansion and reaching dominance is limited by the time required to reach reproductive maturity (admittedly, not altogether independently of nutrient supply). In the alpine zone proper trees do not colonize; however, the same principles apply in regard to successive dominance of species (vegetation types).

## 8.5.1 Arrival and establishment

Suitable mineral substrata, such as glacier forefields, or other similar non-vegetated substrata are often perceived as initially offering conditions for colonization that are not or little space-limited. Accordingly, initial vegetation dynamics is the function of establishment success. Of vascular plants, wind-dispersed species, where present, are the first to colonize (Fig. 8.12). In general, long-distance dispersal via storms/heavy wind, or downwash are the most important vectors, followed by animals, and glacial melt water. Of the propagules that arrive, only very few survive and establish. The reasons for this are broadly related to the safe site concept, initially largely in terms of water availability and supply. For example, Stöcklin and Bäumler (1996) found that of seeds of early colonizing species sown on seasonally dry sites only approximately 0.5–1% emerged (and in only about half the species sown), most of which died by the end of the growing season. On permanently moist sites the emergence of the same species varied between approximately 9–17% and 75%; the rate of death during the growing season was up to a maximum of 32%; however, the

**Box 8.2**  Treeline trees as colonizers of glacier forefields

Tree species, especially conifers (e.g. see Mong 2006 and references therein) are among early colonizers on glacier forefields. The classic, but non-alpine example of tree colonization of glacier foreland is that of Glacier Bay, Alaska (e.g. Fastie 1995), where Sitka spruce (*Picea sitchensis*) appeared, on average 15 years after the retreat of ice, and about 65 years after deglaciation there were ≥10 cone-bearing individuals per hectare. The lack of climatic constraints on the one hand, and copious propagule availability, on the other is shown in the estimated density of shrubs and Sitka spruce (Fig. 1, this box).

In the high mountains, treeline-forming, or accidental tree species may also establish in glacier forelands, especially in those areas where glacier expansion occurred during the Little Ice Age, and which are surrounded by upper montane forest. For example, at Bødalsbreen, in the southern Scandes, Norway, the moraines that deglaciated around 1750 are covered by *Betula pubescens*, as opposed to the young moraines (ice-free since 1930) that have a *Racomitrium canescens–Stereocaulon* moss- and lichen-dominated community (Vetaas 1994). It is noteworthy that despite such physiognomic differences, the species composition (by presence of species alone) can be as much as 79% identical between a cryptogam mat and *Betula* scrub.

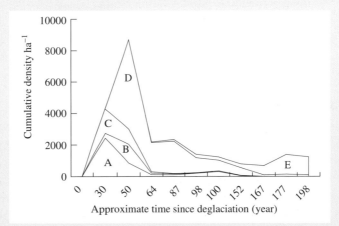

**Fig. 1**  The temporal evolution of the density of some shrub and tree species at Glacier Bay, Alaska (data from Fastie 1995). (A) *Salix sitchensis*; (B) *Salix barclayi*; (C) other *Salix* spp.; (D) *Alnus sitchensis*; (E) *Picea sitchensis*.

first winter killed well over half of the remaining seedlings in most species (for further details see table 6 in Stöcklin and Bäumler 1996). Similarly high rates of mortality were reported from other sites, too (e.g. Galen and Stanton 1999; Erschbamer *et al.* 2001). In addition, as Galen and Stanton

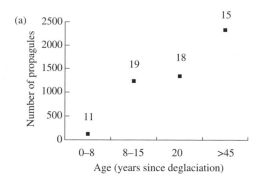

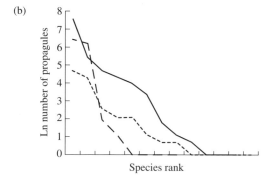

**Fig. 8.12**     Propagule numbers (and number of species) trapped with increasing distance from the snout of Morteratsch Glacier, central Alps, Switzerland (a); and seed rank abundance by species (b) with *plumate* (solid line), *alate* (broken line) and with seeds with no appendices (dotted line). Data from Stöcklin & Bäumler (1996).

(1999) have remarked, safe sites that facilitate emergence may not be safe for the survival of seedlings, giving early establishment a rather stochastic nature.

With the vegetation closing over suitable substratum, the available sites for colonization decrease, but not the overall number of propagules. The propagules available along a series from closed *Carex curvula* sedge heath (2800 m) to open scree vegetation near the upper limit of vascular plants (3050 m, Mount Schrankogel, Austrian Alps, Fig. 8.13) showed that: (1) generalizations are unsafe to make as inter-annual variation can cause about twofold variation in the number of species that produce seeds and up to 10-fold differences in seed output (Ertl 2006), and (2) the potential for 'succession' in terms of available species and safe sites is fundamentally different between recently deglaciated areas and areas at the upper margin of plant life. From the study by Ertl (2006) it appears that in absolute terms, the highest seed output is by the open cushion fields and by the closed *Carex* heath. Interestingly, no propagules of upper montane origin

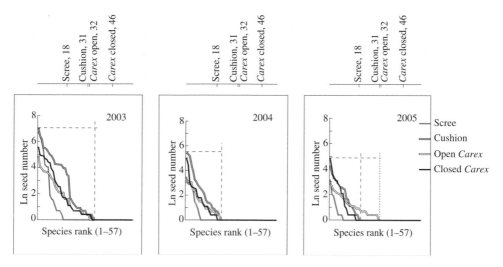

**Fig. 8.13**     Numbers of seeds trapped and contributing species in high alpine closed and open
*Carex curvula* sedge heath, open cushion field and scree at 2800–3050 m asl on
Schrankogel, Austrian Alps. Unpublished data, courtesy of S. Ertl.

There was a c. 10-fold difference between seed numbers trapped in 2003 (the hottest
summer on record in Europe) and 2005 (a year with a short growing season, following
late snowmelt). The number of species that contributed seeds was c. twice as many in
2003 than in the other years. Along the top scale bars are shown the total numbers
of vascular plant species recorded within a 10-m radius circle around the traps. The
number of species contributing to the trapped seeds from outside the 10-m radius
zone were: *Carex* closed, 2; *Carex* open, 3; cushion, 4; scree, 4.

were recorded, i.e. no long distance transport by wind or by other agents
occurred. However, the high dispersal capacity of plumate seeds was evident as seeds that arrived from outside a buffer area of 10-m radius around
the seed traps were all of that kind. The environmental cues that cause
periodic mast seeding events such as in alpine snowgrasses *Chionochloa*
spp. in the New Zealand Alps (Tisch and Kelly 1998), East African *Senecio
keniodendron*, or *Espeletia timotensis* in the northern Andes (Smith and
Young 1994), as well as in arctic tundra plants, such as *Eriophorum vaginatum* (Molau and Shaver 1998) have not been fully consistently explained
to date (see pp. 142–146, Crawford 2008).

The fate of seeds and emerged seedlings is of interest for the turnover
of individuals and species in alpine plant communities. Community type
and year can cause a large variation in the ratio of total species in a stand
to those that produce seeds (e.g. 47–77% in a warm year across four communities; Ertl 2006), in the viability of seeds produced (e.g. Chambers
1995) and the number of seedlings (e.g. 13–64% across four community
types on Niwot Ridge, Colorado Rockies; Forbis 2003). Seedling mortality

may be very high across species in the first year (48–74%), less in the second year (16–32%), and may stabilize later on (Forbis 2003); however, perturbation can much modify survival (Chambers *et al.* 1990; Chambers 1993, 1995). While the population flux in species can be transiently large as a result of seedling flushes the subsequent high rates of seedling deaths, and additional deaths from adults usually result in low overall change in the numbers of individuals (e.g. Nagy and Proctor 1996). This applies to communities that experience relatively little perturbation and are not subject to colonization. In the colonization stage, population fluxes may be very high (e.g. Gibson and Kirkpatrick 1992) and population sizes expand, especially those of aggressive colonizers. In established vegetation, species with a high reproductive capacity, and high seedling survival ability may expand in short transient bursts; however, the overall internal turnover is low. Species have been calculated to contribute established offspring over a wide range, e.g. from about one by 100 adults in *Pinguicula alpina* to 1.35 by a single adult of *Bistorta bistortoides*. However, seedling dynamics can change suddenly in communities that have enjoyed relatively long-term 'stability' and undergo periodic changes as a result of local perturbation, such as that caused by some burrowing animals (Chambers 1995). Such disturbances can prepare a good seedbed that provides an opportunity for mass seedling flushing by stimulating the seed bank (Fig. 8.14).

## 8.5.2 Interactions: perceptions of competition and facilitation

The pattern of vegetation that we perceive is that fashioned by certain combinations of plants along environmental gradients such as water and nutrient availability or space itself. In addition, local perturbations may be frequent and instrumental in certain alpine locales. In broad terms, alpine environments have traditionally been charted out as stressful, following the classification of plant growth strategies (competitor, C; stress tolerator, S; and ruderal, R) by Grime (1979). Such classification places emphasis on general growth responses of plants to available resources whether across or within habitats. For example, if one compares the tall 'weedy' or 'ruderal' vegetation of a sheep fold or a cattle resting area with their highly nutrient-enriched soils in the treeline ecotone, with that of a moderately grazed alpine sedge heath on a well-drained hillside, the C-S-R categorization appears to be an applicable system and covers all three classes. Similar contrasts can be demonstrated along glacier forelands from early colonizing 'ruderals' to late successional 'stress tolerators' (Caccianiga *et al.* 2006). Indeed, such contrasts can be shown at any scale along a long enough environmental gradient, with a suitably large species pool, where species represent wide-ranging characters. Competition versus facilitation are often presented as different manifestations of plant-to-plant interactions along environmental gradients and that the arrangement of species along

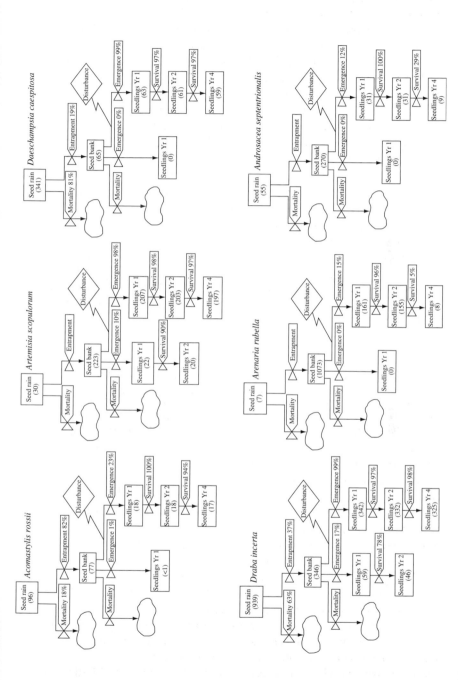

**Fig. 8.14** The impact of small-scale perturbation by pocket gopher (*Thomomys talpoides*)—shown in each panel in a diamond-shaped box as 'Disturbance'—on seedling emergence in the Rocky Mts, Montana (Source: Chambers 1995). The species represent a range of life forms and life histories: *Acomastylis rossii*, clonal herb, 100+ years; *Artemisia scopulorum*, aclonal herb, 50+; *Deschampsia caespitosa*, tussock-forming grass, 10–20; *Draba incerta*, aclonal herb, 5–10; *Arenaria rubella*, aclonal herb, 3–5; *Androsace septentrionalis*, aclonal herb, 2–5.

these continua reflects the net outcome of these interactions (e.g. Choler *et al.* 2001). This can be construed as another way of expressing the C-S(-R) classification by Grime (1979), but where species may be competitors under a set of circumstances and stress tolerators under another set. This may indeed be the case where an environmental impact is large enough to change the prevailing conditions. Perturbations, such as unnatural excess of nutrients can radically transform the nature of allocation patterns and even life history, as was demonstrated for *Cochlearia pyrenaica* ssp. *alpina* after fertilizer addition (Nagy and Proctor 1997). However, such dramatic changes are hardly observed along natural alpine gradients.

For the dynamics of alpine ecosystems the interest is to be able to predict to what extent vegetation dynamics is determined by the abiotic environment as opposed to by biotic interactions. A perennial dilemma is if species-to-species interactions diminish indeed at high environmental constraints (resource limitation) such as in the alpine habitats and what the net outcome of all this is for vegetation dynamics. The answer to the above questions in quantitative terms is rather difficult. Conceptually, it can be measured in an experimental set-up where one considers, for example, annual plants sown at different densities, harvested at different times, and inferences about competition are made by comparing accumulated biomass. However, even-aged monospecific stands are not the norm in nature. None the less, the classic work by de Wit (1963) has been impervious in the way plant competition has been quantified, without questioning its applicability to natural ecosystems (many species and wide age distribution), such as those of the alpine. However, as Taylor *et al.* (1990) so lucidly put it:

In highly fertile habitats which are close to carrying capacity for example, traits such as high nutrient uptake rate and high maximum potential biomass or height will be important in defining competitive ability (...). Competitive ability in a less fertile habitat however, may have little to do with maximum potential biomass or height. Rather, an individual may be a better competitor than its neighbours because it possesses attributes such as proportionately greater root mass,

(See, for example, root mass distribution in a *Carex curvula* dominated sedge heath which makes *Carex curvula* the overriding competitor and matrix species; the high root mass is not a result of high rates of root growth but of long persistence through clonal growth.)

earlier germination time, greater tolerance of low temperatures, more effective association with mycorrhizae, greater attractiveness to common pollinators, more effective allelopathic chemicals, an ability to deplete resources to a lower level, or a higher tolerance threshold of nutrient depletion, among others (Aarssen 1989). In this context, tolerance of competition is regarded primarily as an attribute of competitive ability (...), not as an attribute of stress-tolerance (*sensu* Grime 1979)....However, since resource availability for a plant may be low because of depletion by neighbours (competition) or because of low environmental carrying capacity, the same traits may confer adaptation to both the 'competitor' strategy and the 'impoverishment-tolerator' strategy.

In addition to the question if biomass-related measures are suitable measures of competition, another problem arises in manipulating species interactions by neighbour removals. A large number of workers have used neighbour removals with the aim of quantifying competition between selected target species and the surrounding mix of species in alpine ecosystems (Choler *et al.* 2001; Gerdol *et al.* 2002; Totland and Esaete 2002; Corcket *et al.* 2003; Klanderud 2005; Klanderud and Totland 2005). For example, Choler *et al.* (2001) have demonstrated in an elegant manner that some species, can switch from positive above-ground biomass response to cutting their neighbours to a negative one in relation to the mean altitude distribution range of the target species. Neighbour removals constitute a perturbation and the resulting growth changes of target species are difficult to separate from the impacts of perturbation (such impact in nature can arise from grazing for example). Singling out species, or individuals and instantaneously removing a major part of their growing environment has manifold implications both above-ground and the below-ground. For example, such treatment applied to forest stands would be akin to clear-felling patches around individual trees to demonstrate the benefit to target trees, or thinning by selectively removing species and attributing responses to thinning (space) as competitive release (species) impacts. Remarkably, selective logging in tropical rainforest does indeed result in similar, but unplanned experiments like those made by neighbour removals. The response to such clearing is very much specific to the species: residual trees that belong to canopy species experience better growth as a result of competitive release, but not obligate understorey species, which may even respond negatively. While it can be demonstrated by neighbour removals, or other qualitative or quantitative manipulations that the full or partial removal of neighbouring plants or their parts can result in some cases in increased productivity in alpine habitats, the opposite, i.e. increasing density in natural populations is mostly unfeasible and therefore density-dependent competition effects are difficult to identify with certainty. For self-regulation of density in *Carex curvula* see Grabherr (1989). In a recent welcome clarification on differentiating between the importance versus the intensity of competition, Brooker *et al.* (2005) have offered alternative interpretations of neighbour removal experiments. None the less, the above passage by Taylor *et al.* (1990) remains a reminder of the need for a wider theoretical framework to be developed for advancing the case of plant-to-plant interactions and their impact on vegetation dynamics, especially in alpine habitats.

Competition and facilitation are often portrayed as two ends of a scale with opposite processes, and where competition is not measurable, it is assumed that facilitation exceeds it. Facilitation is passive, i.e. it is not a wilful grouping of species, as opposed to competition that results in the suppression or exclusion of weaker competitors. The mechanics of passive facilitation work by a facilitator forming a nucleus, around which plant and animal species may occur preferentially and in close proximity. This may be the physical presence of a boulder that can trap propagules and provide wind shelter and ameliorated thermal conditions (e.g. Kleier and Lambrinos 2005). A passive facilitator may also be a plant (individual, group, growth form), whose physical presence one way or other ameliorates the chance of success of plant establishment and initial growth. This is a prerequisite for epiphytic cryptogams and facilitates the establishment of foliose and racemose lichens in some alpine vegetation formations. As with vascular plants, alpine cushion plants in particular have been the focus of work on facilitation (Cavieres *et al.* 2002, 2005, 2007). Such initially 'neutral' interspecific assemblages, may, with time, turn into asymmetrical competition between the initial passive facilitator and the beneficiaries (Olofsson *et al.* 1999). Accordingly, the co-occurrence of species, or an increase in frequency of co-occurrences, may or may not be indicative of a mutualistic relationship (Kikvidze *et al.* 2005; Dullinger *et al.* 2007); however, it can serve as a starting point for disentangling the nature of such dynamic associations.

While the exact nature of species-to-species, or individual-to-individual interactions remains to be established, communities with their overall structure, species composition, and realized biomass can be taken as the net outcome of all biotic interactions and abiotic constraints in a given location at a given time, with a given background of history (species pool). Largely unknown in this context is, however, just how overall below-ground microbial processes, and in particular mutualism and parasitism contribute to vegetation dynamics. These are just beginning to unravel (Bardgett *et al.* 2005). While it is generally accepted that mycorrhizal symbiosis is widespread among plants in general (approximately 80%), it is less so in some alpine vegetation types (Smith and Read 1997; Cazares *et al.* 2005; Schmidt *et al.* 2008; Table 8.3). The few quantitative analyses from the various alpine vegetation types from the Alps and arctic mountain ranges suggests that, although mycorrhizal infection is indeed widespread, the proportion of species that form mycorrhizal associations varies widely among habitats and vegetation types (Read and Haselwandter 1981). From the work by Read and Haselwandter (1981, table 3) it appears that the most extreme habitats (subnival and late snowbed) have the lowest level of infection, which is the highest in the closed sedge heath. The authors have explained this by a likely low level of nutrient stress in the snowbed and nival habitats as a result of nutrient flushes at snowmelt (in the range of the total annual atmospheric nitrogen deposition and about 30% of annual soil organic matter nitrogen mineralization; see Bowman 1992), which coincide with intensive root growth. In the meantime, they suggested that the high level of vesicular-arbuscular infection in sedge heath was a reflection of nutrient stress and a high incidence of root contact. Michelsen *et al.* (1998), Nara *et al.* (2003), and

**Table 8.3** The mycorrhizal status of vascular plant species recorded by Cazares *et al.* (2005) between 1800 and 1900 m asl at the Lyman glacier forefront, North Cascades Mountains, Washington

| Family | Number of samples (n>1) | Non-M (%) | AM (%) | EM (%) | ERM (%) | DS (%) |
|---|---|---|---|---|---|---|
| *Monocotyledons* | | | | | | |
| Cyperaceae | 22 | 82 | 4 | 0 | 0 | 18 |
| Asteraceae | 17 | 23 | 76 | 0 | 0 | 35 |
| Juncaceae | 67 | 64 | 6 | 0 | 0 | 34 |
| Poaceae | 20 | 35 | 45 | 0 | 0 | 45 |
| *Dicotyledons* | | | | | | |
| Asteraceae | 17 | 24 | 76 | 0 | 0 | 35 |
| Caryophyllaceae | 21 | 81 | 5 | 0 | 0 | 14 |
| Ericaceae | 78 | 2 | 1 | 0 | 97 | 51 |
| Onagraceae | 34 | 56 | 38 | 0 | 0 | 38 |
| Polygonaceae | 16 | 56 | 19 | 7 | 0 | 31 |
| Rosaceae | 13 | 8 | 92 | 0 | 0 | 54 |
| Salicaceae | 31 | 3 | 16 | 94 | 0 | 84 |
| Saxifragaceae | 54 | 85 | 4 | 0 | 0 | 13 |
| Scrophulariaceae | 67 | 36 | 54 | 0 | 0 | 24 |
| *Gymnosperms* | | | | | | |
| Pinaceae | 27 | 4 | 4 | 96 | 0 | 52 |
| *Pteridophytes* | | | | | | |
| Athyriaceae | 3 | 0 | 0 | 0 | 0 | 100 |
| Cryptogrammaceae | 4 | 0 | 0 | 0 | 0 | 100 |

Non-M, non-mycorrhizal; AM, arbuscular mycorrhizal; EM, ectomycorrhizal; ERM, ericoid mycorrhizal; DS, dark septate endophyte.
For a similar table on some high Andean and Rocky Mt alpine species see Schmidt et al. (2008)

Cazares *et al.* (2005) have reported high incidence of non-infection in non-woody species from various treeline ecotone and alpine heath sites and rates of infection at high altitude sites—especially in the Andes, but also in the Rocky Mountains—much varied, both with regard to vascular taxa and fungal hypha type (Schmidt *et al.* 2008). Showing the presence of mycorrhizal associations is by no means equivalent to proving functional links and causality. For example, as it is reasonably easy to grow plants experimentally with and without infection most knowledge on the impacts of mycorrhizal infection concerns plant growth (biomass). What is lacking, almost entirely, is how mycorrhizal symbionts can affect fitness, population ecology and dynamics, or whether there is a community-level interconnectedness of plants via mycorrhizal hyphae that would allow an exchange of nutrients across individuals (van der Heiden and Sanders 2003). It is known from alpine ecosystems that mycorrhizal fungi are able to enzymatically break down organic substances and pass simple organic molecules on to their hosts (organic nitrogen and phosphorus compounds). In exchange, the plant associates may pass on an estimated 10–50% of their photosynthates to mycorrhizal fungi, bypassing the litter–decomposer route of organic matter addition to the soil. It has also been suggested that the establishment of seedlings and their early development may be improved by the existence of a general hyphal network, connecting individuals of the same or different species. It may be thought that, for example, *Carex curvula* sedge heath (the one with a high degree of infection of its constituent species, and the highest species diversity of all alpine communities in the study by Read and Haselwandter (1981)) is proof of such a hypothesis. However, the closed nature of this vegetation type may in fact be the facilitator of the development of a wide hyphal network in the first place. Most importantly, sedge heaths develop where abiotic conditions are less extreme than in snowbeds and subnival summits (habitats with a low infection rate), i.e. cause and effect need to be put in a wider context beyond the world of mycorrhizae. The exact functional nature of proposed interactions among pathogenic, saprophytic, and mycorrhizal fungi in the alpine is yet unknown, and so is the interaction between bacteria and mycorrhizal fungi. An impressive exception appears to be the seasonal succession of the mycobionts. This has been documented for *Ranunculus adoneus* by Mullen and Schmidt (1993) and Mullen *et al.* (1998). Following snowmelt in late June–early July, dark septate endophytes (that overwinter in the roots) are observed during the period of maximum observed nitrogen uptake by the *Ranunculus* plants. The arbuscules of two types of arbuscular mycorrhizal fungi appear on new roots in August and their highest infection coincides with a period of maximum uptake of phosphorus by the plants. The extent to which this pattern may be generally occurring in alpine plants is not yet known.

Despite this, there is an empirical framework for cataloguing the temporal dynamics of the linkage between above-ground and below-ground processes, proposed by Bardgett *et al.* (2005) at a three-level temporal hierarchy: short-term, <1 year; medium-term, up to approximately 1000 years; long-term, several millennia, with relevant examples for alpine ecosystem dynamics falling in the short- and medium-term. In the shortest term, abiotic factors such as diurnal freeze–thaw cycles, or plant responses to biotic or abiotic factors can greatly vary the abundance and activity of soil micro-organisms through the amount of root exudates (Bardgett 2005)—a factor that may play an important part in tropical alpine ecosystems in regulating nutrient cycling. Similarly to grazing, other biotic (e.g. parasites) and abiotic (mineral nutrients, climate) factors can stimulate an excess of root exudates as a short-term growth compensatory mechanism through enhancing microbial mineralization and availability of nutrients. At the season scale, there is a well-established linkage between microbial activity and plant growth (Box 8.3; Lipson *et al.* 1999). What is less clear is how this linkage influences the variation of plant growth

**Box 8.3** The dynamics of plants and microbial communities in seasonal alpine ecosystems (from Bardgett *et al.* 2005).

The cycling of labile nitrogen pools over seasons relies on intimate, temporal coupling between plants and microbes and their resource demands. The linkage between plant and heterotrophic microbial communities exhibits pronounced seasonal shifts that influence the temporal and spatial variation in diversity of both biotic groups. The supply of limiting resources determining production and species composition is dependent on fluxes of labile carbon and nitrogen between the two groups. Microbial biomass is largely dependent on the carbon supply from plants, whereas plant production is limited by labile forms of nitrogen (small amino acids and inorganic nitrogen), derived from microbial degradation of soil organic matter. The supply rates of these resources change seasonally depending on climatic constraints and the abundance and activity of microbes and plants. Plant and microbial functional group composition also varies spatially, influencing the diversity of the other group.

### Autumn phase

Senescing plants provide a pulse of labile carbon to support microbial growth. Variation in the chemistry of these compounds provides a potential source to promote diversity in the microbial community. For example, litter that is rich in low molecular weight phenolic compounds enhances

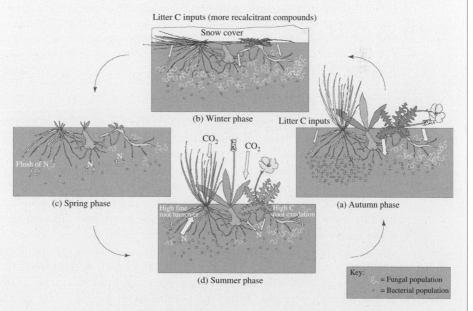

**Fig. 3**    Conceptual model of seasonal dynamics of vascular plant growth and soil microbial processes in an alpine ecosystem (from Bardgett *et al.* 2005).

**Box 8.3** *Continued*

overall microbial biomass, particularly fungi, whereas litter that is rich in carbohydrates and sugars enhances bacterial growth (Bowman and Steltzer 1998).

### Winter phase

Microbial biomass continues to increase in soils warmer than −5°C as carbon and nitrogen in plant litter is consumed and mineralized. Less easily degraded polyphenolic compounds promote dominance by fungal species (Schadt *et al.* 2003).

### Spring phase

Rapid changes in microclimate and the exhaustion of labile carbon compounds lead to turnover of microbial community, with concomitant release of labile nitrogen for plant uptake (Lipson *et al.* 1999).

### Summer phase

Plant uptake of nitrogen to meet growth demands occurs during the early summer (Jaeger *et al.* 1999), followed by a period of carbon sequestration and loss to soil microbes. Some slow-growing phenolic-rich plant species (e.g. the alpine herb *Acomastylis rossi* in the Rocky Mountains) exude carbon into the soil and thereby manipulates microbial immobilization of nitrogen, promoting low nutrient conditions (Bowman *et al.* 2004), whereas other, fast-growing species (e.g. the grass *Deschampsia caespitosa*) exhibit high turnover of fine roots, promoting more fertile conditions (Bowman and Steltzer 1998). Variation in plant-derived carbon substrates influences microbial diversity, whereas microbial activity, biomass, and immobilization of nitrogen influence plant diversity.

and community dynamics between years (Walker *et al.* 1994), or exerts a trend over a longer period of time. It may be postulated that years that differ climatically can favour the growth of certain plant species more than others, perhaps through critical thresholds in mineralization and nutrient availability. The second time-scale proposed by Bardgett *et al.* (2005) is that of succession during which both plants and microbial assemblages become more numerous in species on the one hand, and different from those found in the initial stages, on the other. With the development of complexity, however, increases the number of unanswered questions and probable explanations with regard to cause and effect. A clear trend that occurs is a shift in early colonizing nitrogen-rich litter producing plants to species with persistent litter; the parallel development below ground is characterized by the increasing dominance of fungal decomposers at the expense of bacteria, which are prominent in early stages. The ensuing changes in soil physical and chemical properties and incident changes that may favour some and limit other soil micro-organisms may contribute to the decline of some early colonizing vascular plants; however, this is yet to be demonstrated in alpine succession.

## 8.6 The role of extraneous drivers

It does seem that most alpine and arctic plant communities are in fact very responsive to periodic improvements to their environment, be it by way of a warm growing season, or by enhancements to their micro-environment. This is especially so with regard to seed production (e.g. see Fig. 8.13; Nagy and Proctor 1997). Of the extraneous drivers that may exert intermittent influence on the dynamics of alpine communities at the patch to stand levels, nutrients, temperature, and herbivores are considered here. These themes are revisited under climate change and land use where we consider landscape-scale and long-term impacts. For an overview on the impacts of these drivers in arctic ecosystems see Jonasson *et al.* (2001).

### 8.6.1 Nutrients

Nutrients, land use, and herbivore impact are rather interlinked, as land use in the alpine zone is most often by grazing and changes in the local availability of nutrients can encapsulate land use impacts on the patch to stand scales. However, nutrients arrive from the atmosphere as well by wet and dry deposition and by wind blown dust or insects, or locally they accumulate at the bottom of avalanche tracks as a result of periodic input of soil and high mineralization rates after soil perturbation. Some plant communities create a relatively nutrient-rich environment by either fixing nitrogen, such as the actinomycete *Frankia* in association with *Alnus*, or by having high rates of decomposition and mineralization, such as some tall herb communities.

It can be demonstrated that plant species in alpine ecosystems have distributions that are related to soil nutrient availability (e.g. Ozenda 1985; Atkin and Collier 1992; Ellenberg 1996); therefore, changes in the accessibility of nutrients are likely to alter community composition and structure. On the other hand, it is known that even small changes in atmospheric nutrient input may cause measurable changes in ecosystem properties (Baron *et al.* 2000), and if these changes are large enough to prompt species replacement they can have cascade effects on biogeochemistry (Bowman and Steltzer 1998). While atmospheric input is made at the landscape scale, it may have differential impacts locally, where factors such as topography and climate can much influence deposition rates, and whose relative importance in relation to, for example, local soil chemistry is determined by bedrock and may much vary (Bragazza *et al.* 2003). The actual nitrogen deposition rates are usually below the critical load. Critical nitrogen load is the annual input into an ecosystem below which no perceptible (adverse) impact through eutrophication occurs. This critical load for alpine ecosystems in Switzerland has been calculated at approximately 10 kg nitrogen ha$^{-1}$ year$^{-1}$ (Rihm and Kurz 2001). Given the highly variable and rather weak signal from 'natural experiments', or from those that have tried to mimic levels of nitrogen deposition, most of our understanding about the impacts of changes in nutrient availability is derived from nutrient addition experiments, where the responses of established stands of alpine plant communities were observed (Table 8.4). While these are valid observations in that they report the impacts of a usually one-off high nutrient pulse, they also represent a kind of perturbation. However, such perturbation in nature can indeed be observed at the patch-scale in grazed ecosystems where large herbivores can deposit considerable amounts of urine and faeces, whose impacts may be compared with those of nutrient addition experiments.

The most spectacular responses to nutrient additions have been reported from open habitats, where the addition of nutrients caused large increases in plant cover (Nagy and Proctor 1997; Heer and Körner 2002). In most of the other reported studies nutrient addition increased biomass production, caused shifts in the relative proportion of growth or life forms, or reduced species richness (Table 8.4). Such changes occur regularly, but go mostly unobserved in grazed alpine ecosystems, and form the basis of extraneously driven patch-scale dynamics (that follows, for example, over a range of ages of cowpat left after summer grazing by cattle).

Burrowing animals, such as pocket gophers (*Thomomys talpoides*) in the Colorado Rockies can change nutrient availability and through it local vegetation dynamics. It has been shown that while gopher mound soils are lower in total amounts of nutrients, the available fraction of nitrogen is higher (Litaor *et al.* 1996), which can stimulate better plant growth by the provision of readily available nitrogen (and possibly phosphorus). In addition, such mounds provide ideal seedbed for germination and seedling flush. A similar effect may be expected around burrow entrances of ground squirrel and other burrowing animals.

## 8.6.2 Temperature

Little temperature effect other than that can be observed in alpine habitats as a result of micro-topographic and micro-climatic variation can be expected at the patch- to stand-scale. Temperature is likely to be a driving force of vegetation dynamics indirectly through inter-annual variation in snow cover for example (Edmonds *et al.* 2006). Growing season length can be critical for seed production in alpine annuals as was demonstrated by Reynolds (1984) for two *Polygonum* species. Growing season length (and

**Table 8.4** Responses of alpine organisms and plant communities to nutrient addition

| Nutrient added | Duration | Target species/community | Measured responses | Main impact | Reference |
|---|---|---|---|---|---|
| N, P, N+P, 250 kg ha$^{-1}$ year$^{-1}$ | 2 years | Dry meadow (*Kobresia myosuroides*); wet meadow (*Carex scopulorum*) | Primary production, physiological and developmental constraints; N and P use efficiency | N limited dry meadow PP; N+P limited wet meadow PP; +N, and +N+P increased production and species richness in *Kobresia* dry meadow; shift in structure (+graminoids and herb, −sedge); increase in wet meadow biomass; +graminoids −herbs | Bowman et al. (1993); Bowman (1994) |
| N, P, N+P, 250 kg ha$^{-1}$ year$^{-1}$ | 5 years | Dry meadow (*Kobresia myosuroides*); wet meadow (*Carex scopulorum*) | Diversity and abundance of growth forms/functional types | Dry meadow: +N+P, increase in mycorrhizal and non-mycorrhizal herbs, and grasses; no change in sedges; +P increase in N-fixing herbs; Wet meadow: +N, increase in sedges, +P, +N+P, increase in grasses | Theodose & Bowman (1997) |
| N, 100 kg ha$^{-1}$ year$^{-1}$; P, 50 kg ha$^{-1}$ year$^{-1}$; P, 50 kg ha$^{-1}$ year$^{-1}$ | 3 years | *Vaccinium myrtillus*, *V. uliginosum* heath; *Empetrum hermaphroditum*–*Vaccinium uliginosum* heath | Shoot and leaf growth; above-ground biomass | Decrease in *V. uliginosum* biomass; no effect on *V. myrtillus* or *E. hermaphroditum* | Gerdol et al. (2002) |
| NPK, | 6 years | Fell-field | Shoot growth, meristem activation | Positive | Graglia et al. (1997) |
| NPK + trace elements 100–200 kg ha$^{-1}$ year$^{-1}$ | 3 years | Species near glacial melt | Biomass | 177% biomass increase, mostly by graminoids | Heer & Körner (2002) |

**Table 8.4** *Continued*

| Nutrient added | Duration | Target species/community | Measured responses | Main impact | Reference |
|---|---|---|---|---|---|
| N (and temperature) | | *Betula pubescens* ssp. *tortuosa* | N economy, growth rate | Above 5°C threshold: +N increased growth rate | Karlsson & Nordell (1996) |
| $NH_4NO_3$, 20, 40, 80 kg ha⁻¹ year⁻¹ | 180 days | Nematodes | density | No N effect | Lokupitiya et al. (2000) |
| Ca, NPK, NPKCa, N, 100 kg ha⁻¹ P, 50 kg ha⁻¹ K, 100 kg ha⁻¹, Ca, 1000 kg ha⁻¹ | 2 years | *Cochlearia pyrenaica* ssp. *alpina* and others, incl. *Agrostis vinealis, Festuca rubra* | Cover, flowering, fruiting, population dynamics | +Ca, NPK, NPKCa, increased total cover 3, 5, and 10-fold; population size of *C. pyrenaica* 2, 4, and 8-fold; induced life history changes by depleting original cohorts in +NPK and NPKCa versus halving in control and +Ca | Nagy & Proctor (1997) |
| N 7, 35, 70 kg ha⁻¹ year⁻¹ | 2–3 years | *Cetraria nivalis* heath; *Phyllodoce careula–Juncus trifidus* heath | Species abundance | No effect | Mols et al. (2000) |
| $NH_4NO_3$, 100–200 kg ha⁻¹ year⁻¹; P, 10–20 kg ha⁻¹ year⁻¹; Interaction of nutrient with snowpack | 2 years at high; 1 year no; 3 years low rates | Dry meadow (*Kobresia myosuroides*); mesic meadow (*Deschampsia caespitosa*); snowbed (*Sibbaldia procumbens*) | Above-ground biomass, species richness; abundance of species in *Kobresia* meadow with or without snow | +N and +P, increase in biomass, +N, decrease in species richness | (Seastedt & Vaccaro 2001) |

associated position along a slope gradient) was also paramount for the growth of *Salix herbacea*: 20% difference in season length caused a fivefold difference in shoot growth in the northern Scandes (Wijk 1986). Results from warming experiments do not connect to this, as they usually target specific vegetation types as opposed to boundaries or transitions between vegetation types. Results from warming experiments generally point to increased growth, earlier onset of phenological stages, and more abundant seed production (Table 8.5). These observed changes are often accompanied by reduction in plant density as a result of higher biomass production in the shelter of open-top chambers, which in addition to increasing the temperature, can be seen as major passive facilitators. As most studies have reported single species responses it is difficult to piece together the overall impacts of warmer temperature on the dynamics of species within alpine communities, apart from generalizations that some will be winners and others losers (Erschbamer 2007). As an exception, Hollister *et al.* (2005) have provided an in-depth and insightful analysis of temperature impacts on dry heath and wet meadow plants in Alaska. It appears that, although those species that responded to warming conformed to the generally observed pattern, many species were non-responsive with respect to a number of traits and in many cases the response to warming was strongly correlated with factors other than temperature. In other words, fluctuation in the correlated factors could reinforce or cancel out the impacts of warming on a year-to-year basis. Based on studies such that by Hollister *et al.* (2005), one may begin exploring the potential impacts on vegetation dynamics of individualistic responses in a number of traits by plants to warming.

In addition, a 'natural experiment' reported by Mizuno (2005) who compiled the history of species responses to the retreat of the Tyndall glacier on Mount Kenya is rather informative with regard to interspecific differences in the rates of dispersal and colonization by cryptogams and vascular plants. It appears that giant rosettes (and perhaps analogously, phanerogams in extratopical environments) need a higher temperature threshold (or other environmental cues in the case of *Senecio keniodendron*, a masting species) for propagule production than those species that are able to closely track glacier snouts. Differences of this kind may have important implications for our ability to understand responses to climate warming in alpine community dynamics.

## 8.6.3 Herbivores

Grazing and trampling can have major impacts on the dynamics of alpine vegetation from the patch to the landscape scales (Table 8.6). There is evidence that large browsing animals have the highest impact around their preferred food species (Palmer *et al.* 2003), i.e. by selectively grazing (e.g.

**Table 8.5** Reported temperature impacts on alpine plant species dynamics

| Temperature | Duration | Target species/community | Measured responses | Main impact | Reference |
|---|---|---|---|---|---|
| OTC | 3 years | *Ranunculus glacialis* | Reproductive output | Little temperature impact | Totland & Alatalo (2002) |
| OTC | 4 years | *Ranunculus acris* | Flowering phenology, reproductive success, growth, population dynamics, phenotypic selection | Increased seed output, decreased density | Totland (1999) |
| Observational, OTC | 2 years | *Leontodon autumnalis* var. *taraxaci* | Reproductive success | Reproductive success decreases with time, warmed plants flower earlier and produce more seeds | Totland (1997) |
| OTC | 2 years | *Saxifraga stellaris* | Growth, cost of vegetative propagation, cost of flowering | Flower bud removal increased vegetative propagation, at increased temperature cost of flowering is smaller than at lower temperatures | Sandvik (2001) |
| OTC | – | *Euphrasia frigida* | Growth, seed production, population dynamics | Temperature increased growth and seed production; decreased density in more closed vegetation | Nylehn & Totland (1999) |
| Observational | 3 years | *Betula pubescens* | Soil temperatures, nitrogen uptake by seedlings | Predicted soil temperatures for successful seedling establishment 12°C versus approx. 8.2°C measured | Karlsson & Weih (2001) |
| OTC | 5–7 years | Multispecies in dry heath and wet meadow | Growth, phenology | Increased growth and reproduction, earlier phenological timing where positive response to warming; clear temperature response was the minority. | (Hollister *et al.*, 2005) |
| Modelling | '1000' years | *Pinus mugo* | Increase in cover in response to temperature and dispersal | Resistance to invasion is equally important as increased temperature | Dullinger *et al.* (2004) |

OTC, open top chambers.

**Table 8.6** Grazing experiments and observations in relation to alpine vegetation dynamics

| Grazing treatment | Duration | Target species/community | Measured responses | Main impact | Reference |
|---|---|---|---|---|---|
| Exclusion of lemmings (Lemmus lemmus) | 8 years | Snowbed | Species composition, vegetation structure | Grazing: higher vascular species richness, lower moss cover; Exclosure: lower vascular plant species richness, lower bryophyte species richness, higher Polytrichum moss cover | Virtanen et al. (1997) |
| Exclusion of lemmings and reindeer (Rangifer tarandus) | 5–20 years | Snowbed | Biomass | Increase of moss biomass (principal winter food of lemmings) within exclosures after 5 years; increase in graminoids, change in moss species, and accumulation of litter after 15 years | Virtanen (2000) |
| Transplant Vaccinium myrtillus heath into snowbed; exclosure | 7 years | Vaccinium myrtillus, Deschampsia flexuosa | Cover | Snowbed reduced V. myrtillus cover; grazing reduced cover by 40%; graminoids increased cover; tall herbs appeared in exclosures | Virtanen (1998) |
| Analytical, impacts of intensive sheep grazing | 20 years | páramo | Vegetation cover, soil properties, erosion | Grazing: decrease in species richness, increase in bare ground, erosion, loss of capacity of hydrological regulation | Podwojewski et al. (2002) |
| Enclosure | One-off, twice | Carex alpina, Danthonia cachmyriana, Trachydium roylei | Rate of regrowth | Different preference and clipping height by grazers, regrowth related to clipping height | Negi et al. (1993) |
| Exclosure | 10 years | Gentiana nivalis | Plant height, density | Exclosure: increased vegetation height and reduced bare ground; Gentiana: 5-fold decrease in density after 10 years | Miller et al. (1999) |
| Observational grazing by pika (Ochotona collaris) around and away from scree edge | | Kobresia myosuroides, Oxytropis nigrescens, Erigeron humilis | Leaf demography and length | Reduction of leaf length in Kobresia by current year grazing; historic grazing: Kobresia shorter leaves high rates of production of new leaves; Oxytropis higher rates of new leaf production; Erigeron: fewer new leaves, delay in leaf senescence | McIntire & Hik (2002) |

**Table 8.6** *Continued*

| Grazing treatment | Duration | Target species/community | Measured responses | Main impact | Reference |
|---|---|---|---|---|---|
| Exclosure | 1–3 years | Bofedal (hard cushion wetland) | Standing crop, annual net primary production; species richness | Exclosure reduced species richness, did not affect biomass | Buttolph & Coppock (2004) |
| Exclosure experiments; repeat observation after cessation of grazing | 5–23 years | Alpine heath | Visual estimate of cover types | Decrease in bare ground, increase in vegetation (approx. 1% per year) | Bridle et al. (2001) |
| Exclosures, experimental and simulated grazing | – | Alpine heath | Number of flower heads | Reduction flower head number with increase in grazing intensity; reduction by wallaby grazing, but not by sheep or rabbit | Bridle & Kirkpatrick (2001) |
| Comparative observations | – | Alpine vegetation Kazbegi, Caucasus (long stock grazing history) versus Rocky Mountain National Park (no stock grazing history) | Vegetation composition, growth rates, responses to clipping, digestibility, morphology | No difference in digestibility; no clear difference in morphology or plant family patterns | Bock et al. (1995) |
| Review | | Scandinavian alpine | Species diversity | Increasing intensity of grazing increases species diversity in some vegetation types (e.g. herb meadow, snowbed); in others species diversity decreases (e.g. species poor heath; late snowbeds) | Austrheim and Eriksson (2001) |

Skogland 1980) they create a spatially distinct patch pattern where the rate of vegetation dynamics is high. An added factor is the grazing habit of species through their bite, whether it is deep (e.g. sheep) or shallow (e.g. cattle). Differences in bite depth may affect the speed at which grazed species grow back, as has been shown in a *Danthonia* meadow the Himalayas (Negi *et al.* 1993). Long-term impacts of grazing by rodents and lagomorphs, can condition leaf demographic 'strategies' (e.g. rapid regrowth of shorter leaves, or little regrowth but with an extended leaf life span), as has been shown for species of a *Kobresia* heath, grazed by pikas (*Ochotona collaris*) in the Ruby Range Mountains, Yukon (McIntire and Hik 2002). The rate of dynamics determined by herbivores depends on location (species composition) and grazing history (accidental grazing with no adaptation by species versus regular grazing, such as by pikas around their occupied range on scree heaps). A particular case of alpine vegetation dynamics driven by herbivores is that from the boreal alpine areas where rodents have periodic population peaks and crashes over repeated cycles, several years apart (see Chapter 6). Lemmings in particular have an engineering role in maintaining snowbed communities, which in their absence would become dominated by a tall stand of *Polytrichum* mosses and few vascular plants (Virtanen *et al.* 1997; Virtanen 2000).

Much depends also on the impact grazers exert through trampling. For example, intensive sheep grazing may rapidly lead to a reduction in species richness and breaking up the continuity of the vegetation, such has been observed in the Ecuadorian páramo by Podwojewski *et al.* (2002), while to a lesser extent, the creation of small gaps is beneficial for preparing suitable seed beds for the establishment of species (e.g. Miller *et al.* 1999).

# 8.7 Conclusions

There is a large body of knowledge accumulated, largely from temperate ecosystems on the processes in alpine ecosystems from colonization to patch-scale dynamics in response to biotic and abiotic drivers. It is important to reiterate that dynamics in alpine systems usually involve low rates of perceptible change; for example, the account by Kerner (1863) of a *Carex firma* sedge heath on limestone, is probably as applicable today as it was in the year of its publication. Despite the above studies, we are yet far from being able to formally describe and predict the impact of nutrients, herbivores, and temperature on alpine vegetation dynamics, or on animal assemblages. As temperature and nutrient responses are likely to be species specific, the patch- and stand-scale dynamics of alpine vegetation types very much likely to be continued to be determined by a combination of largely deterministic (habitat, species pool) and stochastic (e.g. incidence of patchiness in grazing, nutrient input) factors, combined with individu-alistic species responses to them. An added uncertainty is the behaviour of the below-ground ecosystem components in their response to abiotic and biotic cues. Under sustained and heavy persistence of extraneous drivers a directional change in for example life forms, or functional types is likely.

# 9 Global change impacts on alpine habitats: climate and nitrogen deposition

## 9.1 Introduction

Global change may be considered to encompass recent (post-1960s) changes in land use and climate, and largely human-induced impacts in atmospheric chemistry (especially inorganic nutrients such as nitrogen compounds and carbon dioxide) and biotic interactions, such as invasions by alien species. The relative importance according to the severity and extent of impact of each of these factors, or drivers of global change on alpine biodiversity has been estimated to follow the order of: land use ≈ climate change > nitrogen deposition > carbon dioxide enrichment > biotic interactions (Walker *et al.* 2001a). We consider the three most important of these. Climate change is the focus of this chapter, together with briefly discussing the potential impacts of nitrogen deposition; land use is dealt with in the following chapter. We start with an outline of past climate changes and their perceived changes in the extent and quality of alpine habitats and vegetation. This is followed by characterizing climatic drivers and enlisting climate change scenarios and observed changes attributed to climate warming. Considerations are given to factors that couple or decouple climate and plant growth at the landscape scale. Finally, the reader is directed to the chapter on land use and conservation.

## 9.2 The environment and its change in space and time

Climate changes in place and time—it has done so in the past and is doing so currently. Depending on the temporal scale we examine, we can detect

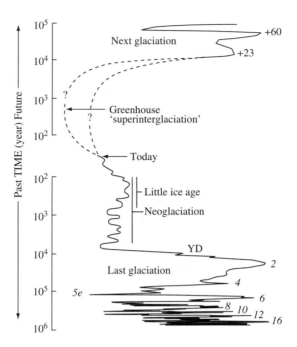

**Fig. 9.1**     Past reconstruction and future projections of temperature fluctuation (S. C. Porter, unpublished, http://faculty.washington.edu/scporter/Rainierglaciers.html)

Past global temperature trends are reconstructed from deep-sea cores and terrestrial glacial analyses (pre-Holocene) and climate proxy data (Holocene). Projections for the next glaciation are based on Milankovitch (Earth-orbital) variations. The dashed portion of the curve that splits into two shows a possible range of warming scenarios based on various assumptions about the rate of greenhouse gas increases in the atmosphere. YD, Younger Dryas—the cold period (stadial) between approximately 12,800 to 11,500 years before present.

directional changes (e.g. short- to medium-term warming or cooling, drying, or wetting) (Greenland *et al.* 2003), or cyclic patterns arising from the Earth's orbital changes (glacial and interglacial epochs) in the longer term. Earth orbital variations have caused a cyclic alternation of cold and warm periods. Currently, we are about 10,000 years after the last glaciation and have another approximately 23,000 years until the next (Fig. 9.1). In the interim, climate is predicted to be heading to warming well beyond today's. The physical consequences of such warming are reasonably straightforward to predict; however, less assured predictions can be made about biological systems, such as alpine ecosystems and their components.

Climate variation can also be detected in space, both over the short- and long-term (Fig. 9.2). The question of being able to predict the impacts of climate change on natural alpine landscapes is whether vegetation, animal, or microbial assemblages track these spatio-temporal changes in climate, or their constituent species dynamically reorganize to form

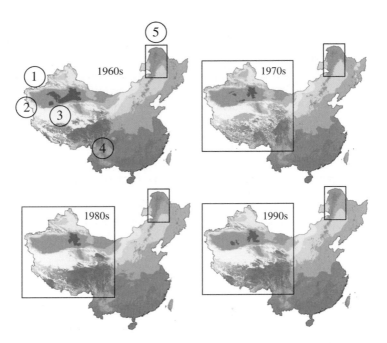

**Fig. 9.2**    The temporal variation of climate in China between 1960 and 2002 (after Yue *et al.* 2005). The depicted polygons (represented in the map as shades of grey) are Holdridge bioclimatic zones, calculated (and spatially modelled) from measured temperature and precipitation values averaged over 10-year periods. Note the changes in spatial distribution, especially evident in the high mountains. Mountain ranges: 1, Tianshan; 2, Kunlunshan; 3, Tibetan Plateau (Qinghai-Xihang); 4, Hengduanshan; 5, Da Higganshan. Observe the changes in the extent of some of the polygons over time.

new assemblages. Air-borne nitrogen compounds may enhance climate impacts by increasing plant growth/soil nitrogen mineralization rates. We are unable to make precise predictions on how species and communities will react to environmental change; however, we have a reasonable capacity to forecast the range of potential climates in the short term (up to approximately 50–100 years). Past climate, and plant and animal community reconstructions lend themselves to draw some lessons about what we may expect to happen to alpine habitats and landscapes in the future.

## 9.2.1 The history of alpine climate

While past climates *per se* are intractable, they do not pass without a trace. A number of indicators can be used to obtain a knowledge of earlier climates: pollen profiles of past epochs preserved in peat and lake sediment, and plant and animal remains in permafrost can inform us about the one time presence (and relative abundance) of some organisms in a locality.

This information can be related to carbon ($^{14}$C) or other isotope dates of the substratum of the origin of the pollen, and plant and other organism remains. By using our knowledge between today's climate and species distribution, and using independent reconstructions of climate, we can 'reconstruct' past climates with a degree of certainty (e.g. Elias 2003; Gosling *et al.* 2008). Reconstructions of climate are made by means of a number of proxies, from geology, palynology, fossil insects, dendrochronology, glacier extents, and pieces of historical evidence (Box 9.1; Joussaume and Guiot 1999).

## 9.2.2 Climate since the retreat of the last glaciation

On the Northern Hemisphere, the climate since the last glacial maximum about 20,000–18,000 years BP has seen the Holocene optimum, with

---

**Box 9.1** Climate proxies

Dendrochronology is based on measuring tree ring width in seasonal climates. Trees grow rapidly in the spring (early wood) and less so later in the season (late wood). This growth pattern results in radial growth in distinct rings, which is easy to observe in some species and less so in others. Tree ring width is a function of weather, i.e. cold springs result in narrow rings, while warm springs produce broad rings. Dendrochronology is particularly useful in a climate, such as that in the boreal zone, where summer drought does not normally interfere with ring width. The time-scale of dating available is species-specific and can reach up to 500 years or more. Combined with isotopic dating, dead wood such as that found lodged in rock crevices in glaciers can be put into a real time frame. Where plant or animal macrofossils are not preserved, palynology may be applied to catalogue past vascular plant genera and sometimes species. It usually allows to reconstruct dominant plants since the Holocene, approximately 10,000 years ago (e.g. Birks 1986; Table 9.1). Other additional historic data may be used such as phenology (date of grape harvest; full series for Bourgogne 1370–2004 in Le Roy Ladurie 2007). The extent of glaciers is another indicator of alpine climate. For example, according to Pfister (1988, cited in Le Roy Ladurie 2007), despite the large variability over the period, the extent of glaciers in the Alps during the Little Ice Age (1303–1859) had been continuously larger than present day.

Freshwater organisms, such as diatoms (Saros *et al.* 2003, Schmidt *et al.* 2004) and insect larvae (Lavoie *et al.* 1997) are also often used. Of terrestrial insects, beetles (Coleoptera) with their hard-wearing chitinuous skeletons preserve well. In addition, beetles (predators and scavengers) are thought to be good indicators of climate, as they do not lag in their colonizing new habitats, unlike treeline trees that may take several hundreds of years to respond to abrupt climate changes (Elias 2003).

temperatures about 1.5°C higher than today that resulted in the Alpine climatic treelines to be at approximately 100–300 m higher than contemporary ones (Table 9.1; Bortenschlager *et al.* 1996, Ali *et al.* 2003). Later cooling established the potential climatic treelines at today's elevations (only preserved where land use has not lowered them) both in Europe (Burga 1991; Amman 1995; Seppa *et al.* 2004; Hall *et al.* 2005) and North America (e.g. Spooner *et al.* 1997).

The last millennium was characterized by a medieval optimum, followed by a typically alpine cooling that largely manifested in increased glacier extent (Little Ice Age, LIA). This in the European Alps was especially marked during three periods: the fourteenth century until 1380, 1570–1630, and 1815–1859. Instrumental temperature data are available for central England 1659–1860, which are often used as a historic reference data set. Overall, annual temperatures during the LIA have been estimated to be between about 0.25°C (Luterbacher *et al.* 2004) and 1°C less than in the twentieth century. Winters were on average 0.5°C cooler, but not summers; exceptionally hot summers occurred in the 1660s, 1680s, 1700s, 1720–1740, 1770s, 1790–1810, and 1820s, even in periods of overall cooling, such as the 1840s, during the last extension of the alpine glaciers (Le Roy Ladurie 2007).

Glacier expansion also occurred in Iceland, Scandinavia, North America; in the Himalayas in the nineteenth century; and in South America between the seventeenth and nineteenth centuries. The impact of the LIA was essentially alpine and purely glaciological; its applicability to draw conclusions about climate–vegetation changes based on glacier expansions and contractions is therefore secondary. For glacier expansion or retraction, the two important factors are temperature (summer ablation) and winter snow (growth).

## 9.2.3 Post-Little Ice Age until the 1980s

Between 1860 and 1910, there were relatively frequent severe winters in Europe; since 1910, there has been a general warming trend, with a cool blip in the 1960s and 1970s. The twentieth century in Europe produced a warming of about 1°C, with different spatial and seasonal patterns (Schönwiese and Rapp 1997). The general warming trend has manifested in less frequent severe winters and more frequent very hot summers. The 1990s had the warmest winters since 1500; the period of 1970–2000 was the warmest three decades over the last millennium (also shown by the longest continuous data series for central England since 1659). Of direct alpine relevance is the exceptionally hot summer of 2003 when the Mont Blanc paths were closed because the thawing of permafrost caused high risks of rock falls. Characteristically, autumns have become milder and longer lasting since 1982, with the warmest being 2006 (not so hot as 1290 when 'trees flushed leaves and flowered before Christmas and in Alsace

**Table 9.1** Vegetation history inferred from pollen and macrofossil analyses at a selection of today's alpine sites. The corresponding climate reconstruction has used additional proxies, for details see references. The figures refer to years before present, unless otherwise stated.

| British Columbia, Canada (Spooner et al. 1997) | Pyhakero, north Finland (Seppa et al. 2004) | Dovre, south Norway (Hall et al. 2005) | Alps (Amman 1995) | Central Otago, New Zealand (McGlone et al. 1997) |
|---|---|---|---|---|
| 9000–7800 Shrubs and herbs | 10,000 ericaceous shrub | 11,000–10,300 Salix dwarf shrubs | 11,000 pioneers, alpine sedge heath | <7500 shrubs and grasses |
| | 9500 Betula pubescens | 10,000 Betula pubescens | 10,000 Larix decidua, Pinus cembra | |
| 7800–4000 Picea; Abies lasiocarpa | 8300 Pinus sylvestris | 8000 Pinus sylvestris | | 7500–3000 trees: Phyllocladus, Podocarpus, Prumnopitys, Nothofagus |
| 5000–4000 Pinus contorta appears | | | | |
| 4000 Disappearance of Picea | | 4400 disappearance of Pinus | | |
| 2000 Disappearance of Abies | 2000 disappearance of Pinus; 1500 disappearance of Betula | Approx. 3100 disappearance of Betula | Approx. 3200–1000 Alnus viridis; anthropogenic forest destruction | 3000–750 Nothofagus menziesii, Dacrydium cupressinum |
| | | | | 750 BP–AD 1860 increase in charcoal, grasses, decrease in trees and scrub |
| Present treeline ecotone with scattered Krummholz of Abies lasiocarpa, Betula glandulosa; Alnus sinuata | Present treeline ecotone and low alpine ericaceous dwarf-shrub heath | Present low alpine heath | Present dwarf-shrub heath (anthropogenic) | Present vegetation at sampling sites: Chionochloa rigida, C. macra tussock grassland |
| Alpine zone 1520–2120 m | Closed forest line approx. 500 m; highest point 711 m | Treeline; highest point >1500 m | | Treeline approx. 1000 m; sample elevations approx. 1400–1550 m |

strawberries were picked in the winter; grape vines had leaves, flower and grapes before 20 January') (Le Roy Ladurie 2001).

# 9.3 Climatic drivers: forecasts and scenarios of future climate changes

The two main climatic drivers of vegetation are temperature and precipitation (see Chapter 4), usually being more emphasis placed on temperature than precipitation. The reason for this, in part, is that temperature changes are easier to predict than corresponding changes in precipitation, and in part, the prevalence of thought that links biological phenomena to energy (heat). This has led to a widespread stance in modelling landscape-scale changes where an equilibrium between vegetation and climate is assumed and thereby predicted changes in temperature are mechanically applied to vegetation (elevation belt displacement, zonal displacement, e.g. McDonald and Brown 1992). However, the importance of precipitation is clear where scenarios that predict temperature impacts alone on vegetation are compared with those that include contrasting precipitation scenarios.

Overall, precipitation changes have indeed been reported from high mountains in the twentieth century and they may be related to observed changes in vegetation (e.g. Cannone et al. 2007). Not only the amount of precipitation, but its temporal availability and form (mist, rain, snow) can have important implications for the functioning of alpine ecosystems. The role of snow-lie in determining habitats (growing season length) and thus shaping vegetation in extra-tropical alpine ecosystems has been emphasized in Chapter 6 (see also Jones and Pomeroy 2001). Walker et al. (1993) have demonstrated at a range of scales from plot to landscape and region how current observed relationship among snowfall, wind patterns, topography, and temperature in relation to vegetation can be used for predicting shift in vegetation in future climates. Gottfried et al. (2002), by using measured temperature data, characterized the differences in relation to temperature and snow duration between 'alpine' and 'subnival' species growing between 2900 and 3100 m (alpine–subnival ecotone) in the Austrian Alps. The authors emphasized the importance of snow duration for 'subnival' species existence in view of potential shortening of snow duration in the future. Snowlie on tropical alpine mountains is transient, and, as such has no comparable importance for habitats and plant communities and animal and microbial assemblages with that in other zonal climates. However, precipitation is important for the survival of organisms and the functioning of ecosystems both in the upper montane forest (e.g. Pounds et al. 1999) and above the treeline. For example, in the light of the sensitivity reported from the East African alpine of giant *Lobelia* species to incidental drought (Young 1994) it might be hypothesized that an increased frequency in drought events, or reduction in precipitation could further affect such sensitive species. Drier climates can lead to more frequent fire events that could transform ecosystems.

Prognoses of temperature increases and precipitation changes for the twenty-first century vary with model, scenario, and region (e.g. IPCC 2007; Settele *et al.* 2009; Fig. 9.3). Predictions agree in that mean annual temperatures will increase, for example, predicted by up to about 7°C in Europe (e.g. Nagy *et al.* 2009a; Fig. 9.4). Precipitation change predictions are less consistent, as some forecast decreases while others predict increases (e.g. see http://cses.washington.edu/data/ipccar4/ for a comparison of models and scenarios).

## 9.3.1 Alpine vegetation and climate

The upper limit to tree growth is an important bioclimatic boundary as it marks the lower edge of the alpine zone (e.g. Körner *et al.* 2003; Körner and Paulsen 2004). Knowing the position of the potential climatic treeline allows the estimation of the upward displacement of this climatic limit in response to forecast rises in temperature. The basis for such calculations is the existence of a temperature gradient with altitude, which, on average, is

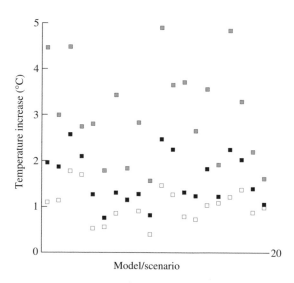

**Fig. 9.3**     The range of predicted climate warming by 20 different combinations of models and IPCC scenarios for the Pacific Northwest, USA (source data: Climate Impact Group, University of Washington, http://cses.washington.edu/data/ipccar4/sum/tempyearly.xls, accessed 5 May 2008). Open squares, 2020; black squares, 2040; grey squares, 2080

about 0.6°C per 100 m (see Chapter 4), i.e. an increase in temperature of 0.6°C could be seen to displace the current lower limit of the alpine climate zone by approximately 100 m upwards.

Forecast changes in temperature for each mountain region in Europe show a dramatic reduction in the extent of today's alpine climate zone (Fig. 9.4). With the exception of the largest massifs, such as the Alps and the southern Scandes, alpine climates would largely be eliminated by an increase of about 3°C in mean annual temperature (see, for example, Pyrenees in Nagy *et al.* (2009a)). As with species and communities that occur in current alpine habitats, instead of mechanistically applying a zonal displacement approach to make predictions about future species richness, and vegetation, animal, and microbial assemblages, it is worth examining at first the lessons one can draw from the palaeoecological literature.

## 9.3.2 Palaeo-climate and vegetation

Recent reconstructions of past climatic changes have shown that climatic changes were abrupt, in contrast to palynological reconstructions (Elias 2003). Elias (2003) has also found that there was a lag in the response by trees growing near the treeline of several 100s to about 1000 years in the Colorado Front Range. Jackson and Overpeck (2000) have provided a conceptual framework that encompasses potential species and community

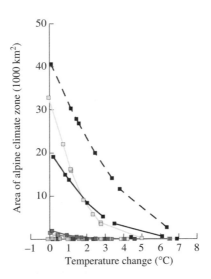

**Fig. 9.4**   Projected decrease in the area of the alpine climate zone in European ranges as a result of predicted climate warming, under a range of scenarios (HAD A1 etc, used by ALARM 2003–2008, http://www.alarmproject.net/alarm/). Dashed black line, Alps treeline taken at 1900 m; solid grey line, Southern Scandes; solid black line, Alps treeline taken as at 2300 m. Observe the small ranges forming a clump at the bottom.

'responses' to climate changes, which, in the authors' view, redraw the niche of species and communities each time they occur (Fig. 9.5). Their framework captures the main arguments used in the climate change impact debate, critically assesses each in turn, based on palaeo-evidence on a time-scale of $10^1$–$10^5$ years. Despite the dearth of specific information about contemporaneous alpine environments, it is a useful reference point for use in assessing potential changes to alpine habitats and communities. The fundamental messages by Jackson and Overpeck (2000) are that (1) climate changes continuously at all scales in time and also in space, and (2) as species differ in their fundamental niches, and they respond differently to different environmental drivers—changes to the environment will result in the 'birth and death' of communities. Climate is thought to exert its influence on individual species *sensu lato*, through changing (or leaving intact) their local environment (*Lebensraum*, habitat, niche), which, if remains within the fundamental niche of a species, the species will not show a reaction to. For convenience, Jackson and Overpeck (2000) have distinguished four main cases whereby species/communities: (1) continue to occupy the same sites as before the change; (2) move along local gradients (e.g. moisture, elevation); (3) migrate, i.e. they may occupy new areas and disappear from formerly occupied ones; and (4) locally become extinct and their range contracts, or conversely, formerly locally occurring ones

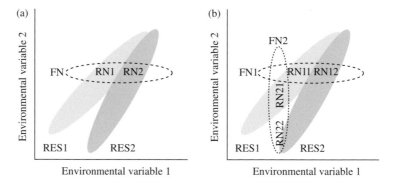

**Fig. 9.5** Schematic illustration of niche changes in species (a) and communities (b) with climate changes (redrawn from Jackson & Overpeck 2001). FN, fundamental niche of a species; RN, potential niche of a species at times 1 and 2; RES, realised environmental space at times 1 and 2.

In (a), a population of a species may occur anywhere in the intersection of FN and RES1. If, after an environmental change the intersection of RES2 no longer contains the population of species that was present in RES1, the population is likely to become extinct. Equally, RES2 may contain sites suitable for populations of the species to establish which will lead to colonisation.

In (b), the potential niches of species 1 and species 2 in RES1 overlap therefore they may co-exist, however, in RES2 they do not overlap, which excludes their co-existence.

greatly extend their distribution (Fig. 9.6; Table 9.2). As with the first contention by Jackson and Overpeck (2000) regarding the changing nature
of climate, it is worth recalling the study by Yue *et al.* (2005) where the
authors explored the short-term temporal and spatial dynamics of bioclimatic conditions in China. China has some very extensive alpine mountain areas (e.g. Tianshan, Kunlunshan, Tibetan Plateau (Qinghai-Xizang),
Hengduanshan, Changbaishan), whose climates range from cold desert
to temperate alpine (Fig. 9.2). Based on measured daily temperature and
precipitation data and by calculating 10-year mean annual bio-temper-
ature, mean total annual precipitation and potential evapotranspiration
ratio, the authors modelled the spatial distribution of the Holdridge life
zones (see Chapter 4) across China between the 1960s and 1990s. There
was an observed shift in the extent (and position) of Holdridge life zones
(not verified in the field) from decade to decade, some of which followed
a trend (a decrease in the extent of the nival zone by 6% per 10 years;
increase in alpine dry tundra by 10% per 10 years), while others did not.
Further relevant examples of climate variation and ecological responses
are presented, based on direct and proxy data collected in Alaskan Long
Term Ecological Research sites at inter-annual (Hobbie *et al.* 2003) and
inter-decadal (Juday *et al.* 2003) scales.

The second contention by Jackson and Overpeck (2000), regarding species and communities in response to changed niche is more difficult to
evaluate. In a central Asian context, none the less, there have been recent
suggestions on the drying and degradation of *Kobresia* heaths (not directly
attributable to the findings of the study by Yue *et al.* (2005)). In contrast,
Cannone *et al.* (2007) linked some of the vegetation changes observed
in the high alpine zone in the European Alps to increased precipitation.

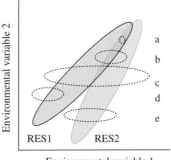

**Fig. 9.6**    Theoretical modes of population response to environmental change, redrawn from
Jackson & Overpeck (2000): a, populations stay in place; b, populations shift along
local habitat gradients; c, populations migrate to newly suitable sites and disappear
from some previously occupied sites; d, disappearance from formerly occupied area,
without colonising new sites; e, formerly rare species colonises new areas extensively.

**Table 9.2** Modes of population and community responses to climate forcing (modes taken from Jackson and Overpeck 2000), with various examples

| Mode | Species | Communities | Notes |
|---|---|---|---|
| (1) Remain in place | Long-lived perennials (up to $10^3$ years), e.g. *Carex curvula* in the Alps (Grabherr *et al.* 1978; Steinger *et al.* 1996) | e.g. *Abies balsamea* upper montane forest/ treeline ecotone established in north-east Appalachian Mountains in early Holocene; maintained throughout the Holocene— stand structure, disturbance changed | Populations may change in density, age and size structure, growth form, recruitment success 'Persistence of particular species' assemblages for >2000–5000 years at sites is rare |
| (2) Move along local gradients | Treeline ecotone filling, treeline advances (Kullman 1998, 2007; Grace *et al.* 2002); mesic grass heath to early snowbed (e.g. Grabherr 2003; Virtanen *et al.* 2003) | Holocene inferences. Non-zonal changes in vegetation, i.e. no displacement of vegetation zones, instead advance of some species led to low-elevation species grow into the upper montane *Abies balsamica* during the Mid-Holocene warm period (see references in Jackson and Overpeck 2000) | Most vegetation models implicitly have a built-in zonal displacement element and lead to such predictions |
| (3) Migration | Colonization of previously unoccupied areas and extinction from former range; rate of migration is determined by rate of change in the environment (e.g. glacier forefields) and ecological processes (dispersal capacity, establishment, reproduction). At retreating end (mortality, lack of recruitment, competition, pathogens and herbivores) | Evidence from paleoecological studies of no analogues of many vegetation types to those reconstructed before the last glacial maximum | Often conceived as zonal changes in plant communities. However, species may have independent rates of colonization and retreat that can lead to the formation of new assemblages, despite that some species are likely to co-migrate |
| (4) Extinction/ contraction | If climate reduces the potential niche of a species it may decline, fragment and survive in scattered local populations. New, favourable change may allow the species to recover and even become widespread (e.g. *Pinus ponderosa*). Some may become extinct (e.g. *Picea critchfieldii* in North America) | Successive interglacial floras do not appear alike, i.e. communities rarely, if at all regroup as previously known | |

In general terms, it seems that rather dramatic changes (e.g. the kinds that are anticipated in a rapidly warming world) would be required for niches to shift and produce either of the four response types proposed by Jackson and Overpeck (2000) to manifest by species. Perhaps the difficulty of being able to test the proposed niche responses is because the framework by Jackson and Overpeck (2000) is biased towards trees, wind-pollinated species and long time-scales, or it might be linked to the shifting nature of climate, or perhaps because past dynamics are reflected in the current altitude and latitude distribution of 'alpine' species. This latter, as far as treeline trees at the pristine treelines of the Colorado Front Range are concerned is favoured by some modern and fossil evidence, according to which treeline ecotones are not in equilibrium with today's climate; instead they are survivors form a previous colonization event (Elias 2003). This may explain cases of treeline inertia (e.g. *Pinus longaeva* at the treeline in the White Mountains and neighbouring ranges has some individuals that are older than 4000 years). While most treeline tree species are able to rapidly establish on open ground, they may well suffer a setback in later years as a result of self-shading, which can tip the balance towards unfavourable soil temperatures that can physically (e.g. through return of permafrost; Crawford 2008, pp. 187–188) or physiologically (e.g. insufficient temperatures for root function; Karlsson and Weih 2001) limit trees. Conversely, some species that inhabit cold environments are limited to them because of their inability to acclimate to warmer growing conditions—see, for example, recent work on arctic and alpine *Ranunculus* spp. by Cooper (2004).

## 9.3.3 The distribution range of alpine species

The alpine zone, by definition is the area above the treeline and is bound by the upper limit to plant life. In this alpine zone, there are various habitats, some with narrow specialist, others with more generalist species. As it is the cumulative response of species in plant communities or animal assemblages to environmental and biotic interactions in a habitat that shape the perceived nature of alpine landscapes, species altitude ranges are important for evaluating their susceptibility to climate change. For example, it may come as a surprise that the over 800-strong flora recorded above the treeline (approximately 2300 m) in the Pyrenees, only 22 species have not been recorded below 2300 m (Gómez *et al.* 2003; but see discussion about definition of alpine species in Chapter 7). Many of the species that have their optimum distribution in the alpine zone also occur below the treeline (which, must be recalled that, is a model in the form of an approximate bioclimatic isoline and not a contour line on a map). The wide distribution range is caused partly by land use that has lowered the treeline and extended the lower limit of alpine grass-sedge heaths, partly by gravity driven dispersal by avalanches, or water. In addition, some species have the plasticity to grow in habitats and communities different from those in the alpine zone, and some have survived in extrazonal enclaves after the last glaciation. An example of the latter is *Poa laxa*, a subnival species on Mount Schrankogel at approximately 3200 m that can be found at about 1500 m, close to the village of Gries in microhabitats, whose climate is more akin to that found on Schrankogel than to the local climate in Gries—an example of Walter's principle of relative habitat constancy (see Breckle 2002, pp. 71–72). Accordingly, wide altitude ranges are not surprising for species that grow in azonal habitats (e.g. rock outcrops or scree). However, singling out the genus *Gentiana* (Gentianaceae) from the Pyrenees and the Alps illustrates that species can also occur across zonal habitats and bioclimatic altitude zones with an altitude span of up to 2000 m. None the less, irrespective of their range size alpine species have their populations centred in the alpine or the treeline ecotone. As follows from the above, a wide altitude distribution does not necessarily imply a corresponding wide temperature range that might be surmised for the gentians found in the alpine zone of the western Pyrenees (Table 9.3), or the Alps (Table 9.4). Even more striking is the example of altitude range that plants reach in the East African mountains, which peak much higher than the

**Table 9.3** Altitude range of gentians (*Gentiana* L.) in the western Pyrenees and the Basque Country (altitude data from Aizpuru *et al.* 2007). Alpine, upper montane, montane indicates upper altitude distribution. Approximate temperature equivalents for altitude ranges are given for information only, without implying that these species do grow across such temperature ranges

| Taxon | Minimum–maximum altitude (m asl) | Altitude range (m) | Approximate temperature equivalent of altitude range (°C) | Potential causes and mechanism |
|---|---|---|---|---|
| *Alpine* | | | | |
| G. verna ssp. verna | 400–2400 | 2000 | 12.0 | |
| G. occidentalis | 500–2300 | 1800 | 10.8 | Phenotypic plasticity, intraspecific genetic variation (fitness), microhabitat similarity (climate, disturbance regime) |
| G. campestris ssp. campestris | (1100) 1300–2350 | 1050 (1250) | 6.3 (7.5) | |
| G. nivalis | 1500–2500 | 1000 | 6.0 | |
| *Upper montane* | | | | |
| G. burseri ssp. burseri | 700–2100 | 1400 | 8.4 | |
| G. ciliata ssp. ciliata | 800–2000 | 1200 | 7.2 | |
| G. lutea ssp. lutea | (750) 1200–1900 | 700 (1150) | 4.2 (6.9) | |
| G. acaulis | (1300) 1600–2200 | 600 (900) | 3.6 (5.4) | |
| *Montane* | | | | |
| G. pneumonanthe | 0–1500 | 1500 | 9.0 | |
| G. cruciata ssp. cruciata | 950–1600 | 750 | 4.5 | |
| G. boryi | (1150) 1300–1700 | 400 (550) | 2.4 (3.3) | |

**Table 9.4** Altitude ranges of the species of the Gentianacea family in Austria. The altitute belts in the table correspond to the following upper limits above sea level: colline, 250–400 (500) m; submontane, 350–500 (700) m; lower montane, 600–800 (900) m; upper montane, up to the closed forest line at 1500–2000 m; subalpine, between the closed forest line and the *Krummholz* at 1800–2100 (2300) m; alpine, 2500–2800 m; subnival, 2800–3100 m (Fischer *et al.* 2005)

| Species | Altitude belt |
|---|---|
| *Gentiana pneumonanthe* | Colline-montane |
| *Gentiana verna* | (Colline-)montane-alpine |
| *Gentianella austriaca* | Colline-subalpine(-alpine) |
| *Gentianella rhaetica* | Colline-alpine |
| *Gentianella bohemica* | Submontane-upper montane |
| *Gentiana asclepiadea* | (Submontane-)montane-subaalpine |
| *Gentianopsis ciliata* | Submontane-subalpine |
| *Gentianella lutescens* | Upper montane-subalpine(-alpine) |
| *Gentianella amarella* | Montane |
| *Gentiana cruciata* | Montane(-subalpine) |
| *Gentiana lutea* | Montane-subalpine(-alpine) |
| *Gentiana nivalis* | (Montane-)subalpine-alpine |
| *Gentiana utriculosa* | Montane-subalpine(-alpine) |
| *Gentiana acaulis* | (Montane-)subalpine-alpine |
| *Gentiana clusii* | Montane-alpine |
| *Gentianella campestris* | Montane-alpine |
| *Gentianella aspera* | Montane-alpine |
| *Gentianella pilosa* | Montane-alpine |
| *Gentianella anisodonta* | Montane-alpine |
| *Gentiana punctata* | (Upper montane-)subalpine-alpine |
| *Gentiana purpurea* | Upper montane-subalpine-(alpine) |
| *Gentiana pannonica* | Upper montane-subalpine(-alpine) |
| *Lomatogonium carinthiacum* | Upper montane-alpine |
| *Gentiana prostrata* | Subalpine-alpine |
| *Gentiana terglouensis* | Subalpine-alpine |
| *Gentiana froelichii* | Subalpine-alpine |
| *Gentiana pumila* | Subalpine-alpine |
| *Comastoma tenellum* | (Subalpine-)alpine |
| *Gentiana brachyphylla* | Subalpine-alpine(-subnival) |
| *Gentiana orbicularis* | Subalpine-alpine(-subnival) |
| *Gentiana bavarica* | Subalpine-subnival |
| *Gentiana frigida* | Alpine |
| *Gentianella engadinensis* | Alpine |
| *Comastoma nanum* | Alpine-subnival |

Alps or Pyrenees. In East Africa, Hedberg (1969) has shown that few species are restricted to the alpine zone (or have narrow ranges), and some are spanning ranges over 2000–3000 m of altitude (Fig. 9.7). The implication of such a wide altitude distribution is that such species are likely to possess a great deal of flexibility to climate variability (perhaps in a way that varies with altitude). Another implication may be that many non-matrix (character-giving) species will have experienced past shifts and are occurring in a variety of habitats, or in outlying habitats. It follows, that no simple relationship can describe absolute species richness losses and gains on the basis of a predicted change in the area of the total alpine climate zone and changes in the area of individual habitats within the alpine zone. Species with a narrow (measured temperature) niche that occur on European mountain summits are currently being characterized for their susceptibility to climate change, as part of the Global Observation Research Initiative in Alpine Environments (GLORIA, http://www.gloria.ac.at).

Total distribution range is only one descriptor of the potential response species may be able to elicit to climate forcing and is important for assessing the survival of species. It must also be borne in mind that species have optima for their functioning and these optima are narrower than the maximum ranges. Optimum ranges form the basis for most modelling studies on predicting the impacts of climate change on alpine species abundances, and on community, habitat, or landscape changes. Prediction based on optimum ranges should reasonably predict plausible ecological responses. However, as admittedly most modelling studies presume an equilibrium between climate and species responses (e.g. Guisan and Zimmermann 2000; Dirnböck *et al.* 2003), they lead to mechanistic predictions of species

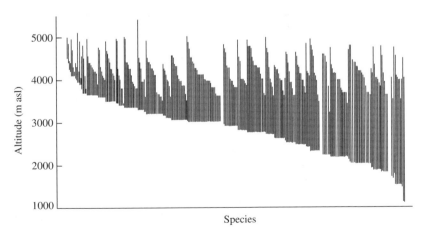

**Fig. 9.7**   The altitude range of species recorded in the alpine zone on the East African mountains (Hedberg 1969). The lower limit of the alpine zone is c. 4000 m, and the treeline ecotone extends upwards from about 3700 m.

increases or decreases (based on habitat suitability, such as measured climatic variables and soil properties) parallel to modelled changes in the environmental descriptors. This may be unwarranted in the light of palaeo data that indicate, for example, that tree colonization may have lagged up to 2000 years behind rapid climate amelioration in the Colorado Front Range at the beginning of the Holocene (Elias 2001). In addition to the abiotic environment, colonization success is also a function of the extent existing vegetation can resist invasion by species favoured by warming, or precipitation change (e.g. Dullinger *et al.* 2004).

Altitude is not the only measure of range. The geographical distribution of species may indicate their environmental tolerance ranges (latitude, temperature; longitude, oceanity; overall size of range, life zones). The sensitivity of plant species to change in climate at a Europe-wide scale has recently been evaluated by modelling plant species' niche properties and their relation to geographical distribution (Thuiller *et al.* 2005). Boreo-alpine (arctic–alpine) species were characterized as having a narrow niche breadth and were predicted to be highly sensitive because of their predominantly northerly distribution, areas predicted to be experiencing dramatic increases in temperature. Temperate alpine species in the study by Thuiller *et al.* (2005), which were also characterized as obligate low temperature species, but with a larger niche breadth than arctic–alpine species, were thought to be able to locally colonize suitable habitats that would become available upslope, above the current snowline. It appears that the altitude range of alpine species might be connected to geographic range, at least at the regional scale (Sklenár and Jørgensen 1999, McGlone *et al.* 2001).

## 9.3.4 Observed changes in alpine vegetation attributed to recent climate warming

Irrespective of whether observed temperature increases represent climate inertia or have been caused by human activities, a number of changes in alpine species distributions and vegetation have been reported recently. The most visibly striking are those caused by the closing up of formerly open upper montane forest stands (e.g. Moiseev and Shiyatov 2003) and by increases in scrub or dwarf-shrub cover (e.g. Cannone *et al.* 2007). Many have also reported treeline altitude increases (e.g. Ninot *et al.* 2008); however, these are more often than not linked to grazing abandonment, and the effects of climate and land use change are often difficult to separate. There are genuine cases where tree species line increases were documented, such as those in the Scandes where seedlings have established in snowbed sites (Kullman 2002).

Above the limit of trees and shrubs, little change has been evident from permanent vegetation plots in zonal grass-sedge heaths (*Kobresia myosuroides*,

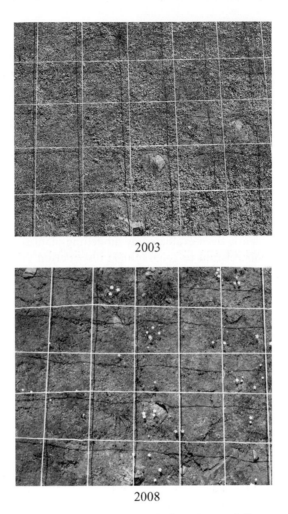

2003

2008

**Fig. 9.8**    On the northern slope the Mte Cinto massif of Corsica small flat areas accumulate shallow soil and bear snowbed vegetation from c. 2200 m upwards. Some of these snowbeds receive seep water from snow melt from upslope. The one depicted in the photographs does not receive irrigation after snow melt and as a result its vegetation drastically changed after the heat of summer 2003—the hottest on record. *Omalotheca supina* that dominated the vegetation in 2003 died out and the formerly closed vegetation is largely open *Sagina pilifera* 5 years later (2008). No visually noticeable change occurred in irrigated snowbeds. (Photo L. Nagy)

*Carex curvula, Sesleria albicans*) over a period spanning between 50 and 100 years (G. Achermann and M. Schütz, pers. comm.) in the Swiss Alps, or in the New Zealand Alps over a 50-year period (Mark 2005). In contrast, snowbed vegetation has been observed to becoming grassier than before in the Alps (Grabherr 2003) and in the Scandes (Virtanen *et al.* 2003). The hot summer in Europe in 2003, eliminated patches of snowbed vegetation that had no water supply from snow melt in the

Corsican high mountains (L. Nagy pers. obs.; Fig. 9.8). Snowbeds with irrigation survived more or less intact. A discernible longer-term impact has been the establishment of *Juniperus communis* ssp. *alpina* at the periphery of some snowbeds, not recorded by Gamisans (1977) in his survey in the late 1960s early 1970s. Recent changes in climate variables and their impact on hydrology and high Andean wetlands in Colombia have been highlighted by Ruiz et al (2008). Decreases in predominantly subnival species cover and concomitant increases in the cover of alpine species was detected over a 10-year period in the European Alps (Pauli *et al.* 2007). Species range extensions have been observed locally in a number of locations in the European Alps (Grabherr *et al.* 1994; Bahn and Körner 2003; Pauli *et al.* 2003; Walther *et al.* 2005).

## 9.3.5 Habitat and vegetation change: landscape- and continent-wide models

The permanence or otherwise of habitats in response to climate change is fundamental in shaping future vegetation and animal assemblages. The above examples illustrate little change so far in response to climate in long-established zonal vegetation; however, azonal habitats, where the over-abundance of certain resources had been prevalent in the past and is diminishing now, such as in some snowbeds, or glacier forefields, (further) rapid and fundamental changes are expected. In that vein, habitat variation within a karstic alpine landscape and forecast changes were examined in an excellent modelling study, supported by empirical data, by Dirnböck *et al.* (2003). On the one hand, the degree of landscape-scale dynamics in local *Pinus mugo* scrub versus vegetation dominated by herbaceous and dwarf-shrub species was estimated, admittedly under the assumption of climatic equilibrium (Fig. 9.9). In addition, the impact of climate change for 85 constituent species were evaluated for their potential niche changes in a warmer and drier climate. A net decrease in potential habitats for two-thirds of the species and a net increase for one-third were predicted within three mountains in the Northern Calcareous Alps, Austria under a mild increase in temperature of 0.7°C and a decrease in annual precipitation of 30 mm; less than 10% were reduced by over 90% as compared with their current distribution. An additional increase of 2°C in temperature (with or without further decrease in precipitation) slightly increased the number of species whose potential habitat decreased; however, importantly the proportion of species whose habitat was reduced by >90% reached half of the total.

Dynamic vegetation models have been developed on the scale of Europe to predict future potential vegetation, or biomes. In these models, treelines were set to be limited by growing day degrees >5°C and winter minimum temperatures (T. Hickler *et al.*, unpublished). The models predict large

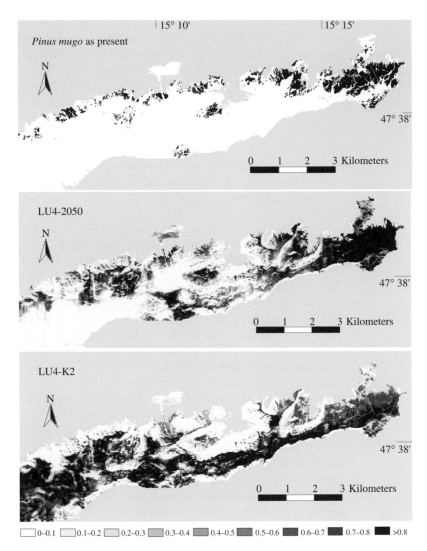

**Fig. 9.9**   The current distribution of and predicted probabilities of a site becoming colonised by *Pinus mugo* under two model assumptions (LU4–2050, grazing abandoned and there is an increase of 0.65°C in temperature and a decrease of 30 mm in precipitation in August; LU4-K2, grazing abandoned and there is an increase of 2.0°C in temperature and a decrease of 30 mm in precipitation in August). Reproduced from Dirnböck et al. (2003)

reductions (Table 9.5) in subnival/arctic desert, alpine/arctic tundra, and upper montane/subarctic forest extents by 2100 in comparison with the baseline predictions (1960–1990) that were calibrated against the Potential Vegetation Map of Europe (Bohn *et al.* 2004). For example, under the Hadley general circulation model climate predictions, both the alpine zone and the upper montane/treeline ecotone forest all but disappear; the

**Table 9.5** The current and predicted percent cover by arctic and alpine vegetation in Europe (forecast figures are from T. Hickler *et al.* unpublished)

| Vegetation type (Bohn *et al.* 2004) | 1961–1990 | 2071–2100 (HAD-A2) | 2071–2100 (PCM-A2) |
|---|---|---|---|
| Polar deserts and subnival vegetation | 0.07 | 0.02 | 0.01 |
| Arctic tundras and alpine vegetation | 3.31 | 0.08 | 0.47 |
| Subarctic, boreal and nemoral montane open woodlands, and subalpine and oro-mediterranean vegetation | 2.81 | 0.50 | 1.11 |
| Total | 6.19 | 0.60 | 1.59 |

HAD, Hadley Circulation Model 3, Met Office, Exeter, UK; PCM, Parallel Climate Model, National Center for Atmospheric Research, Boulder, CO, USA; A2, IPCC medium–high $CO_2$ emissions scenario (tripling of atmospheric $CO_2$ concentrations by 2100 compared with the pre-industrial era).

extent of the alpine vegetation shrinks about 40-fold and that of the upper montane open woods by 7-fold. The corresponding reductions under the cooler and less dry PCM general circulation model (see Table 9.5) are 7-fold and 2.5-fold. The reductions in the area of alpine vegetation suggested by T. Hickler *et al.* (unpublished) are no different from that calculated by Nagy *et al.* (2009a) for the reduction of the potential alpine climate zone, based on the current position of the potential treeline in Europe's high mountains. A yet unborn generation of ecologists will be out there to record a lag or its absence of tree colonization of today's alpine vegetation. A distinct lag, or even standstill in tree colonization of open alpine areas is more than likely where domestic grazers are present, or where wild ungulates abound (Hofgaard 1997).

In summary, climate warming is likely to affect 'alpine' species with a wide altitude distribution to a lesser extent than those whose range is restricted to above the treeline. Conversely, some alpine habitats and their plant communities and animal assemblages could undergo major changes that may lead to the formation of hitherto undescribed communities, depending on species sources available locally (or regionally). On the landscape scale, visual changes are expected from tree colonization; however, the speed at which such colonization events occur will depend on the responsiveness of local treeline species to climate, local substratum conditions (e.g. Körner 2007a), and not the least land use.

## 9.4 Nitrogen input from the atmosphere

Nitrogen from the atmosphere becomes deposited either in a wet or a dry form. Measured data are scarce on nitrogen deposition at alpine altitudes. One exception is the Niwot Ridge Long Term Ecological Research site where nitrogen deposition has been measured since the 1990s (Sievering 2001), and there are incidental reports from short-term measurements in the Alps (Hiltbrunner et al. 2005). An annual mean total of about 6 kg nitrogen ha$^{-1}$ (0.6 g m$^{-2}$) has been reported from an alpine measuring station at 3540 m on Niwot Ridge (3.2 ± 1.0 kg nitrogen ha$^{-1}$ wet deposition; 2.9 ± 0.8 kg nitrogen ha$^{-1}$ dry deposition). The difference in seasonal deposition is interesting, because that during the growing season (1.2 ± 0.1 kg nitrogen ha$^{-1}$ wet; 1.6 ± 0.6 kg nitrogen ha$^{-1}$ dry) becomes available for microbial or plant use immediately, while winter deposition (2.0 ± 0.9 kg nitrogen ha$^{-1}$ wet; 1.3 ± 0.3 kg nitrogen ha$^{-1}$ dry) is released in a major pulse on snow melt, approximately half of which is retained by the vegetation and soil microbes. As we have seen earlier (Chapter 8), nutrient addition experiments at such low doses have not produced significant changes in short-term experiments, and work has suggested that 10 kg nitrogen ha$^{-1}$ should be an appropriate guideline for critical nitrogen load for alpine ecosystems. Should then deposition at rates similar to those reported for Niwot Ridge be a matter of concern? To answer this question long-term cumulative impacts need to be followed, such as for example, Bowman et al. (2006) have done it in a dry *Kobresia* heath on Niwot Ridge. They suggested, on the basis of changes in the Shannon diversity index ($H'$), on the cover of the species that responded most strongly to nitrogen addition over 8 years, and on the scores of the first axis of an ordination analysis that the critical load is somewhere between 4 and 10 kg nitrogen ha$^{-1}$, the former applying to the most sensitive species and the second value indicating community responses. Applying these critical load values to the modelled nitrogen oxide ($NO_x$) deposition data for Europe (Alcamo et al. 2002), it appears that the Pyrenees, Sierra Nevada, Apennines, Corsican high mountains, the Hellenids and the mountains of Crete in the south, and the vast majority of the Scandes in the north of Europe fall below the 10 kg nitrogen ha$^{-1}$ year$^{-1}$ (Fig. 9.10). In contrast, the Scottish Highlands, the very south of the Scandes, the northern and eastern Alps, Dinarids, Carpathians and the Balkan Ranges are exposed to this level of nitrogen deposition. Indeed, total nitrogen deposition (wet $NO_3$ + wet $NH_4$ + dry $NO_2$ + dry $HNO_3$ + dry $NH_3$ + cloud $NO_3$ + cloud $NH_4$), in the Cairngorms, Scotland was in the range of 7.1 – 7.5 kg nitrogen ha$^{-1}$ year$^{-1}$ for 2001–2005 (http://www.nbu.ac.uk/negtap). On the other hand, at the 4 kg nitrogen ha$^{-1}$ year$^{-1}$ exposure level all alpine ranges, with the exception of the Pyrenees, Sierra Nevada, the Corsican high ranges, and the central and northern Scandes are affected (Fig. 9.11). This means that if the modelled figures reflect values on the ground, many temperate and boreal alpine ranges in Europe may receive enough nitrogen from the atmosphere to impact some sensitive species. There is little information on how atmospheric nitrogen input compares with that from urine and faeces by herbivores. Following the calculations by Seagle (2003), one may expect a mean annual input by ungulates that matches the amount derived from the atmosphere. Input by herbivores is different, however, as it is rather patchy

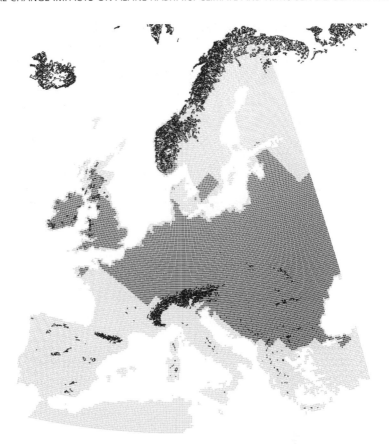

**Fig. 9.10**  Nitrogen deposition in Europe. Data from: ALARM (http://www.alarmproject.net/alarm/). The dark-shaded areas have modelled (e.g. Alcamo *et al.* 2002) annual deposition levels in excess of 10 kg N ha⁻¹ year⁻¹. The outline of nival, alpine, and upper montane zones (black) from the European Potential Vegetation Map (Bohn et al. 2004) are shown for reference.

(and highly concentrated, e.g. a urine hit may be the equivalent of 400 kg nitrogen ha$^{-1}$ year$^{-1}$ (Hamilton III *et al.* 1998) and is especially important in shelter areas, where vegetation may remarkably change as a result (for examples see Erschbamer *et al.* 2003). The quantification of nutrient input into alpine ecosystems by grazing animals and its relative magnitude to that accessed from the atmosphere is yet to be reliably quantified.

Formulating a picture of how nitrogen and climate may interact is difficult. Jonasson *et al.* (2001) have reviewed the results of experimental work carried out in arctic ecosystems on the impacts of nitrogen, water, and warming of ambient air. From that review it emerges that the response varies among habitats/vegetation types, but is consistent within vegetation type. Nutrient addition caused increased net primary production (in part by replacing evergreen dwarf-shrubs with deciduous ones), plant nitrogen, and

**Fig. 9.11**     Nitrogen deposition in Europe. Data from ALARM (http://www.alarmproject.net/alarm/).
The dark areas have modelled (e.g. Alcamo *et al.* 2002) annual deposition levels in excess
of 4 kg N ha⁻¹ year⁻¹. The outline of nival, alpine, and upper montane zones (black) from
the European Potential Vegetation Map (Bohn *et al.* 2004) are shown for reference.

increase in standing biomass in most cases. Warming has caused variable
responses from none to marked increases in biomass, while water had no
effect. Combined application of the treatments yielded conflicting results:
in some cases there were synergistic increases; in contrast, in wet sedge and
tussock tundra temperature × fertilizer responses were negative, as a result
of non-acclimation of root respiration that led to decreased biomass and
nutrient in plant tissues. Model-based general considerations of climate
change and nitrogen deposition interactions are available from Posch (2002)
and Mayerhofer *et al.* (2002). As a closing remark, one must remember again
that one-off additions of nitrogen at concentrations similar to that received
in atmospheric deposition do not prompt any immediate response; however,
continuous input over the long-term may exceed ecosystem thresholds, despite the fact that current scenarios forecast a slight reduction in atmospheric
$NO_x$ in the twenty-first century (Alcamo *et al.* 2002).

# 10 Land use and conservation of alpine landscapes, ecosystems, and species

## 10.1 Introduction

Historically, economic return, health considerations, religious beliefs, and above all population density have been the major shaping forces of human land use in high mountains, as well as in the lowlands. This chapter examines the impact of some of the land-use forms on alpine and upper montane landscapes and habitats. The examples from Europe are contrasted with those from North America, and most notably with those drawn from high-altitude mountain areas with a long history of land use such as the mountains emerging from the extensive plateau of the central Andes. As has been shown, in high mountain ecosystems a vertical zonation causes high biological diversity on a projected area basis and the existence of slopes gives a high-energy status to mountains, i.e. the movement of water and solid material along slopes is much more pronounced than on flatlands. These features make mountain environments sensitive to excessive perturbation, which is often the case with land use. Finally, we discuss conservation management measures that might minimize or ameliorate the impacts of perturbation and contribute to the maintenance of ecosystem functions, continued use, and biological diversity of alpine ecosystems.

## 10.2 Land use from a brief historic perspective

The history of land use may be divided into four broad periods: hunters and gatherers (see for example the recently discovered remains of the Copper Age 'Ötzi' in the South Tyrolean Alps http://en.wikipedia.org/wiki/

Ötzi_the_Iceman), pre-industrial sedentary agriculture, industrial revolution, and post-industrial, the last two being the era of capitalism. Such a classification applies to Europe (e.g. Simmons 2005) but not directly to countries that were colonized and exploited by European powers from the sixteenth century onwards, until the twentieth century. The length of these four periods is unequal and their impacts manifest in different ways. It is generally perceived that hunters and gatherers had light and mostly diffuse impact on the environment; however, some studies have concluded that hunter-gatherers may have used fire to control forest colonization of open mountain slopes after the recession of ice (e.g. Simmons 2005). Sedentary agriculture brought with it the concentration of population in favourable locations. Over millennia, cropping and husbandry of domesticated grazers have transformed vast areas previously covered by natural vegetation, mostly in the lowlands and hills, and lower montane zone. However, in warm climate zones settlements were often established at altitudes where disease risk, such as malaria in southern Europe, or in the Tropics, was less than in the lowlands. The resulting land cover change, and the impact of agricultural intensification, are well illustrated when, for example, the natural vegetation map of Europe (Bohn *et al.* 2004) is compared with today's land cover land-use classes (Nagy *et al.* 2009b). The impacts on the upper montane and alpine zones have been direct, mostly through pastoralism, and indirect through regulations and policies. In the colonies, a major environmental shock occurred soon after the colonists had introduced their European species and cultivation practices, such as in the central Andes (Baied and Wheeler 1993). The industrial revolution brought about fewer direct changes in the mountains than in the lowlands, where an urban workforce could be concentrated. However, major changes did occur in mountains, for example, the replacement of scattered human settlements in the Scottish Highlands by flocks of sheep (Highland Clearances). One of the major impacts that the industrial revolution had directly in Europe and facilitated elsewhere was the development of transport networks. This largely expanded the energy and mining sectors, and, in the relatively recent past, allowed the germination of what today has become a vast tourist industry (for a narrative account on the birth and development of Alpine mountaineering see Fleming (2000)). The 1960s have brought about a very marked change in the use of mountains in the industrialized countries of Europe. A combination of natural migration from the mountains into urban jobs and sometimes government policies aimed at extirpating a 'shameful backward way of life' resulted in an exodus from the mountains, e.g. Pyrenees, Pallaruelo (1994), or the Alps (Ozenda and Borel 2006). Urban migration is the prevailing trend in many parts of the world and has eased land use pressure in some areas (e.g. Grau *et al.* 2007). However, this migration process has reversed recently in some countries, such as Austria, where former productive villages have become tourist or commuter economies. Parallel to this, the tourist industry has undergone an unprecedented explosion (Bätzing 2002).

## 10.2.1 Land-use patterns in agro-pastoral systems

Monasterio (1980) has provided an excellent scheme for the Venezuelan Andes, which can be generalized to illustrate (traditional) land use in relation to bioclimatic factors and land capability (Fig. 10.1). Similar schemes can be drawn up for other mountain regions (e.g. Ellenberg 1979), where land use potential largely depends on temperature and precipitation. The altitude limits of cultivation are sometimes reached or even stretched (e.g. during a favourable spell of climate, population pressure) to increase production such as has been observed in the central Andes (Halloy *et al.* 2006; Young and Lipton 2006), where potato may be grown up to 4600 m (Fig. 10.2). However, in some areas, e.g. the North American ranges, cultivation or herding has never been practised, and in the mountains of Europe after a peak in the nineteenth century, land-use levels have dropped with the intensification of lowland agriculture (e.g. André 1998; Olsson *et al.* 2000). In contrast, in countries with a high population growth, land use has intensified, such as in the Himalayas (Nüsser 2000) or in the high mountain ranges in China (Wu and Tiessen 2002). While the Arctic was altogether spared from cultivation, herding and hunting characterized some of its ranges (McGhee 2006). As Fig. 10.1 indicates, the predominant form of traditional land use in the upper montane and alpine zones of the mountains has been extensive grazing by livestock. In many ranges this long history makes some ecosystems, such as the north Andean páramo, of questionable origin (Balslev 1992), while in ranges where grazing history is relatively short, such as the Australian Snowy Mountains (Williams and Costin 1994), or New Zealand the impacts of grazing on the ecology and management of alpine ecosystems are clearly demonstrable.

## 10.2.2 Land use (change) and its ecological consequences

Land use affects alpine ecosystems at different scales (Table 10.1) from landscape, habitat, to patch (for a recent selection of case studies from mountain regions world-wide see Spehn *et al.* (2006)). The most visual land-use impacts are those that entail treeline ecotone trees and scrub. Sustained use of fire such as that in the páramos (e.g. Hofstede 2001) has lowered and maintained at such lower elevation the woody vegetation, and expanded the area of the landscape dominated by the large caulescent rosettes of *Espeletia*. This ancient means of expanding grazing land has been estimated to have lowered the treeline in many European high mountains by at least 200 m, and its patchy small-scale use has also been reported from the New Guinea Highlands (Smith 1977, 1985), and other tropical ranges. Recent partial recovery of the upper montane forest after the abandonment of grazing can be observed in places (e.g. Hiller and Müterthies 2005); numerous treeline advances have also been reported,

(a)

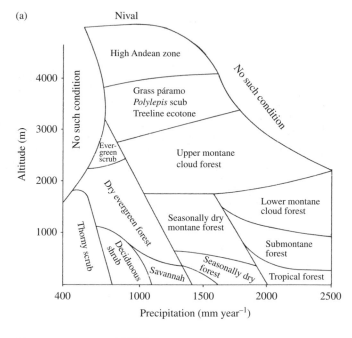

(b)

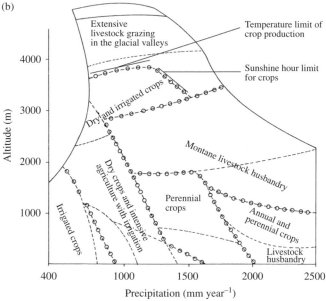

**Fig. 10.1** Schematic representation of (a) natural vegetation formations as a function of mean annual temperature (altitude, vertical axis) and mean annual precipitation (horizontal axis) in the Venezuelan Andes; and (b) agro-ecological potential. Redrawn from Monasterio (1980). The high altitude dry conditions that are absent in Venezuela and other humid tropical countries, dominate in the subtropical dry and arid Andes.

**Fig. 10. 2**     Potato cultivation (middle plane on lower and middle hillside) and heavy grazing by alpaca in the valley bottom in the Andes near Santa Barbara at c. 4400 m. The natural vegetation at this altitude is dominated by tall bunch grasses, scattered specimens of which are visible on the hillside. (Photo: L Nagy)

sometimes uncritically attributed to recent climate amelioration alone (e.g. Meshinev *et al.* 2000).

Habitat-scale impacts arise from differential impacts by management on landscape units, or the habitats they give rise to. Using the example of habitat use by different grazers, some habitats are little affected because of their physical properties (e.g. cliffs and block scree), others are preferred for grazing or browsing (grass and sedge heaths, dwarf-shrub heath), while yet others experience indirect deleterious effects, such as trampling (e.g. mires). Changes may include instances where, for example, a large herbivore species is replaced by another one, whose food preferences and grazing/browsing habits are different (e.g. Negi *et al.* 1993), or when grazing management changes for other reasons. An exemplary case (which also affected landscape patterns) is that in the Andean ranges after the introduction of livestock from Castille from the sixteenth century on (e.g. Gade 1999). In Europe, the post-1960s sharp decline in transhumance have affected the way alpine grazing is managed today. There has been a loss of traditional knowledge, and as a consequence, few shepherds tend large flocks (as opposed to many shepherds and small flocks previously), e.g. in the Pyrenees or the western Alps (Loison *et al.* 2003), which leads to local abandonment of grazing land in some places and a concentration of grazing animals in others. Other ranges have experienced new grazers,

**Table 10.1** Land use and its impacts on alpine habitats at different spatial scales

| Management | Landscape (main physiognomy) | Habitat (secondary physiognomy) | Patch (ecology: species, dynamics) | General |
|---|---|---|---|---|
| Fire, grazing | Treeline lowering | Favours herbaceous versus ligneous plant species | Sheltered patches less affected by fire; wet patches sensitive to trampling | Direct fire effect, browsing of seedlings/ saplings |
| Grazing with small flocks and many shepherds versus large flocks with few shepherds | Erosion scars as a result of heavy or overgrazing | Even grazing pressure versus partial abandonment; partial concentration and locally heavy grazing | Local dynamics caused by urine and faeces deposition | Differences in fodder species preferences, grazing depth, biting versus tearing |
| Abandonment of grazing | Partial upper montane forest recovery | May turn from herbaceous to ligneous | Varied | Secondary succession |
| Hunting and collecting | Hunting may involve occasional use of fire | | Selective collection of medicinal plants | Secondary impact via interfering with wildlife |
| Turf/adobe cutting; Soil extraction | | May create shallow pool habitats | High species turnover | Small-scale habitat transformation |
| Fertilization | Sometimes improved meadows; increasingly used in high altitude agriculture | Mineral nutrients applied via melt-water irrigation of hay meadows | Homogenizing effect | Rare |

such as shifts from sheep to cattle, introduction of horse grazing, or the intensification of reindeer farming. Both intensification in accessible areas and extensification in remote land lead to vegetation change (Tasser and Tappeiner 2002) that manifests in decreased number of plant species, and changes in community structure, both above-ground and below-ground (cf. intermediate disturbance hypothesis). For example, old practices such as melt water fertilization or liming/phosphate addition to melt water and channelling it on to hay meadows/grazing land that had had a local impact on species composition (Grabherr 1997) have been abandoned in the alpine zone, or replaced by synthetic fertilizer addition in upper montane zone secondary grasslands.

Heavy grazing occurs on many alpine pastures from the Scandes (Loffler 2000), to the High Atlas (Bencherifa 1983), the South African Drakensberg (Killick 1997), Tibet (Bauer 1990; Rikhari *et al.* 1993), or the Altiplano

**Fig. 10.3** Erosion in the central Andes as a result of overgrazing by sheep and alpaca. The zonal vegetation of bunch grass dominated grassland has been replaced by small-stature herbs because of burning and long-term grazing at Salma, c. 4800 m. (Photo: L. Nagy)

(Adler and Morales 1999) (Fig. 10.3). The severity of the impact varies from reducing standing phytomass (other than that of giant rosettes) (e.g. Hofstede *et al.* 1995; Hofstede and Rossenaar 1995), through changing the physiognomy of grazed land from tall tussock grass-dominated vegetation to low cushion and rosette cover—such as is the case in the heavily grazed central Andes (Byers 2000)—to modifying functional group (Erschbamer and Mayer 2006) or species composition/richness (Austrheim and Eriksson 2001). When burning accompanies heavy grazing, total above-ground phytomass is further reduced (Fig. 10.4), and on sensitive areas can lead to complete vegetation and soil loss, redrawing large expanses of alpine landscapes. For example, in the New Zealand Alps, excessive pasturing of native snow tussock (*Chionochloa* spp.) by introduced sheep created extensive talus slopes, which have become a typical character of these mountains today (Mark and Dickinson 2003). Grazing affects invertebrates to a largely unknown measure. However, it appears that after the abandonment of grazing on alpine grass- and sedge-heaths there is little difference in above-ground species numbers of flies (Anthomyiidae, Muscidae), while the numbers of flies tend to be higher in abandoned than grazed pastures (Haslett 1996).

Natural habitats in the alpine zone are usually considered free from alien species. Non-native species are most likely to be encountered in those ranges whose elevation is not far above the treeline ecotone, such as the Australian Alps (McDougall *et al.* 2005), and in disturbed habitats, related

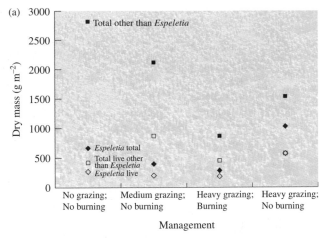

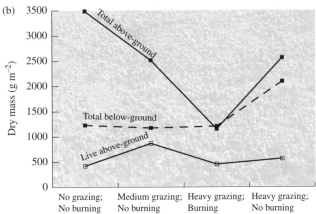

**Fig. 10.4**   The effects of grazing and burning of páramo on (a) above-ground standing litter and biomass in (*Espeletia* vs. others) and on (b) total and live above-ground, and total below-ground phytomass in Colombia (after Hofstede 1995a,b; background: unburnt páramo with a dense growth of *Espeletia* cf. *grandiflora*, Chingaza, Colombia c. 3700 m; photo: H. Pauli).

to infrastructure development and tourism. In addition to current established non-native species in the alpine zone (22 on Mount Kosciuszko), ski resorts, whose gardens are stocked with potential alien alpine escapees and alpine invaders, are a likely source for invasions, particularly in a warmer future climate. It also needs to be borne in mind that highland plateaux in the Andes and Tibet have a large number of naturalized aliens, in close contact with the alpine zone, associated with high-altitude cultivation and animal husbandry. A prominent example is the common stork's bill (*Erodium cicutarium*), a species of European mediterranean origin that has recently been recorded at 5245 m in the high alpine—subnival zone of the central Andes (Schmidt *et al.* 2008).

**Fig. 10.5**    Local impacts of turf cutting in the Andes at c. 4800 m asl near Sibinacocha, Cordillera Vilcanota. Shallow pools are left behind, creating new habitats.

Additional, patch-scale impacts of land use are caused by local organic material-rich soil extraction (Hofstede 2001), turf cutting (Fig. 10.5), sometimes locally extensive collection of medicinal plants (Olsen and Larsen 2003), and by the dynamics caused by animal urine and faeces input, and trampling (see Chapters 8 and 9). The dynamics of re-vegetation after small-scale experimental turf cutting in the Alps in a widespread more or less closed alpine vegetation type is exemplified by the work of Grabherr (2003); Grabherr has shown that approximately 25 years after the removal of *Carex curvula* sedge mats and soil monoliths, and potential propagules, such as stolons and roots, species composition on the bare soil surfaces approximated the original, bar one species. Superficially, this may appear as an example of high community resilience in an alpine ecosystem. However, the species that did not recover was *Carex curvula*, the presence of which defines this community. This example shows, that the species that defines one of the most common community types on siliceous substratum in the European Alps has a low resilience (recovery potential) after catastrophic disturbance. A similar case is observed in the Himalayas, where the accompanying species in a *Kobresia pygmaea* heath require less time to regrow after turf cutting than *Kobresia* (Fig. 10.6).

## 10.2.3 Climate change and land use

Treelines are generally thought to be controlled by thermal conditions (e.g. Körner 2007) and, therefore, warming should release them. However, local modifying factors (e.g. geology, recurrent mass movement events)

**Fig. 10.6**    Turf cutting (bottom of main photograph) and regeneration (insert) in *Kobresia pyg-maea* sedge heath at 4750 m asl near Gokyo, Khumbu Himal, Nepal. Most *Kobresia* heaths in Nepal are grazed by domestic yak. Observe in the insert the difference in structure between vegetation on uncut areas (*Kobresia*) and that regenerating in the hollow. The latter appears more species rich and uneven than the uniform *Kobresia* sward. (Photo: L. Nagy)

and land use may artificially stabilize existing treelines (Young and León 2007). Climate change impacts on treeline advances can manifest where wild or domestic herbivores do not significantly influence recruitment and growth (Cairns and Moen 2004). However, this is rarely the case, as for example, Hofgaard (1997) has shown it in the Scandes and Stutzer (2000) in the Austrian Alps, and little climate change impact can be expected in the way of treeline increases, or in stand density as long as existing grazing practices are in place. Fire suppression at treeline sites can also have an over-riding impact in magnitude over climate warming, such as has been reported from the Canadian Rocky Mountains (Luckman and Kavanagh 2000).

Fire frequency and intensity may be expected to change as a function of precipitation amount and inter-annual variability. Shifts in fire incidences and intensities may redraw the boundaries between upper montane forest and alpine grasslands (Grau and Veblen 2000). An apparent increase in the frequency and intensity of upper montane forest fires on Mount Kilimanjaro has been attributed to a warmer and drier climate during the twentieth century (Hemp 2005, 2006). Such fires, irrespective of origin (natural, unintentional, or intentional) have apparently been lowering the treeline. Importantly, the loss of forest cover reduces interception of occult precipitation (fog) and as such, may have a compound impact on hydrology, and eventually users downhill might be affected by reduced water yields.

Fires, by lowering the treeline, promote shrub and herbaceous species growth and expand the potential area for high-altitude pasturing. A linked question regarding past climate and land use is how the colder period during the Little Ice age may have affected pasture use in the alpine zone of the affected mountains: Did it result in less impact as the winter keep of the animals suffered, or did it result in heavier grazing pressure (such as a modelling study by Thomson (2005) suggested it for Iceland)? And more importantly, how will ongoing and predicted future warming affect future land use in the upper montane and alpine zones worldwide?

## 10.2.4 Socio-economic choice and climate change

Climate change impacts on the productivity of ecosystems and policy decisions to use biofuels increasingly is likely to result in land-use changes in productive ecosystems. In European alpine and upper montane ecosystems with marginal land use potential, little change is likely (Nagy *et al.* 2009c). Adaptation to climate change in transitional economies where traditional high mountain land use and governance are becoming under the influence of modern market forces may have serious ecological and societal consequences (e.g. Young and Lipton 2006).

## 10.2.5 Land-use change impact on alpine climate

A rarely reported case where lowland land use affects alpine climate is from the Colorado Rockies; Stohlgren *et al.* (1998) and Chase *et al.* (1999) found that recently introduced irrigated agriculture had a cooling effect on mountain climate in the summer, in contradiction with general global and regional warming trends. They have found that climate records, conifer seedling distributions, and increased stream flows corroborated evidence and model predictions were indeed supported by physical evidence. The mechanism behind the observed cooling was suggested to be a decoupling between summer adiabatic airflow between the plains and the high mountains and more moisture in the air, causing more cloud and rainfall.

## 10.2.6 Industry: winter tourism and hydropower

Tourism generates vast incomes world-wide, including alpine mountain areas. Poorly managed adventure tourism and uncontrolled infrastructure

**Fig. 10.7**    Alpine tourism is controversal, both in socio-economic and natural resource use terms. Trekking in Nepal generates a large income, however, mostly to the profit of enterpreneurs from the capital or abroad. Trekkers visit in vast numbers, e.g. in October 2008 nearly 10,000 visitors entered the Sagarmatha National Park (Mt Everest region). The management of such a volume of visitors creates numerous problems, one of them being waste disposal. Photograph taken at c. 4600 m asl, Annapurna circuit, Thorong Pedi, Muktinath Himal. (Photo L. Nagy)

development in faraway countries, such as that around Mount Everest does cause, however, damage to alpine ecosystems and landscapes, e.g. through excessive harvesting of alpine woody vegetation for fuel, or overgrazing by pack animals, that leads to erosion (Byers 2005; Figure 10.7). Winter tourism centres around ski developments. The deleterious impacts of the ski industry manifest in excessive and environmentally unfriendly development (ski pistes, and associated infrastructure) at the local scale, and the competing use of water for artificial snow development and hostelling, see e.g. Descamps (2008) for the French Alps. A study by Teich *et al.* (2007) on the economic and ecological impacts of artificial snow in the Swiss Alps has shown that in the Davos region artificial snow production accounted for 0.5% of the total energy consumption, while water used in snow cannons accounted for 25–35% of the total water demand. In terms of ecological impacts, artificial snow is not different from snow that falls and becomes redistributed by wind. Added snow may prolong snow-lie and induce changes in species composition of plants and animals, and if the water source is enriched in minerals it may have further influence through a fertilizer effect. Climate warming is likely to cause an increase in the elevation where snow suitable for skiing (>30 cm depth) lies for at least 100 days between 1 December and 15 April. This will result in the likely closure of low and middle elevation stations in the montane zone and lead to the increased use of alpine stations, with artificial snow generation.

The ecological impacts of skiing are seen in a decrease in species richness and diversity (evenness), standing biomass, and increases in the area covered by open ground. Artificial snow deposition on ski pistes can reverse to some extent the decrease in species numbers and diversity, but pistes in general, regardless of whether they are snow-cannoned or not, are poorer than surrounding areas. The reasons may be connected to the fact that the mechanical compaction of the snow on ski pistes reduces the thermal insulation afforded by intact snow (Rixen *et al.* 2004). In the most serious cases, bare soil becomes extensive, which leads to soil erosion. To counter erosion, alpine seed mixes may be used to stabilize exposed soil surfaces (Krautzer *et al.* 2006). Some animal life (ground beetles, arachnoids) appears little altered on ski pistes, compared with that in intact vegetation, but other insect groups, such as hoverflies (Syrphidae) indicate differences (Haslett 1988).

Körner (2002) emphasized the importance of intact vegetation for soil stability. The author likened it to a diverse array of 'rivets' whereby the root systems of different groups of plants hold the soil in place on alpine slopes. Such an effect is less pronounced in the highest reaches of the alpine zone, where the vegetation is patchy. The root properties of alpine plants investigated on a ski piste at Piz Corvatsch (3303 m asl, Swiss Alps) showed little variation in rooting depth among the 13 investigated species (7.2–20.5 cm mean depth); however, graminoids had larger total lengths (from approximately 0.5 m to 1.5 m) than herbs or prostrate dwarf-shrubs; there were large differences in horizontal extent, too. Large differences in growth and reproductive characters of various native grass species, in view of their potential for using them for re-vegetating ski pistes have been shown by Grabherr *et al.* (1988): tussock-forming species (*Carex curvula, Festuca pseudodura, F. varia, Nardus stricta*) grew radially approximately 1 mm year$^{-1}$, in contrast with the stoloniferous *Agrostis schraderiana* (200–400 mm year$^{-1}$). The grasses all flowered often and copiously, while the sedge *C. curvula* rarely; of the grasses, *F. pseudodura* had an exceptionally high germinability (89%) when compared with the other grasses (approximately 10%). These differences in life strategies much limit which species are suitable for commercial scale re-vegetation of ski pistes.

Hydroelectric power generation has claimed entire glacial valleys through-out the world, one of the latest being the approximately 30 km² dam lake over alpine ground just north-east of Vatnajökull glacier in eastern Iceland (Del Giudice 2008). Such dams obliterate entire alpine ecosystems, with their plant and animal life, and are likely to influence local climate, too. For example, the rare and relict *Carex bicolor* community (Bressoud 1989) that is found along glacier rivers have lost many of its former sites as a result of damming glacial valeys in the Alps. An increase in dam building may be anticipated in the short-term, in an effort to capture more glacial melt water before yields diminish and some glaciers disappear.

## 10.3 Conservation of alpine landscapes and habitats

Land use is one of the most important drivers of change in the cold eco-systems of the alpine and the arctic worlds (Walker *et al.* 2001a). The role of land use is predicted to accelerate, and its impacts on biodiversity to approximate those of climate change and nitrogen deposition taken together, to be the single most important driver in alpine ecosystems worldwide. These land-use impacts combined with feedbacks from climate change, alone or in combination with atmospheric nutrient deposition, may certainly modify land-use preferences over time, which is an import-ant consideration in evaluating land-use options and consequences for sustainable management. Conservation is a management activity that uses a set of tools to achieve its aims from establishing management objectives to their implementation. Worldwide, the objects of conservation concern pertinent to the alpine zone and solutions offered to problems are varied: they reflect local conditions, history, and institutional capacity; however, problems and solutions associated with land use are a recurring theme (Table 10.2).

Examples abound where land use (need or greed) leads to land degrad-ation. In an alpine context, putting grazing animals on an alpine pasture in excess numbers of what it can support will reduce the profit of each shepherd to a point where everybody is a loser and a total system col-lapse, economic as well as ecological, may result—commonly known as the tragedy of the commons (Hardin 1998). In the example, over-grazing will affect not only the pasture ecosystem in question (vegetation and soil loss), but connected systems downstream, and will manifest in the sedi-mentation of water courses, decreased hydrological yield at the watershed level, and increased cost of providing clean potable water to downstream populations. This raises a complex gamut of issues pertinent to valuating alpine habitats and their biota, as exemplified in Table 10.3, where prod-uct and local service values are reasonably easy to express in monetary

**Table 10.2** A selection of alpine conservation problems, objectives and solution reported in the literature

| | Feature/problem | Objective/solution | Reference |
|---|---|---|---|
| Cairngorms, Scotland, UK | The Cairngorms are of national and international importance for geomorphology, woodlands and montane (alpine) environments. Impacts from human land use, atmospheric pollution and climate change, may potentially have widespread effects on the alpine zone | Measures are in place under a range of conservation designations to mitigate historical and recent human impacts and to provide for environmental protection and enhancement. Research is in place or planned to monitor future changes | Gordon et al. (1998) |
| Giant Mountains (Krkonose), Sudetes | Increasing pressure on arctic-alpine vegetation from human activities: (1) visitors; (2) plantations of dwarf-pine stands that disturb recent cryo-pedological processes; (3) air pollution, with signs of climate change | Effective management and conservation of the Giant Mountains | Stursa (1998) |
| European Alps | Maintenance of Golden Eagle (Aquila chrysäetos) populations in the European Alps after its recovery in the twentieth century | Management strategies from models of habitat quality and distribution; scenarios to estimate the potential impact of future human activities on the breeding success and distribution patterns; co-operation of conservationists and land users; environmental education and user-specific public relation activities | Brendel et al. (2002) |
| European Alps | Ibex (Capra ibex ibex): (1) habitat destruction in areas of high population densities, and (2) low genetic variability, possibly a result of inbreeding | | Stuwe and Nievergelt (1991) |
| Grisons, Swiss Alps | Design and implementation of an alpine plant diversity conservation programme in an ecologically, socially, and institutionally diverse environment | Objectives: participatory management and monitoring; reduction in intensive cultivation (fertilizers); cultivation of previously abandoned fields. Design features: rely on existing legislation, limited ecological knowledge and expertise, biodiversity as a common-pool resource | Baumgartner and Hartmann (2001) |

**Table 10.2** *Continued*

| | Feature/problem | Objective/solution | Reference |
|---|---|---|---|
| Valais, Swiss Alps | Local-authority and collective forms of ownership and sustainable use of natural resources: recent trends and emerging management forms of forests, watercourses, and alpine meadows and pastures | | Kissling-Naf *et al.* (2002) |
| Obertauern, Austrian Alps | Grading eliminates plant cover on ski pistes; skiing on non-graded slopes with sedge and rush moors eliminates most of the native species | Restoration of plant cover on graded and non-graded ski runs by using native species for re-vegetating the ski-damaged landscapes of the Alps | Klug-Pümpel and Krampitz (1996) |
| Tibetan Plateau, Haibei | Alpine meadow and shrub are the main pasture types | To promote a sustainable animal production system; improve degraded pasturelands; conserve biodiversity | Zhao and Zhou (1999) |
| Himalayas, Nepal | Commercial collection of alpine medicinal plants (480–2500 t per year) | Conservation and management of alpine medicinal plant species under the community management scheme | Olsen and Larsen (2003) |
| Maloti–Drakensberg Transfrontier Conservation and Development Programme | Alpine and subalpine grassland vegetation for maintenance of catchment integrity and grazing. Problems: range management regimes, spread of alien plants, establishment of agricultural monocultures, poor infrastructure development and maintenance, political engineering of human demographics | To find natural resource and conservation management solutions by integrating biodiversity conservation and socio-economic growth strategies | Zunckel (2003) |
| Mount Kosciuszko, Australian Snowy Mountains | Access tracks and immediate alpine area around the summit of Mount Kosciuszko concentrate a large number of visitors. Soil compaction and erosion; introduction and spread of weeds; faecal contamination of lakes and creeks; increased feral animals; and vegetation clearance | Ensure that the summit and the rest of the Kosciuszko alpine area remains viable for conservation and outdoor recreation. Management solutions: hardening of tracks; provision of toilets; education, including minimum-impact codes; restrictions on activities such as camping in the catchment areas of glacial lakes | Pickering and Buckley (2003) |
| Tasmania, Western Arthur Range | Recreational trampling in undisturbed alpine and subalpine vegetation communities | Concentrating walkers on a minimal number of sites on untracked alpine and subalpine environments | Whinam and Chilcott (2003) |

**Table 10.3** Alpine ecosystem values

| Value category | Item | Example |
| --- | --- | --- |
| Product value | Food | Livestock, game; |
| | Medicine | Traditional medicinal plants |
| | Industrial material | Ores for mining, water for material and energy |
| | Other marketable commodities | Minerals, fodder |
| Function value | Land and watershed protection | Intact, or little disturbed vegetation |
| | Air quality control | Scavenging of radicals and air-borne pollutants |
| | Water quality control | Prevention of erosion; provision/maintenance of hydrological yield |
| Service value | Tourism (eco) and recreation | Mainstream (developed economies) or supplementary (undeveloped and transitory economies) economic activity |
| | Educational object/ material | |
| | Repositories of biological information | Palaeo (pollen, permafrost); modern (extant microbes, plants, and animals) |
| Existence value | Aesthetic/psychological benefits | For inhabitants (those in mountain settlements in particular), tourists |
| | Cross-generation natural and cultural heritage | Cultivated/pastured upper montane and treeline ecotone landscapes and their nature maintained by human activities (transhumance, hay making, tree cutting and burning). General: across mountain ranges; Particular: high mountain cultural and economic centres, e.g. central Andes, Alps |
| | Places of worship | High Andean Inca period shrines of human sacrifice; holy mountains in the Himalayan/Tibetan region (e.g. Kalash, Kawagebo) |

terms (Grêt-Regamey and Kytzia 2007), while taking service value into account is more complex to express across spatial scales downstream; existence values best exemplify the subjective nature of biodiversity conservation (Nunes and van den Bergh 2001). In a conflict, whose resolution led to avoiding the tragedy of the common above, broader environmental and social (e)valuation resulted in the phasing out of grazing on Mount Kosciuszko, Australian Snowy Mountains (Williams and Costin 1994; Costin *et al.* 2000) to prevent erosion and protect water quality and yield. As a corollary, alpine vegetation has since recovered from grazing (Scherrer 2003).

In alpine regions with a long grazing history (e.g. Himalayas, Alps, Pyrenees—see Box 10.1) regulation of grazing rights and management date far back in time. In the Alps, there existed the so-called '*Alpbriefe*', dating back to the fourteenth century, for the owners who shared a

---

**Box 10.1** Transhumance in the Aragon Pyrenees (Pallaruelo 1994)

The valleys of the Aragon Pyrenees, with their habitations and surrounding high altitude grazing land have traditionally been natural and administrative units (Pallaruelo 1994). Exclusive grazing rights to the pastures were granted to the valley communities by successive kings from 1272 onwards into the fourteenth century. Based on these rights, the exploitation of the grazing land was administered by the local communities under strict regulation until the end of the nineteenth century. The establishment of the modern Spanish State then brought state and provincial planning and control over local community rights.

Transhumance existed between the Pre-Pyrenean mountains, the Pyrenean montane and alpine pastures and the Ebro valley plain covering distances of up to 200 km. Available grazing time decreased from approximately 125 days at 1000 m to approximately 60–75 days at 2300 m (the mean altitude of the treeline). The owners of sheep and cattle of the central Pyrenees wintered their animals on the Ebro plain and summered them on the upper montane and alpine pastures.

The Pre-Pyrenean villages kept their livestock in the winter and sent them up the mountain pastures in the summer (*transterminance*). Little change has characterized transhumance from the sixteenth century until the 1960s, when it rapidly declined. The use of the mountain grazing land historically was near saturation, which fell to about 45% by the 1990s. As the owners of the high altitude grazing land, living in the valleys, were the owners of grazing stock, and saturated the carrying capacity of the mountain pastures, only a small number of animals of the Ebro plain origin ever grazed in the Pyrenees. There has been a large drop in sheep numbers between the 1850s and early 1990s, a reduction from about 300,000 to approximately 100,000. The average flock size has increased because of an approximately 10-fold reduction in shepherd numbers. There has been an increase in the number of sheep originating from the Ebro valley, where migrants from the Pyrenean valleys had settled.

---

common grazing area, or *Alm*. These *'Alpbriefe'* characteristically began with highlighting the poor state of the *Alm*. The regulations set out in them included fixing a ceiling to the numbers of animals allowed to graze, and in some cases excluding particular species (e.g. sheep); transgressing rights were not tolerated.

## 10.3.1 Landscapes

In an alpine context, many ecosystems and other features are the subjects of conservation concern. Many of them have resulted from human activities, such as large expanses (but by no means all) of dwarf-shrub heath in north-western Europe, or 'species-rich grasslands' in central European mountains; these bear witness to earlier pastoral and other economic activities at the interface of the alpine and upper montane zones. As well as 'natural' landscape features, 'semi-natural', or sometimes termed 'cultural heritage' features are often considered conservation worthy.

There are constraints on conservation. At one extreme, it is concerned with the preservation of existing landscape, ecosystem, or taxonomic features (or their restoration to some preferred state). Such an approach is usual where current pressure for land is relatively low and economies can bear the associated costs. For example, the Swiss National Park, with large expanses in the alpine zone, has a strict regulation as to the prescribed and excluded activities within its boundaries; similarly, in many mountain wilderness areas of the United States the principle of exclusion applies. At the other extreme, the concept of conservation concerns the maintenance of the productive quality of the environment (capacity to provide goods and services, i.e. economic benefit), already practised in the distant past, such as in the Inca empire in the Andes (Soriano 1981).

Cold environments, such as the alpine, have been less hospitable to humans than productive lowlands and, as a result, many alpine and upper montane environments have largely escaped the transformations that had impacted lowland and lowland hill ecosystems. Despite the relatively little visually striking changes in alpine ecosystems, they have had a long history of management, primarily through pastoral activities; e.g. illustrated by the documented historical use for sheep grazing at near full carrying capacity of the alpine meadows (sedge and grass heaths) in the Aragon Pyrenees until about the 1960s (Pallaruelo 1994; Box 10.1). Clearly, our perception of naturalness of alpine landscapes and ecosystems incorporates elements that are based on observed vegetation structure (treeless, dwarf-shrub, and sedge heaths), and on the knowledge that such vegetation formations are climatically determined to the exclusion of trees; therefore, the long grazing history is perceived as not having contributed to fundamentally changing the natural climate-determined vegetation.

Grabherr (2005) highlighted some practical areas of concern specific to high mountains and their inhabitants: the continuing change in land use (in particular, the disappearance of traditional pastoral and farming practices, followed by land abandonment or by the introduction of industrial agriculture), climate change, and a need for a vision to implement sustainable development that includes set-aside areas for preservation, and an improved balance between set-aside and development.

## 10.3.2 Identifying priorities for conservation of alpine biodiversity: genes, species, ecosystems, and beyond

Of genes, species, and ecosystems, species have been the most often used units of alpine biodiversity counts, reflecting our relying on the long tradition of the work of taxonomists and their compilations of floras and faunas: accounts derived from species distributions, totals for geographic areas and derived observations on rarity, or uniqueness (Väre *et al.* 2003). For plants, this level is complemented by detailed works on the classification of communities in many parts of the world, allowing the establishment of a measure of commonness or rarity at the community level. Less clear is the role of genetic diversity and the way our knowledge on it can be linked to species and communities, largely because our understanding does not go beyond characterizing genetic diversity by using certain markers. We do not know what level of adaptive potential such diversity confers on today's species, or what evolutionary value it has. Such lack of present-day understanding, however, should not lead to the prevention of maintaining as high a level of genetic diversity as possible for the future.

Many general frameworks have been developed for the conservation of biodiversity (e.g. Vane-Wright 1996), which attempt to find the optimum way of determining conservation value, based on biodiversity measures such as richness, rarity, and endemism. The applicability of a general theoretical framework itself is not straightforward, as has been recently shown for alpine plant species in the European Alps and Carpathians. Taberlet *et al.* (pers. comm.) have found that commonly used measures of biodiversity (species richness, rarity, and endemism) did not coincide for the investigated almost 900 alpine vascular plant species in the European Alps, i.e. protecting the areas richest in species would not equally protect rare, or endemic species. The same measures applied to genetic diversity (quantified on a subsample of the species) showed similar incongruence and, in addition, there was a poor spatial match between the measures for species and genes. In conclusion, no single measure can inform us about how best to protect alpine plant diversity.

Conservation-centred management can circumvent such problems, by protecting the whole continuum of extant communities/ecosystems. For example, the European Union Natura 2000 programme has identified 959 sites (that includes alpine and non-alpine ones) for protection in the Scandes, Alps, Pyrenees, Apennines, and additionally the Carpathians, and the Rila–Rodope–Balkan massifs. The achievement of conservation objectives through appropriate management practices is checked against a baseline of 'favourable state' (an example guidance note for alpine heaths in Scotland is shown in Table 10.4).

## 10.3.3 Protected areas network

The Convention on Biological Diversity, the organization charged by the United Nations to provide a consensual world-wide framework for managing biodiversity, approved a Mountain Work Programme in 2004 (http://www.cbd.int/mountain/default.shtml) that provides a broadly defined framework for alpine biodiversity conservation (for a list of relevant

**Table 10.4** Joint Nature Conservation Committee, UK guidance on conservation objectives for alpine and boreal heath for Natura 2000 reporting

| Attributes* | Targets | Method of assessment | Comments |
|---|---|---|---|
| Habitat extent (ha) | No measurable decline in the estimated area of the feature on the site | A baseline map showing the distribution of heath vegetation should be used to assess any changes in extent. Comparison of maps, aerial photographs, changes in boundary observable in the field, or changes in proportional presence at grid of sample points may all be useful | On sites where the objective is to increase the extent of the habitat (e.g. by reducing grazing pressure), favourable condition may not be achieved until there has been a large expansion in heath cover |
| | | Aerial photographs offer a convenient means of rapidly assessing the extent of heath vegetation over large upland areas | |
| Vegetation structure/ composition | Collective cover of typical species at least 66% of total vegetation cover | Visual estimate at the 1 m × 1 m scale | Exclude any patches of obvious snowbed vegetation too small to appear on vegetation maps, and not caused by snow-fencing, from assessment. Exclude bare areas obviously due to exposure |
| | At least one typical species of dwarf-shrub and one typical species of moss/lichen is present | Count at the 1 m × 1 m scale | |
| Grazing species cover | Collective cover of *Agrostis capillaris*, *Agrostis vinealis*, *Anthoxanthum odoratum*, *Deschampsia flexuosa*, *Festuca ovina/vivipara*, *Galium saxatile*, *Poa* spp. (other than arctic-alpine spp.) and *Potentilla erecta* less than 10% of total vegetation cover | Assess at 1 m × 1 m scale | |
| Ground disturbance | Less than 10% of bare mineral soil, bare peat, ground covered by algal mats and/or bare gravel, occurring in sheltered topographical, and microtopographical, locations, with hoof, boot, or vehicle prints present | Visual estimate at the 1 m × 1 m scale | Exclude distinct paths or tracks, bare areas obviously due to exposure, and bare rock |
| | Less than 5% is clearly defined paths or tracks of bare mineral soil, bare peat, ground covered by algal mat, or bare gravel | Visual estimate at the 40 m × 40 m scale | |
| Burning | No burned patches, nor any burnt areas extending more than 25 m into the feature | Assess at whole feature | |

Includes the following NVC (Rodwell 1992) types: H13 *Calluna vulgaris–Cladonia arbuscula* heath, H14 *Calluna vulgaris–Racomitrium lanuginosum* heath, H15 *Calluna vulgaris–Juniperus communis* ssp. *nana* heath, H17 *Calluna vulgaris–Arctostaphylos alpinus* heath, H19 *Vaccinium myrtillus–Cladonia arbuscula* heath, H20 *Vaccinium myrtillus–Racomitrium lanuginosum* heath, H22 *Vaccinium myrtillus–Rubus chamaemorus* heath (in part). Source: http://www.jncc.gov.uk/page-2237.

**Table 10.5** Conservation objectives (primary or complementary) applicable to alpine habitats (from Tokeshi 1999, p. 372) in a number of types of protected areas: scientific reserves, national parks, natural monuments, managed nature reserves, protected landscapes, resource reserves, natural biotic reserves, and multiple use areas

| Conservation objective |
| --- |
| Maintain sample ecosystem in natural state |
| Maintain ecological diversity and environmental regulation |
| Conserve genetic resources |
| Provide education, research and environmental monitoring |
| Conserve watershed, flood control |
| Control erosion and sedimentation |
| Maintain indigenous use or habitation |
| Produce protein from wildlife |
| Produce timber, forage or extractive commodities |
| Provide recreation and tourism service |
| Protect sites and objects of cultural, historical, or archaeological heritage |
| Protect scenic beauty |
| Maintain open options, management flexibility, multiple-use |
| Contribute to rural development |

conventions and legislature see Thompson *et al.* (2005)). The system in place for alpine nature conservation is based on protected areas (Hamilton 2002). Protected areas fall into a number of designations to provide for multiple purposes (Table 10.5). Protected areas are usually in remote areas on low productivity land, not affected or sought after by economic activities (Huggett 2004), attributes that further the case of alpine ecosystem conservation. For example, in Sweden, 5.9% of the land area was protected in 1990, most of which was upper montane birch forest (32%) and alpine heaths (30%) (Nilsson and Gotmark 1992).

## 10.3.4 Monitoring as a tool in alpine conservation management

Grabherr *et al.* (2005) summarized state indicators of the cryosphere, hydrosphere, and terrestrial ecosystems that can be used to detect biotic and abiotic changes triggered by climate, pollution, or land use. The indicators were compiled for application in selected high mountain Biosphere Reserves of the UNESCO Man and the Biosphere programme. The use of appropriate indicators in well-designed monitoring programmes (Yoccoz *et al.* 2001; Legg and Nagy 2006) are essential for informing conservation managers about the state of conservation features in relation to targets (e.g. Table 10.4). Periodic monitoring results allow the implementation of adequate changes in management practices and also help re-evaluate the

**Box 10.2** Ecological and pasture values of alpine and upper montane plant communities in the Pyrenees (Gómez 2008)

The character giving feature in the Ordesa and Monte Perdido National Park, Pyrenees, besides the magnificent geological formations, is the grasslands that occupy about 75% of the alpine belt and the treeline ecotone. They comprise a high number of plant species (Villar *et al.* 1997) and communities (Gómez *et al.* 2003). The grasslands are highly complex in structure and dynamics, because of long-term transhumant pasturing and as a result of recent land use changes. To provide researchers and Park managers with high-resolution (5 m² pixel size) digitized baseline information an ecological and pasture valuation of these grasslands was carried out by the Pyrenean Institute of Ecology, Jaca (http://www.ipe.csic.es/). By assessing individual species, the ecological and forage values of each plant community were determined. In a following step, the value of each of the management units was established. The methodology used took into consideration ecological features such as species and community rareness, European distribution areas, diversity, and presence of Red Data Book entries; and nutritive characteristics (biomass, primary production, digestibility, protein content, and herbivore preferences). Thus, the method identified, for each plant community or management unit, its relative ecological and nutritive values, and provided an assessment of its conservation condition. The repeat application of this standardized protocol allows periodic monitoring of habitats and can form the basis of conservation management decisions.

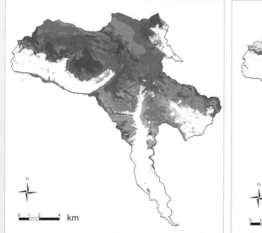 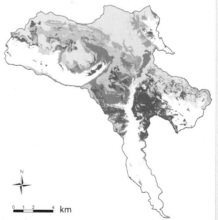

**Fig. 1**    The ecological (left) and pastoral value (right) of the vegetation in the Ordesa and Monte Perdido National Park, Pyrenees, Spain. White, forest; dark, high value; light grey, low value. Note that over large areas ecological and pastoral values separate, while in other areas they overlap. Overlap is likely to indicate grasslands, whose species richness is enhanced by managed grazing.

monitoring programme as a whole. Such a tool is directly applicable to land-use impacts, as has been used in the Cairngorms, Scottish Highlands by Gordon *et al.* (1998). Meanwhile, additional information about changes in local climate drivers and their effects on the cryosphere and hydrosphere allows the implementation of limited mitigation programmes. To underpin the monitoring programmes, a wide selection of methods of long-term surveillance has been tested for a variety of alpine organisms and purposes (e.g. Lesica and Steele 1996; Maudet *et al.* 2002; Pauli *et al.* 2004). An excellent illustration of long-term scientific studies undertaken to aid monitoring for conservation and livestock management in the Ordesa and Monte Perdido National Park, Pyrenees, Spain is the recent publication of ecological and pasture values of alpine and upper montane plant communities (Box 10.2).

# 10.4 Conclusions

Varying degree of land use has affected many alpine ranges world-wide. Global change impacts are continuing to be largely dominated through land use; however, other factors such as climate, regional nitrogen deposition, and locally alien, or so far excluded species may have important additional roles for alpine habitats. The low economic value of alpine ecosystems (except for local industrial exploitation) favours them in protected area selections and helps their formal protection, together with cultural landscapes at the upper montane periphery. Particular attention to conservation via sustainable management is required in alpine ranges adjacent to highland plateaux with high population densities. Indicators of ecosystem state are available for integrated monitoring for conservation management.

# 11  Concluding remarks

Alpine environments are 'cold'; however, in other aspects they encompass a large variability along three main axes: altitude, moisture, and seasonality (Fig. 11.1). There are many combinations possible in an environmental space that is bounded by these variables. Arctic alpine environments have a high seasonality, low precipitation, and negligible air density impact (bottom back left-hand side of Fig. 11.1). At the opposite end of the spectrum, aseasonal tropical alpine environments have very little seasonality, medium to high precipitation, and a marked reduction in air density (top front right-hand in Fig. 11.1). Ordering mountain regions in this way highlights the variety of alpine environments world-wide.

This virtual ordination space allows us to recount what we know about alpine environments, their habitats and biology. With regard to biodiversity, the effort in terms of coverage of alpine areas and organism groups has been uneven. On average, we have a reasonably good knowledge of the vascular plant flora of all the mountain regions (sometimes found in little known local literature). This knowledge can be in the form of very detailed accurate accounts in extensive mountain ranges where there is a long tradition of research. Few mountain range level accounts, if any match that in the Alps (Flora Alpina; Aeschimann *et al.* 2004), the existence of which allowed a recent systematic assessment of plant diversity patterns at the scale of the whole Alps range (INTRABIODIV). This is not the general pattern, however, and in most botanically well-known mountain ranges one has to piece the jigsaw together from country accounts (e.g. Scandinavia), or county- and state-level accounts (e.g. North American ranges). Most of these botanically well-known mountains would occupy the lower right-hand portion of Fig. 11.1. Typically, many areas have sketchy basic accounts at the mountain range level, but sometimes with detailed local floras for certain mountains. Country-level floras cater for their alpine components; however, taxonomic traditions differ with countries and the advantages of derived standard floras (e.g. Flora Europaea) are appreciable. For an

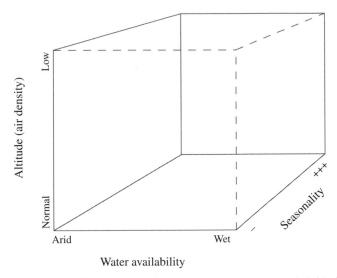

**Fig. 11.1**     Hypothetical ordination framework work for alpine zone mountains, their biodiversity, and habitat diversity patterns and processes.

overview of standard floras of the world see Frodin (2001). There are some excellent alpine floras from the tropics (e.g. New Guinea; van Royen 1980; Johns *et al.* 2006). All encompassing taxonomic treatment of the tropical and subtropical Andean alpine is less complete, despite some families having been revised at the country level recently (e.g. Poaceae Tovar 1993; Renvoize 1998). The less complete list also includes the Himalayas, where the standard floras of varying completeness are split among many countries, and the many other high mountain ranges. At this stage, one may only wish for a world-wide alpine flora account, or at least the initiation of compiling a peer-reviewed database. The treatment of cryptogams is similarly uneven; for example, the Andean páramos are much better known then the puna, which was probably last collected on an extensive scale over 100 years ago. With regard to zoology, the emerging picture is essentially similar: some European and North American boreal and temperate mountains have a detailed taxonomic treatment of most animal groups, but the majority of the other ranges has only been collected by sporadic expeditions. However, as animal taxa are more numerous and varied their account is less complete than that of plants. Some animal groups, such as vertebrates are probably sufficiently known the world over, while many of the soil-dwelling microfauna are yet to be described. Without a doubt the least known of all alpine biodiversity is that of micro-organisms (e.g. Bardgett 2005).

Habitat accounts in terms of vegetation, animal communities, and soil micro-organisms are of a varying degree of completeness and depth. The

Zürich–Monpellier tradition of phytosociology over the past 80 years has produced detailed systematic accounts (supplemented by vegetation maps) of the alpine vegetation, either as part of country or regional works, or specific to mountain ranges, mainly in Europe and in some other regions, such as North Africa (Atlas), Iran (Elburz), or the Japanese mountains; for a recent overview on Holarctic mountains see (Ozenda 2002). In addition, there are excellent accounts from the puna region of the Andes in the Neotropics. Many non-phytosociological habitat-based accounts are also available, e.g. in the tropics (páramo, e.g. Monasterio 1980), and from East Africa (Hedberg 1964; Coe 1967), or North America. Comparative vegetation descriptions are more advanced than flora accounts, as habitats can be characterized by a typical assemblage of species, without an exhaustive list of the flora. Of the vegetation accounts, those with georeferenced locations (maps and detailed sample plots) and with full species enumerations provide useful resources for today's researchers interested in a number of issues, e.g. from comparing temporal changes in species composition and spatial development of vegetation boundaries to modelling species coexistence.

Despite the large body of published zoological literature, few have attempted to give a comprehensive account of the animal life of alpine ranges. An exception is that by Green and Osborne (1994), which has synthesized information on both the vertebrate and invertebrate fauna in each of the Australian alpine habitats. Such detailed accounts are much needed for understanding the animal–plant interactions that are important drivers of the dynamics in many habitat types, and for initiating studies for their quantification. Similarly, a synthetic, habitat-based analysis of what is known about the microbial assemblages, an admittedly arduous task is now probably becoming possible for at least the long-term alpine research site at Niwot Ridge in the Colorado Rocky Mountains. While a picture that will emerge from such an account is probably applicable to other temperate and boreal, to some arctic, and a few Mediterranean alpine habitats, the aseasonal tropics and the non-snow affected subtropics will pose challenges for years to come. The microbial life in aeolian environments and glacier forefields is being investigated in a number of mountain ranges world-wide (e.g. Schmidt *et al.* 2008; Freeman *et al.* 2009)

A burgeoning area of alpine research today concerns revisiting biogeographical theories, based on the distribution of plant and animal species, and underpinning these theories by genetic means. Phylogeographies have been traced and speciation and species distributions during glacial and interglacial epochs investigated (see papers in Abbott 2008). The European Alps is one of the main laboratories of this kind of research, with centres in each alpine country. In the same vein, similar research in the Scandes links alpine phylogeography to that in the arctic, connecting research interest in North America. International collaboration is expanding the geographical horizons of this type of work. One area where applied genetics may have an important contribution to make in the future beyond phylogeography is establishing if links exist between genetic diversity of alpine taxa and its adaptive and evolutionary significance.

Beyond descriptive accounts of taxa and communities lies the domain of understanding the functioning of alpine habitats and their sensitivity to extraneous drivers. The examples to plot in Fig. 11.1 concentrate again in few places. The single most coherent set of long-term studies on the structure and functioning of alpine ecosystems have been carried out in the Niwot Ridge long-term research site (Bowman and Seastedt 2001). That type of work is an excellent example of long-term integrated research that has covered habitats along a topo-sequence (excluding rocky habitats and scree). Patterns of species and biomass distribution, and dynamics of productivity and species abundances across habitats and their linkages with inter-annual climate, animal activity and soil processes have provided the fullest picture available in seasonal alpine environments. In many respects similar work, was carried out in the scope of the International Biosphere Programme and UNESCO Man and the Biosphere programmes in the Alps (Austria, Switzerland, Germany), whose results are available in specific accounts for the study sites; unfortunately, however, not in a comprehensive synthesis. Thorough investigations have been made across the Arctic, with much of the focus on experimental simulation of climate change there (e.g. Callaghan *et al.* 2004) that can be related to alpine habitats in the arctic. A thorough comparatively long-term research engagement in the aseasonal Andes has been in the Merida region of the Venezuelan Andes (Monasterio 1980). The sites used there have a high potential to become the tropical alpine equivalent of the Niwot Ridge knowledge base, with the additional bonus of existing interdisciplinarity that has been employed to adequately address the impact of land use. What is much missed is a long-term integrated programme in dry to arid tropical latitudes that are not affected by seasonal snow accumulation.

Today's view of temperature impacts on alpine ecosystem dynamics has been much influenced by circumstantial evidence, and by drawing parallels between glacier retreats and potential temperature impacts on alpine habitats. However, glaciers have their own dynamics, based on mass balance changes that may bear no relevance to overall alpine vegetation dynamics, with the exception of providing substratum for vegetation succession on exposed moraines. Climate change impacts on functional aspects of alpine habitats are not easy to consistently report in the absence of repeat observations. Repeat observations themselves need to consider the degree of year-to-year variation (Walker *et al.* 1994), before a trend can be proposed (Yoccoz *et al.* 2001). Inter-annual variation can be high for many ecosystem variables, but it is rather negligible for species composition; this has led to a dual-track approach in recording species and sample stands in alpine plant communities, along with continuous logging of rooting zone temperature world-wide for detecting long-term change (GLORIA, http://www.gloria.ac.at). To date, in temperate alpine ecosystems most work has concentrated on individual species responses (e.g. physiological, see Körner 2003), or on the sexual reproduction of alpine plant species (Molau *et al.* 2005), both from purely plant resource allocation point of view or from a combined plant-pollinator system viewpoint (e.g. Totland 1993), and there are few studies that go beyond the lifetime of (one to three) research grant(s). Reproduction and pollination of herbs are important and responsive facets and illustrate how climate variability may be readily observed (and how colonization patterns may be affected); however, from the character of current alpine habitats their dominant graminoid species would need much required studies. These species are clonal (maybe aged >1000 years) and wind pollinated, but most reproduce sexually only occasionally. What we need to know for predicting future alpine vegetation is how these genetically old clonal structures can respond to warming scenarios. In tropical and Southern Hemisphere alpine environments tussock grasses and some character-giving long-lived monocarpic species mast flower periodically. How masting frequency and establishment success may be affected in these ecosystems are important unknowns. In arctic heaths, experimental studies have shown that acute warming (especially with a higher nitrogen supply) can change life-form composition over short time-scales, especially from evergreen to deciduous. Such studies may bear most predictive relevance for boreal alpine areas. We would need studies on snow removal (e.g. A. F. Mark, unpublished), as opposed to snow fencing, to be able to make predictions about snowbed habitats, perceived as likely to most change in a warmer climate. In aseasonal climates changes in water availability are likely to be the main drivers of vegetation and habitat change, and watering and droughting experiments might help complement theoretical consideration.

From a land-use point of view, we may populate Fig. 11.1 with patchy heavy contrasts: of little human impact on most arctic–alpine ecosystems versus serious environmental degradation in Iceland, or relatively little impacted New Guinean alpine ecosystems versus some heavily exploited páramo. We are generally aware where degradation is taking place as a result of land use in high mountains; however, there are few examples where pure and applied ecology have combined to offer practical management advice. A recent publication on the alpine vegetation of the Pyrenees (Fillat *et al.* 2008) has discussed habitats and vegetation from their pasturing potential and ecological value. Importantly, it has provided estimates of carrying capacity for livestock, not only for the individual alpine habitats, but also in view of maintaining the environmental services across the alpine–montane–lowland continuum. Such work would especially be highly relevant for high mountain areas that are densely populated, or the grazing rights are at maximum capacity. Aseasonal environments where snow does not exclude grazing for most of the year may particularly be heavily affected.

By way of closing, we would like to emphasize the role of alpine ecosystems in providing goods and environmental services, most of all water and livelihoods. Equally important is their role in buffering hydrological extremes. The sustained functioning of these systems in heavily used locales lies, to a large extent, in the hands of humans. We need an integrated and unbiased interdisciplinary approach to continue study and use the accumulated knowledge to manage these ecosystems in an unselfish way.

# REFERENCES

Aarssen, L.W. (1989) Competitive ability and species coexistence—a plants-eye view. *Oikos,* **56,** 386–401.

Abbott, R.J. ed. (2008) History, evolution and future of arctic and alpine flora. *Special Issue, Plant Ecology & Diversity,* **1,** 129–349.

Abbott, R.J. & Bochmann, C. (2002) History and evolution of the arctic flora: in the footsteps of Eric Hultén. *Molecular Ecology,* **12,** 299–313.

Adler, P.B. & Morales, J.M. (1999) Influence of environmental factors and sheep grazing on an Andean grassland. *Journal of Range Management,* **52,** 471–481.

Aeschimann, D., Lauber, K., Moser, D.M. & Theurillat, J.-P. (2004) *Flora Alpina.* Haupt, Bern.

Agakhanjanz, O. & Breckle, S.W. (1995) Origin and evolution of the mountain flora in Middle Asia and neighbouring mountain regions. *Arctic and Alpine Biodiversity* (eds C. Körner & F.S. Chapin III), pp. 63–80. Springer, Berlin.

Agakhanyantz, O.E. & Lopatin, I.K. (1978) Main characteristics of the ecosystems of the Pamirs, USSR. *Arctic and Alpine Research,* **10,** 397–407.

Ågren, G.I. & Bosatta, E. (1996) *Theoretical Ecosystem Ecology. Understanding Element Cycles.* Cambridge University Press, Cambridge.

Aizpuru, I., Aseginolaza, C., Uribe-Echebarria, P.M., Urrutia, P. & Zorrakin, I. (2007) *Claves illustradas de la flora del pais vasco y territorios limitrofes.* Eusko Jaurlaritzaren Argitalpen Zerbitzu Nagusia, Vitoria-Gasteiz.

Alcamo, J., Mayerhofer, P., Guardans, R., van Harmelen, T., van Minnen, J., Onigkeit, J., Posch, M. & de Vries, B. (2002) An integrated assessment of regional air pollution and climate change in Europe: findings of the AIR-CLIM Project. *Environmental Science & Policy,* **5,** 257–272.

Alexandrowicz, Z., Margielewski, W. & Perzanowska, J. (2003) European ecological network NATURA 2000 in relation to landslide areas diversity: a case study in the Polish Carpathians. *Ekologia Bratislava,* **22,** 404–422.

Ali, A.A., Carcaillet, C., Guendon, J.L., Quinif, Y., Roiron, P. & Terral, J.F. (2003) The early Holocene treeline in the southern French Alps: new evidence from travertine formations. *Global Ecology & Biogeography,* **12,** 411–419.

Allainé, D. & Yoccoz, N.G. (2003) Rodents in the European Alps: population ecology and potential impacts on ecosystems. *Alpine Biodiversity in Europe* (eds L. Nagy, G. Grabherr, C. Körner & D.B.A. Thompson), pp. 339–349. Springer, Berlin, Heidelberg, New York.

Allen, C.E. & Burns, S.F. (2000) Characterization of alpine soils, Eagle Cap, Wallowa Mountains, Oregon. *Physical Geography*, **21**, 212–222.

Allison, I. & Bennett, J. (1976) Climate and microclimate. *The Equatorial Glaciers of New Guinea* (eds G.S. Hope, J.A. Peterson, I. Allison & U. Radok), pp. 61–80. A.A. Balkema, Rotterdam.

Altschuler, D.L. & Dudley, R. (2006) The physiology and biomechanics of avian flight at high altitude. *Integrative and Comparative Biology*, **46**, 62–71.

Amman, B. (1995) Palaeorecords of plant biodiversity in the Alps. *Arctic and Alpine Biodiversity. Patterns, Causes and Ecosystem Consequences* (eds F.S. Chapin III & C. Körner), pp. 137–149. Springer, Berlin, Heidelberg, New York.

André, M.F. (1998) Depopulation, land-use change and transformation in the French Massif Central. *Ambio*, **27**, 351–353.

Anonymous (1998) Systematik der Böden und der Bodenbildenden Substrate Deutschlands. Systematik der Böden. Systematik der Bodenbildenden Substrate. Gliederung der Periglaziären Lagen *Mitteilungen der Deutschen Bodenkundlichen Gesellschaft*, **Band 86**, 1–180.

Armstrong, D.M., Halfpenny, J.C. & Southwick, C.H. (2001) Vertebrates. *Structure and Function of an Alpine Ecosystem, Niwot Ridge, Colorado* (eds W.D. Bowman & T.R. Seastedt), pp. 128–156. Oxford University Press, Oxford.

Arnalds, A. (1987) Ecosystem disturbance and recovery in Iceland. *Arctic and Alpine Research*, **19**, 508–513.

Arnalds, A. (2004) Carbon sequestration and the restoration of land health—An example from Iceland. *Climatic Change*, **65**, 333–346.

Atkin, O.K. & Collier, D.E. (1992) Relationship between soil-nitrogen and floristic variation in late snow areas of the Kosciusko alpine region. *Australian Journal of Botany*, **40**, 139–149.

Austrheim, G. & Eriksson, O. (2001) Plant species diversity and grazing in the Scandinavian mountains—patterns and processes at different spatial scales. *Ecography*, **24**, 683–695.

Ayenew, T. (2003) Evapotranspiration estimation using thematic mapper spectral satellite data in the Ethiopian rift and adjacent highlands. *Journal of Hydrology*, **279**, 83–93.

Bagchi, S., Namgail, T., Ritchie, M.E. (2006) Small mammalian herbivores as mediators of plant community dynamics in the high-altitude arid rangelands of Trans-Himalaya. *Biological Conservation* **127**:438–442.

Bahn, M. & Körner, C. (2003) Recent increases in summit flora caused by warming in the Alps. *Alpine Biodiversity in Europe* (eds L. Nagy, G. Grabherr, C. Körner & D.B.A. Thompson), pp. 437–441. Ecological studies 167, Springer, Berlin, Heidelberg, New York.

Baied, C.A. & Wheeler, J.C. (1993) Evolution of high Andean puna ecosystems—environment, climate, and culture change over the last 12,000 years in the central Andes. *Mountain Research and Development*, **13**, 145–156.

Baillie, I.C., Tshering, K., Dorji, T., Tamang, H.B., Norbu, C., Hutcheon, A.A. & Baumler, R. (2004) Regolith and soils in Bhutan, Eastern Himalayas. *European Journal Of Soil Science*, **55**, 9–27.

Bale, J.S., Hodkinson, I.D., Block, W., Webb, N.R., Coulson, S.J. & Strathdee, A.T. (1997) Life strategies of arctic terrestrial arthropods. *Ecology of Arctic Environments* (eds S.J. Woodin & M. Marquiss), pp. 137–165. Blackwell, Oxford.

Ballare, C.L., Rousseaux, M.C., Searles, P.S., Zaller, J.G., Giordano, C.V., Robson, T.M., Caldwell, M.M., Sala, O.E. & Scopel, A.L. (2001) Impacts of solar ultraviolet-B radiation on terrestrial ecosystems of Tierra del Fuego (southern Argentina)—an overview of recent progress. *Journal of Photochemistry and Photobiology B-Biology,* **62,** 67–77.

Balslev, H. (1992) *Páramo: an Andean Ecosystem under Human Influence.* Academic Press, London.

Bannister, P., Maegli, T., Dickinson, K.J.M., Halloy, S.R.P., Knight, A., Lord, J.M., Mark, A.F. & Spencer, K.L. (2005) Will loss of snow cover during climatic warming expose New Zealand alpine plants to increased frost damage? *Oecologia,* **144,** 245–256.

Bardgett, R.D. (2005) *The Biology of Soil.* Oxford University Press, Oxford.

Bardgett, R.D., Bowman, W.D., Kaufmann, R. & Schmidt, S.K. (2005) A temporal approach to linking aboveground and belowground ecology. *Trends in Ecology and Evolution,* **20,** 634–641.

Bardou, E. & Delaloye, R. (2004) Effects of ground freezing and snow avalanche deposits on debris flows in alpine environments. *Natural Hazards and Earth System Sciences,* **4,** 519–530.

Barinov, M.G. (1988) *Mesostructure of Photosynthetic Apparatus of Plants from Different Climatic Zones.* Komi Sientific Center of Urals Branch, Academy of Sciences of the USSR, Syvktyvkar.

Baron, J.S., Rueth, H.M., Wolfe, A.M., Nydick, K.R., Allstott, E.J., Minear, J.T. & Moraska, B. (2000) Ecosystem responses to nitrogen deposition in the Colorado Front Range. *Ecosystems,* **3,** 352–368.

Barry, R.G. (1992) *Mountain Weather and Climate.* Methuen, New York.

Bätzing, W. (2002) *Die Alpen. Gesichte und Zukunft einer europäischen Kulturlandschaft.* Verlag C.H. Beck, München.

Bauer, J.J. (1990) The analysis of plant-herbivore interactions between ungulates and vegetation on alpine grasslands in the Himalayan region of Nepal. *Plant Ecology,* **90,** 15–34.

Baumgartner, J. & Hartmann, J. (2001) The design and implementation of sustainable plant diversity conservation program for alpine meadows and pastures. *Journal of Agricultural and Environmental Ethics,* **14,** 67–83.

Baumler, R. (2001) Pedogenic studies in aeolian deposits in the high mountain area of eastern Nepal. *Quaternary International,* **76–7,** 93–102.

Bayer, R.J. (1997) *Antennaria rosea* (Asteraceae)—a model group for the study of the evolution of polyploid agamic complexes. *Opera Botanica,* **0,** 53–65.

Beall, C.M. (2006) Andean, Tibetan, and Ethiopian patterns of adaptation to high-altitude hypoxia. *Integrative and Comparative Biology,* **46,** 18–24.

Beck, E. (1994) Cold tolerance in tropical alpine plants. *Tropical Alpine Environments* (eds P.W. Rundel, A.P. Smith & F.C. Meinzer), pp. 77–110. Cambridge University Press, Cambridge.

Beck, E. (1994) Turnover and conservation of nutrients in the pachycaul *Senecio keniodendron. Tropical Alpine Environments. Plant Form and Function* (eds P.W. Rundel, A.P. Smith & F.C. Meinzer), pp. 215–221. Cambridge University Press, Cambridge.

Beekman, A.M. & Verweij, P.A. (1987) Structure and nutrient status of a paramo bunchgrass vegetation in relation to soil and climate. M.Sc. Thesis, University of Amsterdam, Amsterdam.

Beerling, D.J., Chaloner, W.G., Huntley, B., Pearson, J.A. & Tooley, M.J. (1993) Stomatal density responds to the glacial cycle of environmental-change. *Proceedings of the Royal Society of London Series B—Biological Sciences,* **251,** 133–138.

Bell, R.A., Athey, P.V. & Sommerfeld, M.R. (1986) Cryptoendolithic algal communities of the Colorado Plateau. *Journal of Phycology*, **22**, 429–435.

Belnap, J. & Lange, O.L. (2003) *Biological Soil Crusts: Structure, Function, and Management*. Springer, Heidelberg, Berlin, New York.

Bencherifa, A. (1983) Land-use and equilibrium of mountain ecosystems in the high atlas of western Morocco. *Mountain Research and Development*, **3**, 273–279.

Berg, N.H. (1986) Blowing snow at a Colorado alpine site: measurements and implications. *Arctic and Alpine Research*, **18**, 147–161.

Billings, W.D. (1974) Adaptations and origins of alpine plants. *Arctic and Alpine Research*, **6**, 129–142.

Billings, W.D. (2000) Alpine vegetation. *North American Terrestrial Vegetation* (eds M.G. Barbour & W.D. Billings), pp. 537–572. Cambridge University Press, Cambridge.

Birkenland, P.W., Berry, M.E. & Swanson, D.K. (1991) Use of soil catena field data for estimating relative ages of moraines. *Geology*, **19**, 281–283.

Birkenland, P.W., Shroba, R.R., Burns, S.F., Price, A.B. & Tonkin, P.J. (2003) Integrating soils and geomorphology in mountains—an example from the Front Range of Colorado. *Geomorphology*, **55**, 329–344.

Birks, H.J.B. (1986) Late-Quaternary biotic changes in terrestrial and lacustrine environments, with particular reference to north-west Europe. *Handbook of Holocene Palaeoecology and Palaeohydrology* (ed B.E. Bergland), pp. 3–65. Wiley, New York.

Bleeker, P. (1980) The alpine soils of the New Guinea high mountains. *Alpine Flora of New Guinea* (ed P. van Royen), pp. 59–74. Cramer-Verlag, Vaduz.

Bliss, L.C. (1985) Alpine. *Physiological Ecology of North American Plant Communities* (eds B.F. Chabot & H.A. Mooney), pp. 41–65. Chapman and Hall, New York, London.

Blumer, P. & Diemer, M. (1996) The occurrence and consequences of grasshopper herbivory in an alpine grassland, Swiss Central Alps. *Arctic and Alpine Research*, **28**, 435–440.

Bocher, J. & Nachman, G. (2001) Temperature and humidity responses of the arctic-alpine seed bug Nysius groenlandicus. *Entomologia Experimentalis et Applicata*, **99**, 319–330.

Bock, J.H., Jolls, C.L. & Lewis, A.C. (1995) The effects of grazing on alpine vegetation—a comparison of the central Caucasus, Republic of Georgia, with the Colorado Rocky Mountains, USA. *Arctic and Alpine Research*, **27**, 130–136.

Bockheim, J.G., Munroe, J.S., Douglass, D. & Koerner, D. (2000) Soil development along an elevational gradient in the southeastern Uinta Mountains, Utah, USA. *Catena*, **39**, 169–185.

Bohn, U., Gollub, G., Hettwer, C., Neuhauslova, Z., T., a., Sclueter, H. & Weber, H. (2004) Map of the natural vegetation of Europe. Scale 1: 2 500 000. Part I. Explanatory text with CD-ROM. Bundesamt für Naturschutz, Bonn.

Bortenschlager, S., Oeggl, K. & Wahlmueller, N. (1996) IGCP Project 158b: Austria—regional syntheses. *Palaeoecological Events During the Last 15 000 Years: Regional Syntheses of Palaeoecological Studies of Lakes and Mires.* (eds B.E. Berglund, H.J.B. Birks & M. Ralska-Jasiewiczowa), pp. 667–685. Wiley, Chichester.

Bowman, W.D. & Seastedt, T.R. (2001) *Structure and Function of an Alpine Ecosystem, Niwot Ridge, Colorado.* Oxford University Press, Oxford.

Bowman, W.D. & Steltzer, H. (1998) Positive feedbacks to anthropogenic nitrogen deposition in Rocky Mountain Alpine tundra. *Ambio, 27,* 514–517.

Bowman, W.D. (1992) Inputs and storage of nitrogen in winter snowpack in an alpine ecosystem. *Arctic and Alpine Research, 24,* 211–215.

Bowman, W.D. (1994) Accumulation and use of nitrogen and phosphorus following fertilization in 2 alpine tundra communities. *Oikos, 70,* 261–270.

Bowman, W.D. (2001) Introduction. Historical perspective and significance of alpine ecosystem studies. *Structure and Function of an Alpine Ecosystem, Niwot Ridge, Colorado* (eds W.D. Bowman & T.R. Seastedt), pp. 3–12. Oxford University Press, Oxford.

Bowman, W.D., Gartner, J.R., Holland, K. & Wiedermann, M. (2006) Nitrogen critical loads for alpine vegetation and terrestrial ecosystem response: Are we there yet? *Ecological Applications, 16,* 1183–1193.

Bowman, W.D., Steltzer, H., Rosenstiel, T.N., Cleveland, C.C. & Meier, C.L. (2004) Litter effects of two co-occurring alpine species on plant growth, microbial activity and immobilization of nitrogen. *Oikos, 104,* 336–344.

Bowman, W.D., Theodose, T.A., Schardt, J.C. & Conant, R.T. (1993) Constraints of nutrient availability on primary production in 2 alpine tundra communities. *Ecology, 74,* 2085–2097.

Bradshaw, A.D. (1993) Introduction: understanding the fundamentals of succession. *Primary Succession on Land* (ed J.W. Miles, D.W.H.), pp. 1–3. Blackwell, Oxford.

Bragazza, L., Gerdol, R. & Rydin, H. (2003) Effects of mineral and nutrient input on mire bio-geochemistry in two geographical regions. *Journal of Ecology, 91,* 417–426.

Braun Wilke, R.H. (2001) *Carta de Aptitud Ambiental de la Provincia de Jujuy.* San Juan de Jujuy.

Breckle, S.W. & Wucherer, W. (2006) Vegetation of the Pamir (Tajikistan): land use and desertification problems. *Land use Change and Mountain Biodiversity* (eds E. Spehn, M. Lieberman & C. Körner), pp. 225–237. CRC Taylor & Francis, Boca Raton, London, New York.

Breckle, S-W. (2006) *Flora and Vegetation of Afghanistan.* Internet document http://www.ag-afghanistan.de/files/breckle_flora.pdf, accessed 16 December 2008.

Breckle, S.-W. (2002) *Walter's Vegetation of the Earth. The Ecological Systems of the Geo-Biosphere.* Springer, Berlin, Heidelberg, New York.

Brendel, U.M., Eberhardt, R. & Wiesmann, K. (2002) Conservation of the Golden Eagle (*Aquila chrysäetos*) in the European Alps—a combination of education, cooperation, and modern techniques. *Journal of Raptor Research, 36,* 20–24.

Brent, J., Sinclair, M., Worland, R. & Wharton, D.A. (1999) Ice nucleation and freezing tolerance in New Zealand alpine and lowland weta, *Hemideina* spp. (Orthoptera; Stenopelmatidae). *Physiological Entomology, 24,* 56–63.

Bressoud, B. (1989) Contributions a la connaissance du Caricion atrofusco-saxatilis dans les Alpes. *Phytocoenologia, 17,* 145–270.

Bridle, K.L. & Kirkpatrick, J.B. (2001) Impacts of grazing by vertebrate herbivores on the flower stem production of tall alpine herbs, Eastern Central Plateau, Tasmania. *Australian Journal of Botany, 49,* 459–470.

Bridle, K.L., Kirkpatrick, J.B., Cullen, P. & Shepherd, R.R. (2001) Recovery in alpine heath and grassland following burning and grazing, Eastern Central Plateau, Tasmania, Australia. *Arctic, Antarctic and Alpine Research, 33,* 348–356.

Brittain, J.E. & Milner, A.M. (2001) Ecology of glacier-fed rivers: current status and concepts. *Freshwater Biology, 46,* 1571–1578.

Brooker, R., Kikvidze, Z., Pugnaire, F.I., Callaway, R.M., Choler, P., Lortie, C.J. & Michalet, R. (2005) The importance of importance. *Oikos,* **109,** 63–70.

Broomhall, S.D., Osborne, W.S. & Cunningham, R.B. (2000) Comparative effects of ambient ultraviolet-B radiation on two sympatric species of Australian frogs. *Conservation Biology,* **14,** 420–427.

Brown, C.S., Mark, A.F., Kershaw, G.P. & Dickinson, K.J.M. (2006) Secondary succession 24 years after disturbance of a New Zealand high-alpine cushionfield. *Arctic Antarctic and Alpine Research,* **38,** 325–334.

Buchner, O. & Neuner, G. (2003) Variability of Heat Tolerance in Alpine Plant Species Measured at Different Altitudes. *Arctic, Antarctic, and Alpine Research,* **35,** 411–420.

Burga, C.A., Klötzli, F. & G., G. (2004) *Gebirge der Erde. Landschaft, Klima, Pflanzenwelt.* Ulmer, Stuttgart.

Burns, S.F. (1980) *Alpine Soil Distribution and Development, Indiana Peaks, Colorado Front Range.* University of Colorado, Boulder.

Burrows, C.J. (1990) *Processes of Vegetation Change.* Unwin Hyman, London.

Buse, A., Hadley, D. & Sparks, T. (2001) Arthropod distribution on an alpine elevational gradient: the relationship with preferred temperature and cold tolerance. *European Journal of Entomology,* **98,** 301–309.

Buttolph, L.P. & Coppock, D.L. (2004) Influence of deferred grazing on vegetation dynamics and livestock productivity in an Andean pastoral system. *Journal of Applied Ecology,* **41,** 664–674.

Byers, A. (2005) Contemporary human impacts on alpine ecosystems in the Sagarmatha (Mt. Everest) National Park, Khumbu, Nepal. *Annals of the Association of American Geographers,* **95,** 112–140.

Byers, A.C. (2000) Contemporary landscape change in the Huascaran National Park and buffer zone, Cordillera Blanca, Peru. *Mountain Research and Development,* **20,** 52–63.

Cabrera, A.L. (1976) *Regiones Fitogeográficas de Argentina.* Editorial ACME, Buenos Aires.

Caccianiga, M., Luzzaro, A., Pierce, S., Ceriani, R.M. & Cerabolini, B. (2006) The functional basis of a primary succession resolved by CSR classification. *Oikos,* **112,** 10–20.

Caine, N. (2001) Geomorphic systems of Green Lakes Valley. *Structure and Function of an Alpine Ecosystem, Niwot Ridge, Colorado* (eds W.D. Bowman & T.R. Seastedt), pp. 45–74. Oxford University Press, Oxford.

Cairns, D.M. & Moen, J. (2004) Herbivory influences tree lines. *Journal of Ecology,* **92,** 1019–1024.

Caldwell, M.M. (1968) Solar ultraviolet radiation as an ecological factor for alpine plants. *Ecological Monographs,* **38,** 243–268.

Caldwell, M.M. (1979) Plant life and ultraviolet radiation: some perspective in the history of the Earth's UV climate. *BioScience,* **29,** 520–525.

Callaghan, T.V., Bjorn, L.O., Chernov, Y., Chapin, T., Christensen, T.R., Huntley, B., Ims, R.A., Johansson, M., Jolly, D., Jonasson, S., Matveyeva, N., Panikov, N., Oechel, W., Shaver, G., Elster, J., Jonsdottir, I.S., Laine, K. & Taulavuori, K. (2004) Responses to projected changes in climate and UV-B at the species level. *Ambio,* **33,** 418–435.

Cannone, N., Sgorbati, S. & Guglielmin, M. (2007) Unexpected impacts of climate change on alpine vegetation. *Frontiers in Ecology and the Environment,* **5,** 360–364.

Cardozo, C.H. & Schnetter, M.L. (1976) Estudios ecologicos en el Paramo de Cruz Verde, Colombia. III. La biomasa de tres asociaciones vegetales y la productividad de *Calamagrostis effusa* (H.B.K.) Steud. y *Paepalanthus columbiensis* Ruhl. en comparacion con la concentracion de clorofila. *Caldasia,* **11,** 85–91.

Cassagne, N., Remaury, M., Gauquelin, T. & Fabre, A. (2000) Forms and profile distribution of soil phosphorus in alpine Inceptisols and Spodosols (Pyrenees, France). *Geoderma,* **95,** 161–172.

Cavieres, L., Arroyo, M.T.K., Penaloza, A., Molina-Montenegro, M. & Torres, C. (2002) Nurse effect of *Bolax gummifera* cushion plants in the alpine vegetation of the Chilean Patagonian Andes. *Journal of Vegetation Science,* **13,** 547–554.

Cavieres, L.A. & Arroyo, M.T.K. (1999) Bancos de semillas en Phacelia secunda J.F. Gmelin (Hydrophyllaceae): variacion altitudinal en los Andes de Chile central (33" S). *Revista Chilena De Historia Natural,* **72,** 541–556.

Cavieres, L.A. & Arroyo, M.T.K. (2001) Persistent soil seed banks in Phacelia secunda (Hydrophyllaceae): experimental detection of variation along an altitudinal gradient in the Andes of central Chile (33 degrees S). *Journal of Ecology,* **89,** 31–39.

Cavieres, L.A., Badano, E.I., Sierra-Almeida, A. & Molina-Montenegro, M.A. (2007) Microclimatic modifications of cushion plants and their consequences for seedling survival of native and non-native herbaceous species in the high Andes of central Chile. *Arctic, Antarctic and Alpine Research,* **39,** 229–236.

Cavieres, L.A., Badano, E.I., Sierra-Almeida, A., Gomez-Gonzalez, S. & Molina-Montenegro, M.A. (2005) Positive interactions between alpine plant species and the nurse cushion plant *Laretia acaulis* do not increase with elevation in the Andes of central Chile. *New Phytologist,* **169,** 59–69.

Cazares, E., Trappe, J.M. & Jumpponen, A. (2005) Mycorrhiza-plant colonization patterns on a subalpine glacier forefront as a model system of primary succession. *Mycorrhiza,* **15,** 405–416.

Cernusca, A. (1976) Bestandesstruktur, Bioklima und Energiehaushalt von alpinen Zwergstrauchbeständen. *Oecologia Plantarum,* **11,** 71–102.

Chambers, J.C. (1995) Disturbance, life-history strategies, and seed fates in alpine herbfield communities. *American Journal of Botany,* **82,** 421–433.

Chang, D.H.S. & Gauch, H.G. (1986) Multivariate analysis of plant communities and environmental factors in Ngari, Tibet. *Ecology,* **67,** 1568–1575.

Chapin III, F.S. & Shaver, G.R. (1985) Arctic. *Physiological Ecology of North American Plant Communities* (eds B.F. Chabot & H.D. Mooney), pp. 16–40. Chapman and Hall, New York.

Chase, T.N., Pielke, R.A., Kittel, T.G.F., Baron, J.S. & Stohlgren, T.J. (1999) Potential impacts on Colorado Rocky Mountain weather due to land use changes on the adjacent Great Plains. *Journal of Geophysical Research—Atmosphere,* **104,** 16673–16690.

Chernov, Y.I. (1995) Diversity of the arctic terrestrial fauna. *Arctic and Alpine Biodiversity: Patterns, Causes and Ecosystem Consequences* (eds F.S. Chapin III & C. Körner), pp. 81–95. Springer, Berlin, Heidelberg, New York.

Choler, P., Michalet, R. & Callaway, R.M. (2001) Facilitation and competition on gradients in alpine plant communities. *Ecology,* **82,** 3295–3308.

Clements, F.E. (1928) *Plant Succession and Indicators. A Definitive Edition of Plant Succession and Plant Indicators.* Hafner Press, New York.

Coe, M.J. (1967) *The Ecology of the Alpine Zone of Mount Kenya.* W. Junk, The Hague.

Comes, H.P. & Kadereit, J.W. (2003) Spatial and temporal patterns in the evolution of the flora of the European alpine system. *Taxon,* **52,** 451–462.

Cooper, D.J. (1989) Geographical and ecological relationships of the arctic-alpine vascular flora and vegetation, Arrigetch Peaks region, central Brooks Range, Alaska. *Journal of Biogeography,* **16,** 279–295.

Cooper, E.J. (2004) Out of sight, out of mind: thermal acclimation of root respiration in arctic *Ranunculus. Arctic, Antarctic, and Alpine Research,* **36,** 308–313.

Corcket, E., Callaway, R.M. & Michalet, R. (2003) Insect herbivory and grass competition in a calcareous grassland: results from a plant removal experiment. *Acta Oecologica—International Journal of Ecology,* **24,** 139–146.

Costin, A., Gray, M., Totterdell, C. & Wimbush, D. (2000) *The Kosciuszko Alpine Flora (Field Edition).* CSIRO, Collingwood.

Costin, A.B. (1954) *A Study of the Ecosystems of the Monaro Region, New South Wales.* Government Printer, Sydney.

Costin, A.B., Wimbush, D.J., Barrow, M.D. & Lake, P. (1969) Development of soil and vegetation climaxes in the Mount Kosciuszko area, Australia. *Plant Ecology,* **18,** 273–288.

Crawford, R.M.M. (2008) *Plants at the Margin. Ecological Limits and Climate Change.* Cambridge University Press, Cambridge.

Crawford, R.M.M., Chapman, H.M. & Smith, L.C. (1995) Adaptation to variation in growing season length in Arctic populations of *Saxifraga oppositifolia* L. *Botanical Journal of Scotland,* **41,** 177–192.

Crisci, J.V. (2001) The voice of historical biogeography. *Journal of Biogeography,* **28,** 157–168.

Crisci, J.V., Cigliano, M.M., Morrone, J.J. & Roig-Junent, S. (1991) Historical biogeography of southern South America. *Systematic Zoology,* **40,** 152–171.

Crouch, H.J. (1993) Plant distribution patterns and primary succession on a glacier foreland: a comparative study of cryptogams and higher plants. *Primary Succession on Land* (ed J.W. Miles, D.W.H.), pp. 133–145. Blackwell, Oxford.

Dahl, E. (1951) On the relation between the summer temperature and the distribution of alpine vascular plants in Fennoscandia. *Oikos,* **3,** 22–52.

Dahl, E. (1998) *The Phytogeography of Northern Europe (British Isles, Fennoscandia and Adjacent Areas).* Cambridge University Press, Cambridge.

Dai, W.H. & Huang, Y. (2006) Relation of soil organic matter concentration to climate and altitude in zonal soils of China. *Catena,* **65,** 87–94.

Davie, T.J.A., Fahey, B.D. & Stewart, M.K. (2006) Tussock grasslands and high water yield: a review of the evidence. *Journal of Hydrology (Wellington North),* **45,** 83–93.

de Wit, C.J. (1963) Competition among plants. *Agronomy Journal,* **55,** 1–31.

DeChaine, E.G. & Martin, A.P. (2005) Marked genetic divergence among sky island populations of *Sedum lanceolatum* (Crassulaceae) in the Rocky Mountains. *American Journal of Botany,* **92,** 477–486.

Del Giudice, M. (2008) Power Struggle. The people of Iceland awaken to a stark choice: exploit wealth of clean energy or keep their landscape pristine. *National Geographic,* **213,** 62–89.

del Moral, R. (1993) Mechanisms of primary succession on volcanoes: a view from Mount St Helens. *Primary Succession on Land* (ed J.W. Miles, D.W.H.), pp. 79–100. Blackwell, Oxford.

Descamps, P. (2008) La montagne victime des sports d'hiver. *Le Monde Diplomatique*, **55**, 20–21.

Descroix, L. & Mathys, N. (2003) Processes, spatio-temporal factors and measurements of current erosion in the French Southern Alps: A review. *Earth Surface Processes and Landforms*, **28**, 993–1011.

Diaz, A., Pefaur, J.E. & Durant, P. (1997) Ecology of South American paramos with emphasis on the fauna of the Venezuelan paramos. *Ecosystems of the World 3 Arctic and Alpine Tundra* (ed F.E. Wielgolaski), pp. 263–309. Elsevier, Amsterdam.

Dickoré, W.B. & Miehe, G. (2002) Cold spots in highest mountains of the world—diversity patterns and gradients in the flora of the Karakorum. *Mountain Biodiversity. A Global Assessment* (eds C. Koerner & E. Spehn), pp. 129–147. Parthenon, Boca Raton, London.

Dickoré, W.B. (1991) Zonation of flora and vegetation of the Northern declivity of the Karakoram/Kunlun mountains (SW Xinjiang China) *GeoJournal*, **25**, 265–284.

Diemer, M., Körner, C. & Prock, S. (1992) Leaf life spans in wild perennial herbaceous plants: a survey and attempts at a functional interpretation. *Oecologia*, **89**, 10–16.

Dierssen, K. (1996) *Vegetation Nordeuropas*. Ulmer, Stuttgart.

Dillon, M.E., Frazier, M.R. & Dudley, R. (2006) Into thin air: physiology and evolution of alpine insects. *Integrative and Comparative Biology*, **46**, 49–61.

Dirksen, V. & Dirksen, O. (2007) Vegetation dynamics and ecological consequences of the 1996 eruption in Karymsky Volcanic Center, Kamchatka. *Journal of Volcanology and Seismology*, **1**, 164–174.

Dirnböck, T., Dullinger, S. & Grabherr, G. (2003) A regional impact assessment of climate and land-use change on alpine vegetation. *Journal of Biogeography*, **30**, 401–417.

Dolédec, S., Chessel, D., terBraak, C.J.F. & Champely, S. (1996) Matching species traits to environmental variables: A new three-table ordination method. *Environmental and Ecological Statistics*, **3**, 143–166.

Douglas, G.W. & Bliss, L.C. (1977) Alpine and high subalpine plant communities of the North Cascades Range, Washington and British Columbia. *Ecological Monographs*, **47**, 113–150.

Dullinger, S., Dirnböck, T. & Grabherr, G. (2004) Modelling climate change-driven treeline shifts: relative effects of temperature increase, dispersal and invasibility. *Journal of Ecology*, **92**, 241–252.

Edmonds, T., Lunt, I.D., Roshier, D.A. & Louis, J. (2006) Annual variation in the distribution of summer snowdrifts in the Kosciuszko alpine area, Australia, and its effect on the composition and structure of alpine vegetation. *Austral Ecology*, **31**, 837–848.

Edwards, J.S. (1972) Arthropod fallout on Alaskan snow. *Arctic and Alpine Research*, **4**, 167–176.

Egger, B. (1994) Végétation et station alpines sur serpentine près de Davos. Veröffentlichungen des Geobotanischen Institutes der Eidgenössische Technische Hochschule Stiftung Rübel in Zürich, **117**, 1–275.

Egli, M., Fitze, P. & Mirabella, A. (2001) Weathering and evolution of soils formed on granitic, glacial deposits: results from chronosequences of Swiss alpine environments. *Catena*, **45**, 19–47.

Egli, M., Mirabella, A., Sartori, G. & Fitze, P. (2003) Weathering rates as a function of climate: results from a climosequence of the Val Genova (Trentino, Italian Alps). *Geoderma*, **111**, 99–121.

Egli, M., Wernli, M., Kneisel, C. & Haeberli, W. (2006) Melting glaciers and soil development in the proglacial area Morteratsch (Swiss alps): I. Soil type chronosequence. *Arctic, Antarctic, and Alpine Research*, **38**, 499–509.

Ehrich, D., Gaudeul, M., Assefa, A., Koch, M.A., Mummenhoff, K., Nemomissa, S., Consortium, I., Brochmann, C., Af Ehrich, D., Gaudeul, M., Assefa, A., Koch, M.A., Mummenhoff, K., Nemomissa, S. & Intrabiodiversity Consortium (2007) Genetic consequences of Pleistocene range shifts: contrast between the Arctic, the Alps and the East African mountains. *Molecular Ecology*, **16**, 2542–2559.

Elias, S.A. (2001) Paleoecology and late Quaternary environments of the Colorado Rockies. *Structure and Function of an Alpine Ecosystem, Niwot Ridge, Colorado* (eds W.D. Bowman & T.R. Seastedt), pp. 285–303. Oxford University Press, Oxford.

Elias, S.A. (2003) Millennial and century climate changes in the Colorado alpine. *Climate Variability and Ecosystem Response at Long-term Ecological Research Sites* (eds D. Greenland, D.G. Goodin & R.C. Smith), pp. 370–383. Oxford University Press, Oxford.

Ellenberg, H. (1979) Man's influence on tropical mountain ecosystems in South-America—2nd Tansley lecture. *Journal of Ecology*, **67**, 401–416.

Ellenberg, H. (1996) *Vegetation Ecology of Central Europe*. Cambridge University Press, Cambridge.

Emanuelsson, U. (1984) Ecological effects of grazing and trampling on mountain vegetation in northern Sweden. PhD Thesis, University of Lund, Lund.

Embrechts, J. & Tavernier, R. (1987) Soil climate regimes of Cameroon and their relation to the distribution of major crops and grazing systems. *Proceedings of the Third International Forum on Soil Taxonomy and Agrotechnology Transfer. Benchmark Soils Project Technical Report No. 10* (eds H. Ikawa & G.Y. Tsuji), pp. 134–153. University of Hawaii, HITAR.

Erschbamer, B. & Mayer, R. (2006) *Grazing in the Biosphere Reserve Gurgler Kamm*. Institute of Botany, University of Innsbruck, Innsbruck.

Erschbamer, B. (1989) Vegetation on avalanche paths in the alps. *Vegetatio*, **80**, 139–146.

Erschbamer, B. (2007) Winners and losers of climate change in a central alpine glacier foreland. *Arctic, Antarctic and Alpine Research*, **39**, 237–244.

Erschbamer, B., Bitterlich, W. & Raffl, C. (1999) Die Vegetation als Indikator für die Bodenbildung im Gletschervorfeld des Rotmoosferners (Obergurgl, Ötztal, Nordtirol). *Bericht der naturwissenschaftlich-medizinischer Verein Innsbruck*, **86**, 107–122.

Erschbamer, B., Kneringer, E. & Schlag, R.N. (2001) Seed rain, soil seed bank, seedling recruitment, and survival of seedlings on a glacier foreland in the Central Alps. *Flora*, **196**, 304–312.

Erschbamer, B., Virtanen, R. & Nagy, L. (2003) The impacts of vertebrate grazers on vegetation on European high mountains. *Alpine Biodiversity in Europe* (eds L. Nagy, G. Grabherr, C. Körner & D.B.A. Thompson), pp. 377–396. Springer, Berlin, Heidelberg, New York.

Ertl, S. (2006) *Three years of seed rain studies in the Austrian Alps: seed rain studies in the alpine to nival vegetation zone at Mount Schrankogel, Stubaier Alps, Tyrol.* pp. 1–7. University of Vienna, Vienna.

FAO (2001) Global Forest Resources Assessment 2000. *Main Report.* FAO, Rome.

FAO-UNESCO (1992) UNEP Gridded FAO/UNESCO Soil Units. Digital Raster Data on a 2-minute Cartesian Orthonormal Geodetic (lat/long) 10800 x 5400 grid. *Global Ecosystems Database Version 2.0. One single-attribute spatial layer. 58,343,747 bytes in 6 files. [first published in 1984].* NOAA National Geophysical Data Center, Boulder, CO.

Fastie, C.L. (1995) Causes and ecosystem consequences of multiple pathways of primary succession at Glacier Bay, Alaska. *Ecology,* **76,** 1899–1916.

Fedorova, I.T. & Volkova, Y.A. (1990) World vegetation cover map. Global analog map of 1:80,000,000 scale in Russian polyconical projection. HTML publication on CD-ROM, unpublished.

Fedorova, I.T., Volkova, Y.A. & Varlyguin, D.L. (1993) Legend to the world vegetation cover map. Unpublished.

Ferreyra, M., Cingolani, A., Ezcurra, C. & Bran, D. (1998) High-Andean vegetation and environmental gradients in northwestern Patagonia, Argentina. *Journal of Vegetation Science,* **9,** 307–316.

Fiedler, K., Brehm, G., Hilt, N., Süssenbach, D. & Häuser, C.L. (2008) Variation of diversity patterns across moth families along a tropical elevational gradient. *Gradients in a Tropical Mountain Ecosystem* (eds E. Beck, J. Bendix, I. Kottke, F. Makeschin & R. Mosandl), pp. 167–179. Springer, Berlin, Heidelberg, New York.

Fillat, F., Garcia-Gonzalez, R., Gomez, D. & Reine, R. (2008) *Pastos del Pirineo.* Consejo Superior de Investigaciones Científicas, Madrid.

Fischer, M.A., Adler, W. & Oswald, K. (2005) *Exkursionsflora für Österreich, Liechtenstein und Südtirol, 2nd edition.* Land Oberösterreich, Biologiezentrum der OÖ Landesmuseen, Linz.

Fischer, M.A., Adler, W. & Oswald, K. (2005) *Exkursionsflora für Österreich, Liechtenstein und Südtirol, 2nd edition.* Land Oberösterreich, Biologiezentrum der OÖ Landesmuseen, Linz.

Fisk, M.C., Brooks, P.D. & Schmidt, S.K. (2001) Nitrogen cycling. *Structure and Function of an Alpine Ecosystem, Niwot Ridge, Colorado* (eds W.D. Bowman & T.R. Seastedt), pp. 237–253. Oxford University Press, Oxford.

Fisk, M.C., Schmidt, S.K. & Seastedt, T.R. (1998) Topographic patterns of above- and belowground production and nitrogen cycling in alpine tundra. *Ecology,* **79,** 2253–2266.

Fitzpatrick, E.A. (1997) Arctic soils and permafrost. *Ecology of Arctic Environments* (eds S.J. Woodin & M. Marquiss), pp. 1–39. Blackwell, Oxford.

Flanagan, P.W. & Veum, A.K. (1974) Relationships between respiration, weight loss, temperature and moisture in organic residues in tundra. *Soil Organisms and Decomposition in Tundra* (eds A.J. Holding, O.W. Heal, S.F. MacLean & P.W. Flanagan). Tundra Biome Steering Committee, Stockholm.

Fleming, F. (2000) *Killing dragons. The conquest of the Alps.* Granta, London.

Fogg, G.E. (1998) *The Biology of Polar Habitats.* Oxford University Press, Oxford.

Forbis, T.A. (2003) Seedling demography in an alpine ecosystem. *American Journal of Botany,* **90,** 1197–1206.

Foster, B.L. & Tilman, D. (2000) Dynamic and static views of succession: Testing the descriptive power of the chronosequence approach. *Plant Ecology,* **146,** 1–10.

Frangi, J.L., Barrera, M.D., Richter, L.L. & Lugo, A.E. (2005) Nutrient cycling in *Nothofagus pumilio* forests along an altitudinal gradient in Tierra del Fuego, Argentina. *Forest Ecology and Management,* **217,** 80–94.

Freeman, K.R., Pescador, M.Y., Reed, S.C., Costello, E.K., Robeson, M.S. & Schmidt, S.K. (2009) Soil $CO_2$ flux and photoautotrophic community composition in high-elevation, "barren" soil. Environmental Microbiology (In Press).

Fridley, J.D., Grime, J.P. & Bilton, M. (2007) Genetic identity of interspecific neighbours mediates plant responses to competition and environmental variation in a species-rich grassland. *Journal of Ecology,* **95,** 908–915.

Friedel, H. (1956) Die alpine Vegetation des obersten Mölltals (Hohe Tauern)—Erläuterungen zur Vegetationskarte der Umgebung der Pasterze (Großglockner). *Wissenschaftliche Alpenvereinshefte (Universitätsverlag Wagner, Innsbruck),* **16,** 1–153.

Frodin, D.G. (2001) *Guide to the Standard Floras of the World.* Cambridge University Press, Cambridge.

Füreder, L. (1999) High alpine streams: cold habitats for insect larvae. *Cold-Adapted Organisms. Ecology, Physiology, Enzymology and Molecular Biology* (eds R. Margesin & F. Schinner), pp. 181–196. Springer, Heidelberg Berlin New York.

Gade, D.W. (1999) *Nature and culture in the Andes.* The University of Wisconsin Press, Wisconsin.

Galen, C. & Stanton, M.L. (1999) Seedling establishment in alpine buttercups under experimental manipulations of growing-season length. *Ecology,* **80,** 2033–2044.

Galen, C., Shore, J.S. & Deyoe, H. (1991) Ecotypic divergence in alpine *Polemonium viscosum*—genetic structure, quantitative variation, and local adaptation. *Evolution,* **45,** 1218–1228.

Gamisans, J. (1999) *La végétation de la Corse.* Édisud, Aix-en-Provence.

Gamisans, J. (1977) *La végétation des montagnes corses.* Phytocoenologia **4,** 35-131.

Gansert, D. (2004) Treelines of the Japanese Alps—altitudinal distribution and species composition under contrasting winter climates. *Flora,* **199,** 143–156.

Garay, I., Sarmiento-Monasterio, L. & Monasterio, M. (1983) Le paramo désertique: éléments biogènes, peuplemenets des microarthropodes et startégies de survie de la végétation. *International Colloquium of Soil Biology* (eds P. Lebrun, H.M. Andre, A. de Medts, C. Gregoire-Wibo & G. Waughty), pp. 127–134. Ottignies, Louvain-la-Neuve.

García-González, R., Marinas, A. (2008) Bases ecológicas para la ordenación de territorios pastorales. Pastos del Pirineo (eds F. Fillat, R. Garcia-González, D. Gómez & R. Reiné), pp. 229–253. CSIC - Diputación de Huesca, Madrid.

Gauthier, P., Lumaret, R. & Bedecarrats, A. (1998) Genetic variation and gene flow in Alpine diploid and tetraploid populations of *Lotus* (*L. alpinus* (D.C.) Schleicher/*L. corniculatus* L.). I. Insights from morphological and allozyme markers. *Heredity,* **80,** 683–693.

Gerdol, R., Brancaleoni, L., Marchesini, R. & Bragazza, L. (2002) Nutrient and carbon relations in subalpine dwarf shrubs after neighbour removal or fertilization in northern Italy. *Oecologia,* **130,** 476–483.

Germino, M.J. & Smith, W.K. (2000) Differences in microsite, plant form, and low-temperature photoinhibition in alpine plants. *Artic, Antarctic, and Alpine Research,* **32,** 388–396.

Gerrard, A.J. (1990) *Mountain Environments.* Belhaven Press, London.

Ghalambor, C.K., Huey, R.B., Martin, P.R., Tewksbury, J.J. & Wang, G. (2006) Are mountain passes higher in the tropics? Janzen's hypothesis revisited. *Integrative and Comparative Biology,* **46,** 5–17.

Gibson, N. & Kirkpatrick, J.B. (1992) Dynamics of a Tasmanian cushion heath community. *Journal of Vegetation Science,* **3,** 647–654.

Gigon (1999) Positive Interaktionen in enem alpinen Blumenpolster. *Berichte der Reinhard Tüxen Gesellschaft,* **11,** 321–330.

Gómez, D. (2008) Métodos para el estudio de los pastos, su caracterización ecológica y valoración. Pastos del Pirineo Madrid (eds F. Fillat, R. Garcia-González, D. Gómez & R. Reiné, editors), pp. 75–109. CSIC – Diputación de Huesca, Madrid.

Gómez, D., Sesé, J.A. & Villar, L. (2003) The vegetation of the alpine zone in the Pyrenees. *Alpine Biodiversity in Europe* (eds L. Nagy, G. Grabherr, C. Körner & D.B.A. Thompson), pp. 85–92. Springer, Berlin, Heidelberg, New York.

Gonzalez, G. & Novoa, C. (1989) Partage de l'espace entre le lagopède Lagopus mutus pyrenaicus et la perdix grise Perdix perdix hispaniensis dans le massif du Carlit (Pyrénées Orientales) en fonction de l'altitude et de l'exposition. *Revue Ecologic (Terre Vie),* **44,** 347–360.

González, J.A. (1985) The water potential of some plants and the hydric stress in high mountain environment. *Lilloa,* **36,** 167–172.

Gorchakovsky, P.L. (1989) Horizontal and altitudinal differentiation of the vegetational cover of the Ural mountains. *Pirineos,* **133,** 33–54.

Gordon, J.E., Thompson, D.B.A., Haynes, V.M., Brazier, V. & Macdonald, R. (1998) Environmental sensitivity and conservation management in the Cairngorm Mountains, Scotland. *Ambio,* **27,** 335–344.

Gottfried, M., Pauli, H. & Grabherr, G. (1998) Prediction of vegetation patterns at the limits of plant life: a new view of the alpine-nival ecotone. *Arctic and Alpine Research,* **30,** 207–221.

Gottfried, M., Pauli, H., Reiter, K. & Grabherr, G. (2002) Potential effects of climate change on alpine and nival plants in the Alps. *Mountain Biodiversity—A Global Assessment* (eds C. Körner & E.M. Spehn), pp. 213–223. Parthenon Publishing, London, New York.

Grabherr, G. (1979) Variability and ecology of the alpine dwarf shrub community Loiseleurio-Cetrarietum. *Vegetatio,* **41,** 111–120.

Grabherr, G. (1989) On community structure in high alpine grasslands. *Vegetatio,* **83,** 223–227.

Grabherr, G. (1995) Alpine vegetation in a global perspective. *Vegetation Science in Forestry* (eds E.O. Box, R.K. Peet, T. Masuzawa, I. Yamada, K. Fujiwara & P.F. Maycock), pp. 441–451. Kluwer, Dordrecht.

Grabherr, G. (1997) The high-mountain ecosystems of the Alps. *Ecosystems of the World. 3. Polar and Alpine Tundra* (ed F.E. Wielgolaski), pp. 97–121. Elsevier, Amsterdam.

Grabherr, G. (1997) Vegetations- und Landschaftsgeschichte als Grundlage für Natur- und Landschaftsschutz. *Bericht der Reinhold-Tüxen-Gesellschaft,* **9,** 37–48.

Grabherr, G. (2003) Alpine vegetation dynamics and climate change—a synthesis of long-term studies and observations. *Alpine Biodiversity in Europe* (eds

L. Nagy, G. Grabherr, C. Körner & D.B.A. Thompson), pp. 399–409. Springer, Berlin, Heidelberg, New York.

Grabherr, G. (2005) Priorities for the conservation and management of the natural heritage in Europe's high mountains. *Mountains of Northern Europe. Conservation, Management, People and Nature* (eds D.B.A. Thompson, M.F. Price & C.A. Galbraith), pp. 371–376. TSO Scotland, Edinburgh.

Grabherr, G., Gottfried, M. & Pauli, H. (1994) Climate effects on mountain plants. *Nature*, **369**, 448.

Grabherr, G., Gottfried, M. & Pauli, H. (2000) Hochgebirge als "hot spots" der Biodiversität—dargestellt am Beispiel der Phytodiversität. *Bereichten der Reinhardt Tuexen Gesellschaft*, **12**, 101–112.

Grabherr, G., Gottfried, M., Gruber, A. & Pauli, H. (1995) Patterns and current changes in alpine plant diversity. *Arctic and Alpine Biodiversity. Patterns, Causes and Ecosystem Consequences* (eds F.S. Chapin III & C. Körner), pp. 167–181. Springer, Berlin.

Grabherr, G., Gurung, A.B., Dedieu, J.P., Haeberli, W., Hohenwallner, D., Lotter, A.F., Nagy, L., Pauli, H. & Psenner, R. (2005) Long-term environmental observations in mountain biosphere reserves: recommendations from the EU GLOCHAMORE project. *Mountain Research and Development*, **25**, 376–382.

Grabherr, G., Mähr, E. & Reisigl, H. (1978) Nettoprimärproduktion in einem Krummseggenrasen (Caricetum curvulae) der Ötztaler Alpen, Tirol. *Oeclogia Plantarum*, **13**, 227–251.

Grabherr, G., Mair, A. & Stimpfl, H. (1988) Vegetationsprozesse in alpinen Rasen und die Chance einer echten Renaturierung von Schipisten und anderer Erosionsflächen in alpinen Hochlagen. *Ingenieurbiologie—Erosionsbekämpfung im Hochgebirge*, **3**, 93–114.

Grabherr, G., Nagy, L. & Thompson, D.B.A. (2003) Overview: an outline of Europe's alpine areas. *Alpine Biodiversity in Europe* (eds L. Nagy, G. Grabherr, C. Körner & D.B.A. Thompson), pp. 3–12. Springer Verlag, Berlin, Heidelberg, New York.

Grace, J., Berninger, F. & Nagy, L. (2002) Impacts of climate change on the tree line. *Annals of Botany*, **90**, 537–544.

Gradstein, S.R. (1995) Diversity of hepaticae and Anthocerotae in montane forests of the tropical Andes. *Biodiversity and Conservation of Neotropical Montane Forests* (eds S.P. Churchill, H. Balsev, E. Forero & J.L. Luteyn), pp. 321–334. New York Botanical Garden, New York.

Graglia, E., Jonasson, S., Michelsen, A. & Schmidt, I.K. (1997) Effects of shading, nutrient application and warming on leaf growth and shoot densities of dwarf shrubs in two arctic-alpine plant communities. *Ecoscience*, **4**, 191–198.

Grau, H.R. & Veblen, T.T. (2000) Rainfall variability, fire and vegetation dynamics in neotropical montane ecosystems in north-western Argentina. *Journal of Biogeography*, **27**, 1107–1121.

Grau, H.R., Gasparri, N.I., Morales, M., Grau, A., Aráoz, E., Carilla, J. & Gutiérrez, J. (2007) Regeneración ambiental en el Noroeste argentino. Oportunidades para la conservación y restauración de ecosistemas. *Ciencia Hoy*, **17**, 42–56.

Green, K. & Osborne, W.S. (1994) *Wildlife of the Australian Snow-Country*. Reed, Sydney.

Greenland, D. & Losleben, M. (2001) Climate. *Structure and Function of an Alpine Ecosystem, Niwot Ridge, Colorado* (eds W.D. Bowman & T.R. Seastedt), pp. 15–31. Oxford University Press, Oxford.

Greenland, D. (1991) Surface-energy budgets over alpine tundra in summer, Niwot Ridge, Colorado Front Range. *Mountain Research and Development,* **11,** 339–351.

Greenland, D. (1993) Spatial energy budgets in alpine tundra. *Theoretical and Applied Climatology,* **46,** 229–239.

Greenland, D., Goodin, D.G. & Smith, R.C. (2003) *Climate Variability and Ecosystem Response at Long-term Ecological Research Sites.* Oxford University Press, Oxford.

Grêt-Regamey, A. & Kytzia, S. (2007) Integrating the valuation of ecosystem services into the Input–Output economics of an Alpine region. *Ecological Economics,* **63,** 786–798.

Grime, J.P., Hodgson, J.G. & Hunt, R. (2007) Comparative Plant Ecology. A Functional Approach to Common British Species Chapman & Hall.

Grime, P. (1979) *Plant Strategies and Vegetation Processes.* Wiley, Chichester.

Grishin, S.Y., Krestov, P. & Okitsu, S. (1996) The subalpine vegetation of Mt. Vysokaya, central Sikhote-Alin. *Plant Ecology,* **127,** 155–172.

Grubb, P.J. (1977) Control of forest growth and distribution on wet tropical mountains: with special reference to mineral nutrition. *Annual Review of Ecology and Systematics,* **8,** 83–107.

Gruber, F. (1980) Die Verstaubung der Hochgebirgsböden im Glocknergebiete. *Untersucchungen an alpinen Böden in den Hohen Tauern 1974-1978. Stoffdynamik und Wasserhaushalt* (ed H. Franz), pp. 69–91. Österreichen Akademie Wissentschaftliche Veröffentlichungen, Wien.

Gruber, S., Hoelzle, M. & Haeberli, W. (2004) Permafrost thaw and destabilization of Alpine rock walls in the hot summer of 2003. *Geophysical Research Letters,* **31,** L13504.

Guisan, A. & Zimmermann, N.E. (2000) Predictive habitat distribution models in ecology. *Ecological Modelling,* **135,** 147–186.

Guliashvili, V.Z., Mahatadze, L.B. & Prilipko, L.I. (1975) *Rastitelnost Kavkaza [Vegetation of the Caucasus].* Nauka, Moscow.

Hadley, K.S. (1987) Vascular alpine plant distributions within the central and southern Rocky Mountains, U.S.A. *Arctic and Alpine Research,* **19,** 242–251.

Hahn, S.C., Oberbauer, S.F., Gebauer, R., Grulke, N.E., Lange, O.L. & Tenhunen, J.D. (1996) Vegetation structure and aboveground carbon and nutrient pools in the Imnavait Creek watershed. *Landscape Function and Disturbance in Arctic Tundra* (eds J.F. Reynolds & J.D. Tenhunen), pp. 109–128. Springer, Berlin, Heidelberg, New York.

Halfpenny, J. & Heffernan, M. (1992) Nutrient input to an alpine tundra—an aeolian insect component. *Southwestern Naturalist,* **37,** 247–251.

Hall, G., Larsen, J., Eide, W., Peglar, S.M., John, H. & Birks, B. (2005) Holocene environmental history and climate of Ratasjoen, a low-alpine lake in south-central Norway. *Journal of Paleolimnology,* **33,** 129–153.

Hall, K. & Lamont, N. (2003) Zoogeomorphology in the Alpine: some observations on abiotic—biotic interactions. *Geomorphology,* **55,** 219–234.

Hall, K., Boelhouwers, J. & Driscoll, K. (1999) Animals as erosion agents in the alpine zone: some data and observations from Canada, Lesotho, and Tibet. 436–446. *Arctic, Antarctic, and Alpine Research,* **31,** 436–446.

Halliday, G., Kliim-Nielsen, L. & Smart, I.H.M. (1974) Studies on the flora of the North Blosseville Kyst and on the hot springs of Greenland. *Meddelelser om Grønland,* **199,** 1–48.

Halloy, S. & González, J.A. (1993) An inverse relation between frost survival and atmospheric-pressure. *Arctic and Alpine Research,* **25,** 117–123.

Halloy, S. (1991) Islands of life at 6000 m altitude: the environment of the highest autotrophic communities on earth (Socompa Volano, Andes). *Arctic and Alpine Research,* **23,** 247–262.

Halloy, S., Seimon, A., Yager, K. & Tupayachi, A. (2006) Multidimensional (climatic, biodiversity, socioeconomic, and agricultural) context of changes in land use in the Vilcanota watershed, Peru. *Land Use Change and Mountain Biodiversity* (eds E. Spehn, M. Lieberman & C. Körner), pp.319–333 CRC Press, Boca Raton

Halloy, S.R.P. & Mark, A.F. (1996) Comparative leaf morphology spectra of plant communities in New Zealand, the Andes and the European Alps. *Jounal of the Royal Society of New Zealand,* **26,** 41–78.

Halloy, S.R.P. & Mark, A.F. (2003) Climate-change effects on alpine plant biodiversity: a New Zealand perspective on quantifying the threat. *Arctic, Antarctic and Alpine Research,* **35,** 248–254.

Halloy, S.R.P. (2002) Variations in community structure and growth rates of high-Andean plants with climatic fluctuations. *Mountain Biodiversity. A Global Assessment* (eds C. Körner & E. Spehn), pp. 225–237. Parthenon, Boca Raton, London, New York.

Hamilton III, E.W., Giovannini, M.S., Moses, S.A., Coleman, J.S. & McNaughton, S.J. (1998) Biomass and mineral element responses of a Serengeti short-grass species to nitrogen supply and defoliation: compensation requires a critical [N]. *Oecologia,* **116,** 407–418.

Hamilton, L.S. (2002) Conserving mountain biodiversity in protected areas. *Mountain biodiversity. A Global Assessment* (eds C. Körner & E. Spehn), pp. 295–306. Parthenon, London.

Handa, I.T., Körner, C. & Hättenschwiler, S. (2005) A test of the treeline carbon limitation hypothesis by in situ $CO_2$ enrichment and defoliation. *Ecology,* **2005,** 1288–1300.

Hardin, G. (1998) The tragedy of the commons. *Debating the Earth: The Environmental Politics Reader* (eds J. Dryzek & D. Schlosberg). Oxford University Press, Oxford.

Haslett, J.R. (1988) Assessing the quality of alpine habitats—hoverflies (Diptera, Syrphidae) as bio-indicators of skiing pressure on alpine meadows in Austria. *Zoologische Anzeiger,* **220,** 179–184.

Haslett, J.R. (1996) *Functional ecology and conservation of invertebrates in mountain environments: insect communities associated with vegetation of grazed and abandoned alpine pastures.* Zoological Institute, University of Salzburg, Salzburg.

Hassan, R., Scholes, R. & Ash, N. (2005) *Ecosystems and human well-being. Volume one: current state and trends. Findings of the Condition and Trends Working Group of the Millennium Ecosystem Assessment.* Island Press, Washington.

Hauman, L. (1933) Esquisse de la vegetation des hautes altitudes sur le Ruwenzori. *Bulletin de l'Academie Royale Scientifique de la Belgique, Climat Sciences series* 5, **19,** 602–616.

Hayes, J.P. & O'Connor, C.S. (1999) Natural selection on thermogenic capacity of high-altitude deer mice. *Evolution,* **53,** 1280–1287.

Hedberg, I. & Hedberg, O. (1979) Tropical-alpine life-forms of vascular plants. *Oikos,* **33,** 297–307.

Hedberg, O. (1964) Features of afroalpine plant ecology. *Acta Phytogeographica Suecica,* **49,** 1–144.

Hedberg, O. (1969) Evolution and speciation in a tropical high mountain flora. *Botanical Journal of the Linnean Society,* **1,** 135–148.

Hedberg, O. (1970) Evolution of the Afroalpine flora. *Biotropica,* **2,** 16–23.

Hedberg, O. (1986) Origins of the Afroalpine flora. *High Altitude Tropical Biogeography* (eds F. Vuilleumier & M. Monasterio), pp. 443–468. Oxford University Press, Oxford.

Hedberg, O. (1992) Afroalpine vegetation compared with páramo: convergent adaptatios and divergent differentiation. *Páramo. An Andean Ecosystem Under Human Influence* (eds H. Balslev & J.L. Luteyn), pp. 15–29. Academic Press, London, San Diego, New York.

Hedberg, O. (1997) High-mountain areas of tropical Africa. *Polar and Alpine Tundra* (eds F.E. Wielgolaski & D.W. Goodall), pp. 185–197. Elsevier, Amsterdam.

Heer, C. & Körner, C. (2002) High elevation pioneer plants are sensitive to mineral nutrient addition. *Basic and Applied Ecology,* **3,** 39–47.

Hemp, A. (2005) Climate change-driven forest fires marginalize the impact of ice cap wasting on Kilimanjaro. *Global Change Biology,* **11,** 1013–1023.

Hemp, A. (2006) The impact of fire on diversity, structure, and composition of the vegetation on Mt. Kilimanjaro. *Land Use Change and Mountain Biodiversity* (eds E. Spehn, M. Lieberman & C. Körner), pp. 51–68. Taylor & Francis, CRC Press, Boca Raton, London, New York.

Henttonen, H. & Kaikusalo, A. (1993) Lemming movements. *The Biology of Lemmings* (eds N.C. Stenseth & R.A. Ims), pp. 157–186. Academic Press, London.

Henttonen, H., Kaikusalo, A., Tast, J. & Viitala, J. (1977) Interspecific competition between small rodents in subarctic and boreal ecosystems. *Oikos,* **29,** 581–590.

Hestmark, G., Skogesal, O. & Skullerud, O. (2007) Early recruitment equals long-term relative abundance in an alpine saxicolous lichen guild. *Mycologia,* **99,** 207–214.

Heywood, V.H. (1995) *Global Biodiversity Assessment.* Cambridge University Press, Cambridge.

Hickler, T., Vohland, K., Costa, L., Miller, P.A., Smith, B., Feehan, J., Kühn, I., Cramer, W., and M.T. Sykes (2009) Vegetation on the move – where do conservation strategies have to be re-defined? Atlas of Biodiversity *Risks – from Europe to the Globe, from Stories to Maps* (eds J. Settele et al.), in press. Pensoft, Sofia & Moscow.

Hiller, B. & Müterthies, A. (2005) Humus forms and reforestation of an abandoned pasture at the alpine timberline (Upper Engadine, Central Alps, Switzerland). *Mountain ecosystems. Studies in Treeline Ecology* (eds G. Broll & B. Keplin), pp. 203–218. Springer, Berlin.

Hiltbrunner, E., Schwikowski, M. & Körner, C. (2005) Inorganic nitrogen storage in alpine snow pack in the Central Alps (Switzerland). *Atmospheric Environment,* **39,** 2249–2259.

Hinzmann, L.D., Kane, D.L., Benson, C.S. & Everett, K.R. (1996) Energy balance and hydrological processes in an arctic watershed. *Landscape Function and Disturbance in Arctic Tundra* (eds J.F. Reynolds & J.D. Tehunen), pp. 131–154. Springer, Berlin, Heidelberg, New York.

Hnautiuk, R.J. (1978) The growth of tussock grasses on an equatorial high mountain and on two sub-antarctic islands. *Geoecological Relations Between the Southern Temperate Zone and the Tropical Mountains. Erdwissenschaftliche Forschung 11* (eds C. Troll & W. Lauer), pp. 159–188. Franz Steiner Verlag, Wiesbaden.

Hnautiuk, R.J.J., Smith, J.M.B. & McVean, D.N. (1976) *Mt. Wilhelm Studies. 2. The Climate of Mt. Wilhelm*. Australian National University Press, Canberra.

Hobbie, J.E., Bettez, N., Deegan, L.A., Laundre, J.A., MacIntyre, S., Oberbauer, S., O'Brien, W.J., Shaver, G. & Slavik, K. (2003) Climate forcing at the Arctic LTER site. *Climate Variability and Ecosystem Response at Long-Term Ecological Research Sites* (eds D. Greenland, D.G. Goodin & R.C. Smith), pp. 74–91. Oxford University Press, Oxford.

Hoch, G. & Körner, C. (2003) The carbon charging of pines at the climatic treeline: a global comparison. *Oecologia*, **135**, 10–21.

Hofgaard, A. (1997) Inter-relationships between treeline position, species diversity, land use and climate change in the central Scandes Mountains of Norway. *Global Ecology and Biogeography Letters*, **6**, 419–429.

Hofstede, R. (2001) El impacto de las actividades humanas en el páramo. *Los paramos del Ecuador. Particularidades, Problemas y Perspectivas* (eds V.P. Mena, G. Medina & R. Hofstede), pp. 161–185. Abya Yala, Quito.

Hofstede, R., Segarra, P. & Mena, V.P. (2003) *Los Páramos del Mundo. Proyeto Atlas Mundial de los Páramos* Peatland Initiative/NC-IUCN/EcoCiencia, Quito.

Hofstede, R.G.M. & Rossenaar, A. (1995) Biomass of grazed, burned and undisturbed páramo grasslands, Colombia. 2. Root mass and aboveground/belowground ratio. *Arctic and Alpine Research*, **27**, 13–18.

Hofstede, R.G.M. (1995) *Effects of Burning and Grazing on a Colombian Paramo Ecosystem*. PhD Thesis, University of Amsterdam, Amsterdam.

Hofstede, R.G.M., Mondragon, M.X. & Rocha, C.M. (1995) Biomass of grazed, burned, and undisturbed páramo grasslands, Colombia. 1. Aboveground vegetation. *Arctic and Alpine Research*, **27**, 1–12.

Holderegger, R., Herrmann, D., Poncet, B., Gugerli, F., Thuiller, W., Taberlet, P., Gielly, L., Rioux, D., Brodbeck, S., Aubert, S., Manel, S. (2008) Land ahead: using genome scans to identify molecular markers of adaptive relevance. *Plant Ecology & Diversity* **1**, 273–283.

Holderegger, R. (2006) The meaning of genetic diversity. *Landscape Ecology*, **21**, 797–807.

Holdrige, L.R. (1967) *Life Zone Ecology*. Tropical Science Center, San José.

Hollister, R.D., Webber, P.J. & Bay, C. (2005) Plant response to temperature in Northern Alaska: Implications for predicting vegetation change. *Ecology*, **86**, 1562–1570.

Holten, J.I. (2003) Altitude ranges and spatial patterns of alpine plants in Northern Europe. *Alpine Biodiversity in Europe* (eds L. Nagy, G. Grabherr, C. Korner & D.B.A. Thompson), pp. 173–184. Springer, Berlin, Heidelberg, New York.

Holtmeier, F.K. (1989) Ökologie und Geographie der oberen Waldgrenze. *Bericht der Reinhold-Tüxen-Gesellschaft*, **1**, 15–45.

Holzner, W,, Huebl, E. (1988) Vergleich zwischen Flora und Vegetation der sub-alpin – alpinen Styfe in der japonischen Alpen und in den Alpen Europas. *Veröffentlichungen des Geobotanischen Institutes der ETH, Stiftung Rübel, Zürich* **98**, 299–329.

Hope, G.S. & Peterson, J.A. (1975) Glaciation and vegetation in the high New Guinean mountains. *Quaternary Studies* (eds R.P. Suggate & M.M. Creswell), pp. 155–162. Royal Society of New Zealand, Wellington.

Hope, G.S. (1976) Vegetation. *The Equatorial Glaciers of New Guinea* (eds G.S. Hope, J.A. Peterson, I. Allison & U. Radok), pp. 112–172. A.A. Balkema, Rotterdam.

Hope, G.S. (1980) Historical influences on the New Guinea flora. *The Alpine Flora of New Guinea* (ed P. van Royen), pp. 223–248. J. Cramer, Vaduz.

Horsfield, D. & Thompson, D.B.A. (1996) The uplands: guidance on terminology regarding altitudinal zonation and related terms. *Information and Advisory Note No. 26*. Scottish Natural Heritage, Battleby.

Hörandl, E., Cosendai, A.-C., Temsch. E.M. (2008) Understanding the geographic distributions of apomictic plants: a case for a pluralistic approach. *Plant Ecology & Diversity* **1**:309–320.

Hubbell, S.P. (2001) *The Unified Neutral Theory of Biodiversity and Biogeography.* Princeton University Press, Princeton, Oxford.

Huggett, R.J. (2004) *Fundamentals of Biogeography.* Routledge, London, New York.

Hughes, C. & Eastwood, R. (2006) Island radiation on a continental scale: exceptional rates of plant diversification after uplift of the Andes. *Proceedings of the National Academy of Sciences,* **103,** 10334–10339.

Hülber, K., Gottfried, M., Pauli, H., Reiter, K., Winkler, M. & Grabherr, G. (2006) Phenological responses of snowbed species to snow removal dates in the Central Alps: Implications for climate warming. *Arctic Antarctic and Alpine Research,* **38,** 99–103.

Humboldt, F.H.A. (1817) *De Distributiona Geographica Plantarum Secundum Coeli Temepriem et Altitudinem Montium Prolegomena.* Lutetiae Parisiorum, Paris.

Huntington, H.P. (2000) Ecology. *Arctic Flora and Fauna. Status and Conservation* (ed H.P. Huntington), pp. 17–49. CAFF—Edita, Helsinki.

Hutchinson, V.H. (1982) *Physiological Ecology of the Telmatobiid Frogs of Lake Titicaca.* pp. 357–361. National Geographic Society, Washington, D.C.

Ikeda, H. & Setoguchi, H. (2007) Phylogeography and refugia of the Japanese endemic alpine plant, *Phyllodoce nipponica* Makino (Ericaceae). *Journal of Biogeography,* **34,** 169–176.

Ingraham, N.L. & Mark, A.F. (2000) Isotopic assessment of the hydrologic importance of fog deposition on tall snow tussock grass on southern New Zealand uplands. *Austral Ecology,* **25,** 402–408.

IPCC (2007) Climate Change 2007: The Physical Science Basis. *Contribution of Working Group I to the Fourth Assessment Report of the Intergovernmental Panel on Climate Change* (eds S. Solomon, D. Qin, M. Manning, Z. Chen, M. Marquis, K.B. Averyt, M. Tignor & H.L. Miller). Cambridge University Press, Cambridge, New York.

Ishizuka, K. (1974) Mountain vegetation. *The Flora and Vegetation of Japan* (ed M. Numata), pp. 173–210. Kodansha, Tokyo.

Islebe, G.A. & Velázquez, A. (1994) A comparison of methods for predicting vegetation type affinity among mountain ranges in Megamexico: a phytogeographical scenario. *Vegetatio,* **156,** 3–18.

Jackson, S.T. & Overpeck, J.T. (2000) Responses of plant populations and communities to environmental changes of the late Quaternary. *Paleobiology*, **26**, 194–220.

Jaeger, C.H., Monson, R.K., Fisk, M.C. & Schmidt, S.K. (1999) Seasonal partitioning of nitrogen by plants and soil microorganisms in an alpine ecosystem. *Ecology*, **80**, 1883–1891.

Jeník, J. & Stursa, J. (2003) Vegetation of the Giant Mountains, Central Europe. *Alpine Biodiversity in Europe* (eds L. Nagy, G. Grabherr, C. Koerner & D.B.A. Thompson). Springer Verlag, Berlin Heidelberg New York, 47–51.

Jeník, J. (1998) Biodiversity of the Hercynian mountains in central Europe. *Pirineos*, **151/152**, 83–99.

Jerosch, M. (1903) *Geschichte und Herkunft der schweizerische Alpenflora.* Engelmann, Leipzig.

Jing, S., Solhoy, T., Wang, H.F., Vollan, T.I. & Xu, R.M. (2005) Differences in soil arthropod communities along a high altitude gradient at Shergyla Mountain, Tibet, China. *Arctic, Antarctic and Alpine Research*, **37**, 261–266.

Jochimsen, M. (1970) Die Vegetationsentwicklung auf Moränenböden in Abhängigkeit von einigen Umweltfaktoren. *Alpin-Biologische Studien, geleitet von H. Janetschek und H. Pitschmann, Veröffentlichung der Universität Innsbruck*, **46**, 1–22.

Johns, R.J., Edwards, P.J., Utteridge, T.M.A. & Hopkins, H.C.F. (2006) *A Guide to the Alpine and Subalpine Flora of Mount Jaya.* Royal Botanic Garden, Kew, London.

Johnston, S.W. (2001) The influence of aeolian dust deposits on alpine soils in south-eastern Australia. *Australian Journal of Soil Research*, **39**, 81–88.

Jomelli, V. & Francou, B. (2000) Comparing the characteristics of rockfall talus and snow avalanche landforms in an Alpine environment using a new methodological approach: Massif des Ecrins, French Alps. *Geomorphology*, **35**, 181–192.

Jonasson, S., Chapin III, F.S. & Shaver, G.R. (2001) Biogeochemistry in the Arctic: patterns, processes and controls. *Global Biogeochemical Cycles in the Climate System* (eds E.-D. Schultze, M. Heimann, A. Harrison, E. Holland, J. Lloyd, I.C. Prentice & D. Schimel), pp. 139–150. Academic Press, San Diego.

Jones, H.G. & Pomeroy, J.W. (2001) The ecology of snow and snow-covered systems: summary and relevance to Wolf Creek, Yukon. *Snow Ecology: An Interdisciplinary Examination of Snow-covered Ecosystems* (eds H.G. Jones, J.W. Pomeroy, D.A. Walker & R.W. Hoham), pp. 1–14. Cambridge University Press, Cambridge.

Jørgensen, P.M., Ulloa Ulloa, C., Madsen, J.E. & Valencia, R. (1995) A floristic analysis of the high Andes of Ecuador. *Biodiversity and Conservation of Neotropical Montane Forests* (eds S.P. Churchill, H. Balsev, E. Forero & J.L. Luteyn), pp. 221–237. New York Botanical Garden, New York.

Joussaume, S. & Guiot, J. (1999) Reconstruire les chauds et froids de l'Europe. *La Recherche*, **321**, 54–59.

Juday, G.P., Barber, V., Rupp, S., Zasada, J. & Wilmking, M. (2003) A 200-year perspective of climate variability and response of white spruce in interior Alaska. *Climate Variability and Ecosystem Response at Long-term Ecological Research Sites* (eds D. Greenland, D.G. Goodin & R.C. Smith), pp. 226–250. Oxford University Press, Oxford.

Kadereit, J.W., Griebeler, E.M. & Comes, H.P. (2004) Quaternary diversification in European alpine plants: pattern and process. *Philosophical Transactions of the Royal Society of London Series B—Biological Sciences*, **359**, 265–274.

Kallio, P. & Lehtonen, J. (1975) On the ecocatastrophe of birch forest caused by *Oporinia autumnata* (Bkh.) and the problem of reforestation. *Fennoscandian Tundra Ecosystems 2* (ed F.E. Wielgolaski), pp. 175–180. Springer, Berlin, Heidelberg, New York.

Karlsson, J., Jonsson, A. & Jansson, M. (2005) Productivity of high-latitude lakes: climate effect inferred from altitude gradient. *Global Change Biology,* **11,** 710–715.

Karlsson, P.S. & Jacobson, A. (2001) Onset of reproduction in *Rhododendron lapponicum* shoots: the effect of shoot size, age, and nutrient status at two subarctic sites. *Oikos,* **94,** 279–286.

Karlsson, P.S. & Nordell, K.O. (1996) Effects of soil temperature on the nitrogen economy and growth of mountain birch seedlings near its presumed low temperature distribution limit. *Écoscience,* **3,** 183–189.

Karlsson, P.S. & Weih, M. (2001) Soil Temperatures near the Distribution Limit of the Mountain Birch (*Betula pubescens* ssp. *czerepanovii*): Implications for Seedling Nitrogen Economy and Survival. *Arctic, Antarctic and Alpine Research,* **33,** 88–92.

Kaufmann, R. (2001) Invertebrate succession on an alpine glacier foreland. *Ecology,* **82,** 2261–2278.

Kebede, M., Ehrich, D., Taberlet, P., Nemomissa, S., Brochmann, C., Af Kebede, M., Ehrich, D., Taberlet, P., Nemomissa, S. & Brochmann, C. (2007) Phylogeography and conservation genetics of a giant lobelia (*Lobelia giberroa*) in Ethiopian and Tropical East African mountains. *Molecular Ecology,* **16,** 1233–1243.

Kerner von Marilaun, A. (1863) *Das Pflanzenleben der Donaulander.* Wagner, Innsbruck.

Kessler, M. (1998) Forgotten forests of the high Andes. *Plant Talk,* **15,** 25–28.

Kessler, M. (2002) The elevational gradient of Andean plant endemism: varying influences of taxon-specific traits and topography at different taxonomic levels. *Journal of Biogeography,* **29,** 1159–1165.

Kiffney, P.M., Clements, W.H. & Cady, T.A. (1997) Influence of ultraviolet radiation on the colonization dynamics of a Rocky Mountain stream benthic community. *Journal of the North American Benthological Society,* **16,** 520–530.

Killick, D.J.B. (1978) The Afro-alpine region. *Biogeography and Ecology of Southern Africa* (ed M.J.A. Werger), pp. 515–560. Dr. W. Junk bv Publishers, The Hague.

Killick, D.J.B. (1997) Alpine tundra of southern Africa. *Ecosystems of the World. 3. Polar and Alpine Tundra* (eds F.E. Wielgolaski & D.W. Goodall), pp. 199–209. Elsevier, Amsterdam.

Kimball, K.D. & Weihbrauch, D.M. (2000) Alpine vegetation communities and alpine—treeline ecotone boundary in New England as biomonitors for climate change. *USDA Forest Service Proceedings RMPS-P-15,* **3,** 93–101.

Kimball, S., Wilson, P. & Crowther, J. (2004) Local ecology and geographic ranges of plants in the Bishop Creek watershed of the eastern Sierra Nevada, California, USA. *Journal of Biogeography,* **31,** 1637–1657.

Kirkbride, M.P. & Dugmore, A.J. (2005) Late Holocene solifluction history reconstructed using tephrochronology. *Cryospheric Systems: Glaciers and Permafrost. Geological Society of London Special Publication 242* (eds C. Harris & J.B. Murton), pp. 145–155. Geological Society of London, London.

Kirkpatrick, J.B. & Bridle, K.L. (1999) Environment and floristics of ten Australian alpine vegetation formations *Australian Journal of Botany,* **47,** 1–21.

Kirkpatrick, J.B. (2002) Factors influencing the spatial restriction of vascular plant species in the alpine achipelagos of Australia. *Mountain Biodiversity. A Global Assessment* (eds C. Körner & E. Spehn), pp. 155–164. Parthenon, London.

Kissling-Naf, I., Volken, T. & Bisang, K. (2002) Common property and natural resources in the Alps: the decay of management structures? *Forest Policy and Economics*, **4**, 135–147.

Klanderud, K. & Totland, R. (2005) The relative importance of neighbours and abiotic environmental conditions for population dynamic parameters of two alpine plant species. *Journal of Ecology*, **93**, 493–501.

Klanderud, K. (2005) Climate change effects on species interactions in an alpine plant community. *Journal of Ecology*, **93**, 127–137.

Klatzel, F. (2001) Natural History Handbook for the Wild Side of the Everest: the Eastern Himalaya and Makalu-Barun Area. The Mountain Institute, Kathmandu.

Kleier, C. & Lambrinos, J.G. (2005) The importance of nurse associations for three tropical alpine life forms. *Arctic Antarctic and Alpine Research*, **37**, 331–336.

Klimeš, L. (2003) Life-forms and clonality of vascular plants along an altitudinal gradient in E Ladakh (NW Himalayas). Basic and Applied Ecology 4, 317–328.

Klötzli, F. (2004) Kilimanjaro—Berg der Pracht, Berg der Götter. *Gebirge der Erde. Landschaft, Klima, Pflanzenwelt* (eds C.A. Burga, F. Klötzli & G. Grabherr), pp. 380–390. Ulmer, Stuttgart.

KlugPümpel, B. & Krampitz, C. (1996) Conservation in alpine ecosystems: the plant cover of ski runs reflects natural as well as anthropogenic environmental factors. *Bodenkultur*, **47**, 97–117.

Komárková, V. & Webber, P.J. (1978) An alpine vegetation map of Niwot Ridge, Colorado. *Arctic and Alpine Research*, **10**, 1–29.

Komárková, V. (1979) *Alpine Vegetation Types of the Indian Peaks area, Front Range, Colorado Rocky Mountains*. J. Cramer, Vaduz.

Köppen, W. (1931) *Grundriss der Klimakunde*. Walter de Gruyter, Berlin, Leipzig.

Körner, C. & Paulsen, J. (2004) A world-wide study of high altitude treeline temperatures. *Journal of Biogeography*, **31**, 713–732.

Körner, C. (2000) Why are there global gradients in species richness? Mountains might hold the answer. *Trends in Ecology & Evolution*, **15**, 513–514.

Körner, C. (2002) Mountain biodiversity, its causes and function: an overview. *Mountain Biodiversity: A Global Assessment* (eds C. Körner & E. Spehn), pp. 3–20. Pergamon, London.

Körner, C. (2003) *Alpine Plant Life. Functional Plant Ecology of High Mountain Ecosystems*. Springer, Berlin.

Körner, C. (2003) *Alpine Plant Life. Functional Plant Ecology of High Mountain Ecosystems*. Springer, Berlin.

Körner, C. (2007a) Climatic treelines: conventions, global patterns, causes. *Erdkunde*, **61**, 316–324.

Körner, C. (2007b) The use of 'altitude' in ecological research. *Trends in Ecology and Evolution*, **22**, 569–574.

Körner, C. (2008) Winter crops: growth at low temperature may hold the answer for alpine treeline formation. *Plant Ecology & Diversity*, **1**, 3–11.

Körner, C., Paulsen, J. & Pelaez-Riedl, S. (2003) A bioclimatic characterisation of Europe's alpine areas. *Alpine Biodiversity in Europe* (eds L. Nagy, G. Grabherr, C. Körner & D.B.A. Thonpson), pp. 13–28. Springer, Berlin, Heidelberg, New York.

Körner, C., Wieser, G. & Guggenberger, H. (1980) Der Wasseraushalt ines alpinen Rasens in den Zentralalpen. *Veroeffentlichungen des Osterreichen MaB-Hochgebirgsprogramma Hohe Tauern. band 3. Untersuchungen an alpinen Boden in den Hohen Tauern 1974–1978 Stoffdynamik und Wasseraushalt,* pp. 243–264. Universistatsverlag Wagner, Innsbruck.

Kottek, M., Gieser, J., Beck, C., Rudolf, B. & Rubel, F. (2006) World map of the Köppen-Geiger climate classification. *Meteorologische Zeitschrift,* **15,** 259–263.

Krautzer, B., Wittmann, H., Peratoner, G., Graiss, W., Partl, C., Parente, G., Venerus, S., Rixen, C. & Streit, M. (2006) *Site-Specific High Zone Restoration in the Alpine Region.* HBLFA, Ramberg-Gumpenstein, Irding.

Kueppers, L.M., Southon, J., Baer, P. & Harte, J. (2004) Dead wood biomass and turnover time, measured by radiocarbon, along a subalpine elevation gradient. *Oecologia,* **141,** 641–651.

Kullman, L. (1998) Tree-limits and montane forests in the Swedish Scandes: Sensitive biomonitors of climate change and variability. *Ambio,* **27,** 312–321.

Kullman, L. (2002) Rapid recent range-margin rise of tree and shrub species in the Swedish Scandes. *Journal of Ecology,* **90,** 68–77.

Kullman, L. (2007) Tree line population monitoring of *Pinus sylvestris* in the Swedish Scandes, 1973–2005: implications for tree line theory and climate change ecology. *Journal of Ecology,* **95,** 41–52.

Kumar, L., Skidmore, A.K. & Knowles, E. (1997) Modelling topographic variation in solar radiation in a GIS environment. *International Journal of Geographical Information Science,* **11,** 475–497.

Kürschner, H., Herzschuh, U. & Wagner, D. (2005) Phytosociological studies in the north-eastern Tibetan Plateau (NW China)—A first contribution to the subalpine scrub and alpine meadow vegetation *Botanische Jahrbücher,* **126,** 273–315.

Kurz, P. (1987) *Erosionsschutz der natürlichen Vegetation und künstlicher Begrünungen in alpinen Hochlagen dargestellt an Beispiel "Pfannhorn" in Südtirol.* Ph.D., University of Innsbruck, Innsbruck.

Lacoul, P. & Freedman, B. (2006) Relationships between aquatic plants and environmental factors along a steep Himalayan altitudinal gradient. *Aquatic Botany,* **84,** 3–16.

Landolt, E. (1954) Die Artengruppe des Ranunculus montanus Willd. in den Alpen und im Jura (zytologisch-systematische Untersuchungen). *Berichten der Schweitzerischen Botanischen Gesellschaft,* **64,** 9–83.

Landolt, E. (1956) Die Artengruppe des Ranunculus montanus Willd. in den Pyrenäen und anderen europäischen Gebirgen westlich der Alpen. *Berichten der Schweizerischen Botanischen Gesellschaft,* **66,** 92–117.

Larcher, W. (1977) Ergebnisse des IBP-Projekts "Zwergstrauchheide Patscherkofel". Sitzungsberichten der Österreichischen Akademie der Wissenschaften. *Math.-Naturwiss. Kl., Abt. I,* **186,** 302–371.

Larcher, W. (2003) *Physiological Plant Ecology.* Springer, Berlin, Heidelberg, New York.

Lauer, W. (1978) Timberline studies in central Mexico. *Arctic and Alpine Research,* **10,** 383–396.

Lavoie, C., Elias, S.A. & Payette, S. (1997) Holocene fossil beetles from a treeline peatland in subarctic Quebec. *Canadian Journal of Zoology,* **75,** 227–236.

Lawrence, D.B., Schoenike, R.E., Quispel, A. & Bond, G. (1967) The Role of Dryas Drummondii in Vegetation Development Following Ice Recession at Glacier

Bay, Alaska, with Special Reference to Its Nitrogen Fixation by Root Nodules. *The Journal of Ecology*, **55**, 793–813.

Le Roy Ladurie, E. (2001) *Histoire humaine et comparée du climat*. Fayard, Paris.

Leemans, R. & Cramer, W. (1992) IIASA Database for Mean Monthly Values of Temperature, Precipitation, and Cloudiness on a Global Terrestrial Grid. Digital Raster Data on a 30 minute Cartesian Orthonormal Geodetic (lat/long) 360 × 720 grid. *Global Ecosystems Database Version 2.0.*, pp. Thirty-six independent single-attribute spatial layers. 15,588,254 bytes in 77 files. [first published in 1991] NOAA National Geophysical Data Center, Boulder, CO.

Leemans, R. (1992) Global Holdridge Life Zone Classifications. Digital Raster Data on a 0.5-degree Cartesian Orthonormal Geodetic (lat/long) 360 × 720 grid. *Global Ecosystems Database Version 2.0*. Boulder, CO: NOAA National Geophysical Data Center. Two independent single-attribute spatial layers. 537,430 bytes in eight files. [first published in 1989]. NOAA National Geophysical Data Center, Boulder, CO.

Legg, C.J. & Nagy, L. (2006) Why most conservation monitoring is, but need not be, a waste of time. *Journal of Environmental Management*, **78**, 194–199.

Lehtonen, J. & Heikkinen, R.K. (1995) On the recovery of mountain birch after Epirrita damage in Finnish Lapland, with a particular emphasis on reindeer grazing. *Ecoscience*, **2**, 349–356.

Lesica, P. & Steele, B.M. (1996) A method for monitoring long-term population trends: an example using rare arctic-alpine plants. *Ecological Applications*, **6**, 879–887.

Letouzey, R. (1968) *Étude Phytogéographique du Cameroun*. Editions Paul Lechevalier, Paris.

Leuschner, C. (2000) Are high elevations in tropical mountains arid environments for plants? *Ecology*, **81**, 1425–1436.

Levin, D.A. (2002) *The Role of Chromosomal Change in Plant Evolution*. Oxford University Press, Oxford.

Lewis Smith, R.I. (1993) The role of bryophyte propagule banks: case-study of an Antarctic fellfield soil. *Primary Succession on Land* (eds J. Miles & D.W.H. Walton), pp. 55–78. Blackwell, Oxford.

Lewis, N.K. (1998) Landslide-driven distribution of aspen and steppe on Kathul mountain, Alaska. *Journal of Arid Environments*, **38**, 421–435.

Ley, R.E., Williams, M.W. & Schmidt, S.K. (2004) Microbial population dynamics in an extreme environment: Controlling factors in talus soils at 3750m in the Colorado Rocky Mountains. *Biogeochemistry* **68**: 313–335.

Libermann-Cruz, M. (1986) Microclima y distribucion de Polylepis tarapacana en el Parque Nacional Sajama, Bolivia. *Documents Phytosociologiques N. S.*, **X**, 235–272.

Lieth, H., Berlekamp, J., Fuest, S. & Riediger, S. (1999) *Climate Diagram World Atlas*. Backhuys, Leiden.

Lindgren, Å. (2007) *Effects of Herbivory on Arctic and Alpine Vegetation*. PhD Thesis, University of Stockholm, Stockholm.

Lipson, D.A., Schmidt, S.K. & Monson, R.K. (1999) Links between microbial population dynamics and nitrogen availability in an alpine ecosystem. *Ecology*, **80**, 1623–1631.

List, R.J. (1971) *Smithsonian Meteorological Tables*. Smithsonian Institution Press, Washingto DC.

Litaor, M.I., Mancinelli, R. & Halfpenny, J.C. (1996) The influence of pocket gophers on the status of nutrients in Alpine soils. *Geoderma*, **70**, 37–48.

Lods-Crozet, B., Castella, E., Cambin, D., Ilg, C., Knispel, S. & Mayor-Simeant, H. (2001) Macroinvertebrate community structure in relation to environmental variables in a Swiss glacial stream. *Freshwater Biology*, **46**, 1641–1661.

Loffler, J. (2000) High mountain ecosystems and landscape degradation in northern Norway. *Mountain Research and Development*, **20**, 356–363.

Loison, A., Toigo, C. & Gaillard, J.-M. (2003) Large herbovores in European alpine ecosystems: current status and challenges for the future. *Alpine Biodiversity in Europe* (eds L. Nagy, G. Grabherr, C. Körner & D.B.A. Thompson), pp. 351–375. Springer, Berlin, Heidelberg, New York.

Lokupitiya, E., Stanton, N.L., Seville, R.S. & Snider, J.R. (2000) Effects of increased nitrogen deposition on soil nematodes in alpine tundra soils. *Pedobiologia*, **44**, 591–608.

Longton, R.E. & Holdgate, M.W. (1967) Temperature relationships of Antarctic vegetation. *Philosophical Transactions of the Royal Society of London. Series B, Biological Sciences*, **252**, 237–250.

Longton, R.E. (1997) The role of bryophytes and lichens in polar ecosystems. *Ecology of Arctic Environments* (eds S.J. Woodin & M. Marquiss), pp. 69–96. Blackwell, Oxford.

Loope, L.L. & Medeiros, A.C. (1994) Biotic interactions in Hawaiian high elevation ecosystems. *Tropical Alpine Environments. Plant Form and Function* (eds P.W. Rundel, A.P. Smith & F.C. Meinzer), pp. 337–354. Cambridge University Press, Cambridge.

Luckman, B. & Kavanagh, T. (2000) Impact of climate fluctuations on mountain environments in the Canadian Rockies. *Ambio*, **29**, 371–380.

Luebert, F. & Gajardo, R. (2005) Vegetación alto andina de Parinacota (norte del Chile) y una sinopsis de la vegetación de la Puna meridional. *Phytocoenologia*, **35**, 79–128.

Luterbacher, J., Dietrich, D., Xoplaki, E., Grosjean, M. & Wanner, H. (2004) European seasonal and annual temperature variability, trends, and extremes since 1500. *Science*, **303**, 1499–1503.

Luteyn, J.L., Cleef, A.M., Rangel, O.C. & Balslev, H. (1992) Plant diversity in páramo: towards a checklist of páramo plants and a generic flora. *Páramo. An Andean Ecosystem under Human Influence* (eds J.L. Luteyn, & H. Balslev), pp. 71–84. Academic Press, London San Diego New York.

Lynch, C.D. & Watson, J.P. (1992) The distribution and ecology of *Otomys slogetti* (Mammalia: Rodentia) with notes on its taxonomy. *Navos. Nas. Mus. Bloemfontein*, **8**, 141–158.

Maire, R. (1924) *Etudes sur la Vegetation et la Flore du Grand Atlas et du Moyen Atlas Marocains*. L'Institut Scientifique Cherifien, Rabat.

Makarov, M.I., Glaser, B., Zech, W., Malysheva, T.I., Bulatnikova, I.V. & Volkov, A.V. (2003) Nitrogen dynamics in alpine ecosystems of the northern Caucasus. *Plant Soil*, **256**, 389–402.

Makarov, M.I., Malysheva, T.I., Haumaier, L., Alt, H.G. & Zech, W. (1997) The forms of phosphorus in humic and fulvic acids of a toposequence of alpine soils in the northern Caucasus. *Geoderma*, **80**, 61–73.

Mangen, J.M. (1993) *Ecology and Vegetation of Mt Trikora, New Guinea (Irian Jaya, Indonesia)*. Ministère des Affaires Culturelles, Luxembourg.

Mark, A.F. & Adams, N.M. (1995) *New Zealand Alpine Plants.* Godwit Press, Auckland.

Mark, A.F. & Dickinson, K.J.M. (1997) New Zealand alpine ecosystems. *Ecosystems of the World 3. Polar and Alpine Tundra* (ed F.E. Wielgolaski), pp. 311–345. Elsevier, Amsterdam.

Mark, A.F. & Dickinson, K.J.M. (2003) Temporal responses over 30 years to removal of grazing from a mid-altitude snow tussock grassland reserve, Lammerlaw Ecological Region, New Zealand. *New Zealand Journal of Botany,* **41,** 655–667.

Mark, A.F. (1994) Patterned ground activity in a southern New Zealand high-alpine cushionfield. *Arctic and Alpine Research,* **26,** 270–280.

Mark, A.F. (2005) Tempo and mode of vegetation dynamics over 50 years in a New Zealand alpine cushion/tussock community. *Journal of Vegetation Science,* **16,** 227–236.

Mark, A.F., Dickinson, K.J.M., Allen, J., Smith, R. & West, C.J. (2001) Vegetation patterns, plant distribution and life forms across the alpine zone in southern Tierra del Fuego, Argentina. *Austral Ecology,* **26,** 423–440.

Márquez, E.J., Rada, F. & Fariñas, M.R. (2006) Freezing tolerance in grasses along an altitudinal gradient in the Venezuelan Andes. *Oeclogia,* **150,** 393–397.

Martinez-Rica, J.P. (2003) Diversity of alpine vertebrates in the Pyrenees and Sierra Nevada, Spain. *Alpine Biodiversity in Europe* (eds L. Nagy, G. Grabherr, C. Körner & D.B.A. Thompson), pp. 367–375. Springer, Berlin, Heidelberg, New York.

Matthews, J.A. (1992) The ecology of recently-deglaciated terrain. A geoecological approach to glacier forelands and primary succession. Cambridge University Press, Cambridge.

Matzinger, N., Andretta, M., Van Gorsel, E., Vogt, R., Ohmura, A. & Rotach, M.W. (2003) Surface radiation budget in an Alpine valley. *Quarterly Journal of the Royal Meteorological Society,* **129,** 877–895.

Maudet, C., Miller, C., Bassano, B., Breitenmoser-Wursten, C., Gauthier, D., Obexer-Ruff, G., Michallet, J., Taberlet, P. & Luikart, G. (2002) Microsatellite DNA and recent statistical methods in wildlife conservation management: applications in Alpine ibex [Capra ibex (ibex)]. *Molecular Ecology,* **11,** 421–436.

May, D.E. & Webber, P.J. (1982) Spatial and temporal variation of of vegetation and its productivity on Niwot Ridge, Colorado. *Ecological Studies in the Colorado Alpine: A Festschrift for John W. Marr* (ed J.C. Halfpenny), pp. 35–62. University of Colorado, Institute of Arctic and Alpine Research, Boulder.

Mayerhofer, P., de Vries, B., den Elzen, M., van Vuuren, D., Onigkeit, J., Posch, M. & Guardans, R. (2002) Long-term, consistent scenarios of emissions, deposition, and climate change in Europe. *Environmental Science & Policy,* **5,** 273–305.

Mazzoleni, S., Rego, F., Giannino, F. & Legg, C.J. (2004) Ecosystem modelling: vegetation and disturbance. *Environmental Modelling: Finding Simplicity in Complexity* (eds J. Wainwright & M. Mulligan), pp. 171–186. Wiley & Sons, Ltd, London.

McDonald, K.A. & Brown, J.H. (1992) Using montane mammals to model extinctions due to global change. *Conservation Biology,* **6,** 409–415.

McDougall, K.L., Morgan, J.W., Walsh, N.G. & Williams, R.J. (2005) Plant invasions in treeless vegetation of the Australian Alps. *Perspectives in Plant Ecology Evolution and Systematics,* **7,** 159–171.

McGhee, R. (2006) *The Last Imaginary Place. A Human History of the Arctic World*. Oxford University Press, Oxford.

McGlone, M.S., Duncan, R.P. & Heenan, P.B. (2001) Endemism, species selection and the origin and distribution of the vascular plant flora of New Zealand. *Journal of Biogeography,* **28,** 199–216.

McGlone, M.S., Moar, N.T. & Meurk, C.D. (1997) Growth and vegetation history of alpine mires on the Old Man Range, Central Otago, New Zealand. *Arctic and Alpine Research,* **29,** 32–44.

McIntire, E.J.B. & Hik, D.S. (2002) Grazing history versus current grazing: leaf demography and compensatory growth of three alpine plants in response to a native herbivore (*Ochotona collaris*). *Journal of Ecology,* **90,** 348–359.

McVean, D. & Ratcliffe, D. (1962) *The Plant Communities of the Scottish Highlands*. HMSO, Edinburgh.

McVean, D.N. (1969) Alpine vegetation of the central Snowy Mountains of New South Wales. *Jounal of Ecology,* **57,** 67–86.

McVean, D.N. (1974) Mountain climates of the southwest Pacific. *Altitudinal Zonation in Malesia. Transactions of the Third Aberdeen-Hull Symposium on Malesian Ecology* (ed J.R. Flenley), pp. 48–57. University of Hull, Hull.

Melcher, P.J., Goldstein, G., Meinzer, F.C., Minyard, B., Giambelluca, T.W. & Loope, L.L. (1994) Determinants of thermal balance in the Hawaiian giant rosette plant, *Argyroxiphium sandwicense. Oecologia,* **98,** 412–418.

Mena, V.P. & Medina, G. (2001) La biodiversidad de los páramos en el Ecuador. *Los Páramos del Ecuador* (eds V.P. Mena, G. Medina & R. Hofstede), pp. 27–52. Abya Yala, Quito.

Mena, V.P., Medina, G. & Hofstede, R. (2001) *Los Páramos del Ecuador. Particularidades, Problemas y Perspectivas*. Abya Yala/Proyecto Paramo, Quito.

Merzlyakova, I. (2002) The mountains of Central Asia and Kazakhstan. *The Physical Geography of Northern Eurasia* (ed M. Shahgedanova), pp. 377–402. Oxford University Press, Oxford.

Meshinev, T., Apostolova, I. & Koleva, E. (2000) Influence of warming on timberline rising: a case study on Pinus peuce Griseb. in Bulgaria. *Phytocoenologia,* **30,** 431–438.

Meusel, H., Jager, E. & Weinert, E. (1965) *Vergleichende Chorologie der zentraleuropaischen Flora*. Gustav Fisher Verlag, Jena.

Meyer, E. & Thaler, K. (1995) Animal diversity at high altitudes in the Austrian Central Alps. *Arctic and Alpine Biodiversity* (eds F.S. Chapin, III & C. Körner), pp. 97–108. Springer, Berlin, Heidelberg, New York.

Michelsen, A., Quarmby, C., Sleep, D. & Jonasson, S. (1998) Vascular plant 15N natural abundance in heath and forest tundra ecosystems is closely correlated with presence and type of mycorrhizal fungi in roots. *Oecologia,* **115,** 406–418.

Miehe, G. (1997) Alpine vegetation types of the central Himalaya. *Ecosystems of the World. Polar and Alpine Tundra* (ed F.E. Wielgolaski), pp. 161–184. Elsevier, Amsterdam.

Miehe, G. (2004) Himalaya. *Gebirge der Erde. Landschaft, Klima, Pflanzenwelt* (eds C.A. Burga, F. Kloetli & G. Grabherr), pp. 325–348. Ulmer, Stuttgart.

Miehe, S. & Miehe, G. (1994) *Ericaceous Forests and Heathlands in the Bale Mountains of South Ethiopia. Ecology and Man's Impact*. Traute Warnke Verlag, Reinbek bei Hamburg.

Mikhailov, Y.E. & Olschwang, V.N. (2003) High altitude invertebrate diversity in the Ural Mountains. *Alpine Biodiversity in Europe* (eds L. Nagy, G. Grabherr, C. Körner & D.B.A. Thompson), pp. 259–280. Springer, Berlin, Heidelberg, New York.

Miles, J. & Walton, D.W.H. (1993) *Primary succession on land.* Blackwell, Oxford.

Miller, G.R. & Cummins, R.P. (2003) Soil seed banks of woodland, heathland, grassland, mire and montane communities, Cairngorm Mountains, Scotland. *Plant Ecology,* **168,** 255–266.

Miller, G.R., Geddes, C. & Mardon, D.K. (1999) Response of the alpine gentian *Gentiana nivalis* L. to protection from grazing by sheep. *Biological Conservation,* **87,** 311–318.

Milner, A.M., Brittain, J.E., Castella, E. & Petts, G.E. (2001) Trends of macroinvertebrate community structure in glacier-fed rivers in relation to environmental conditions: a synthesis. *Freshwater Biology,* **46,** 1833–1847.

Mizuno, K. (1998) Succession processes of alpine vegetation in response to glacial fluctuations of Tyndall Glacier, Mt. Kenya, Kenya. *Arctic and Alpine Research,* **30,** 340–348.

Mizuno, K. (2005) Glacial fluctuation and vegetation succession on Tyndall Glacier, Mt Kenya. *Mountain Research and Development,* **25,** 68–75.

Moiseev, P.A. & Shiyatov, S.G. (2003) Vegetation dynamics at the treeline ecotone in the Ural highlands, Russia. *Alpine Biodiversity in Europe* (eds L. Nagy, G. Grabherr, C. Körner & D.B.A. Thompson), pp. 423–435. Springer, Berlin, Heidelberg, New York.

Molau, U. & Shaver, G.R. (1998) Controls on seed production and seed germinability in *Eriophorum vaginatum. Global Change Biology,* **3 (Suppl.1),** 80–88.

Molau, U. (2003) Overview: patterns in diversity. *Alpine Biodiversity in Europe* (eds L. Nagy, G. Grabherr, C. Koerner & D.B.A. Thompson), pp. 125–132. Springer, Berlin, Heidelberg, New York.

Molau, U., Nordenhall, U. & Eriksen, B. (2005) Onset of flowering and climate variability in an alpine landscape: a 10-year study from Swedish Lapland. *American Journal of Botany,* **92,** 422–431.

Mols, T., Paal, J. & Fremstad, E. (2000) Response of Norwegian alpine communities to nitrogen. *Nordic Journal of Botany,* **20,** 705–712.

Monasterio, M. & Reyes, S. (1980) Diversidad ambiental y variacion de la vegetation en los paramos de los Andes venezoelanos. *Estudios Ecologicos en los Páramos Andinos* (ed M. Monasterio), pp. 47–91. Universidad de los Andes, Merida.

Monasterio, M. (1980) *Estudios Ecologicos en los Páramos Andinos.* Universidad de los Andes, Merida.

Monasterio, M. (1980) Las formaciones vegetales de los paramos de Venezuela. *Estudios Ecologicos en los Páramos Andinos* (ed M. Monasterio), pp. 93–158. Universidad de los Andes, Merida.

Monasterio, M. (1980) Los páramos andinos como region natural. Caracteristicas biogeograficas generales y afinidades con otras regiones andinas. *Estudios Ecologicos en los Páramos Andinos* (ed M. Monasterio), pp. 15–27. Universidad de los Andes, Merida.

Monasterio, M. (1980) Poblamiento humano y uso de la tierra en los altos Andes de Venezuela. *Estudios Ecologicos en los Páramos Andinos* (ed M. Monasterio), pp. 170–198. Universidad de los Andes, Merida.

Mong, C.E. (2006) Establishment of *Pinus wallichiana* on a Himalayan glacier foreland: Stochastic distribution or safe sites? *Arctic, Antarctic and Alpine Research*, **38**, 584–592.

Monsoon, R.K., Rosenstiel, T.N., Forbis, T.A., Lipson, D.A. & Jaeger, C.H. (2006) Nitrogen and carbon storage in alpine plants. *Integrative and Comparative Biology*, **46**, 35–48.

Montilla, M., Herrera, R.A. & Monasterio, M. (1992) Micorrizas vesiculo-arbusculares en parcelas que se encuentran en sucesion—regeracion en los Andes tropicales. *Suelo y Planta*, **2**, 59–70.

Moreira-Muñoz, A. & Muñoz-Schick, M. (2007) Classification, diversity, and distribution of Chilean Asteraceae: implications for biogeography and conservation. *Diversity and Distributions*, **13**, 818–828.

Morris, W.F. & Doak, D.F. (1998) Life history of the long-lived gynodioecious cushion plant *Silene acaulis* (Caryophyllaceae), inferred from size-based population projection matrices. *American Journal of Botany*, **85**, 784–793.

Morton, J.K. (1972) Phytogeography of the West African Mountains. *Taxonomy, Phytogeography and Evolution* (ed D.H. Valentine), pp. 221–236. Academic Press, London, New York.

Moser, D., Dullinger, S., Englisch, T., Niklfeld, H., Plutzar, C., Sauberer, N., Zechmeister, H.G. & Grabherr, G. (2005) Environmental determinants of vascular plant species richness in the Austrian Alps. *Journal of Biogeography*, **32**, 1117–1127.

Mueller-Dombois, D. & Frosberg, F.R. (1998) *Vegetation of the Tropical Pacific Islands*. Springer, Berlin, Heidelberg, New York.

Mullen, R.B. & Schmidt, S.K. (1993) Mycorrhizal infection, phosphorus uptake, and phenology in *Ranunculus adoneus*: Implications for the functioning of mycorrhizae in alpine systems. *Oecologia*, **94**, 229–234.

Mullen, R.B., Schmidt, S.K. & Jaeger, C.H. (1998) Nutrient uptake during snowmelt by the snow buttercup, *Ranunculus adoneus*. *Arctic and Alpine Research*, **30**, 121–125.

Nagy, L. & Proctor, J. (1996) The demography of *Lychnis alpina* L. on the Meikle Kilrannoch ultramafic site. *Botanical Journal of Scotland*, **48**, 155–166.

Nagy, L. & Proctor, J. (1997) Plant growth and reproduction on a toxic Alpine ultramafic soil: adaptation to nutrient limitation. *New Phytologist*, **137**, 267–274.

Nagy, L. (2003) The high mountain vegetation of Scotland. *Alpine Biodiversity in Europe* (eds L. Nagy, G. Grabherr, C. Körner & D.B.A. Thompson), pp. 39–46. Berlin, Heidelberg, New York.

Nagy, L. (2006) European high mountain (alpine) vegetation and its suitability for indicating climate change impacts. *Biology and Environment*, **106B**, 335–341.

Nagy, L., Grabherr, G., Körner, C. & Thompson, D.B.A. (2003) *Alpine biodiversity in Europe*. Springer, Berlin.

Nagy, L., Ohlemüller, R., Pauli, H., Gottfried, M., Grabherr, G. (2009a) Climate change impact on the future extent of the alpine climate zone. *Atlas of Biodiversity Risks. From Europe to the Globe and from Stories to Maps*. (eds J. Settele et al.), p. 22. Pensoft Publishers, Sofia.

Nagy, L., Gottfried, M., Grabherr, G., Pauli, H. (2009b) The potential interactions of climate, land use change, and nitrogen deposition on shaping alpine biodiversity by 2100. *Atlas of Biodiversity Risks. From Europe to the Globe and from Stories to Maps* (eds Settele J et al.), in press. Pensoft Publishers, Sofia.

Nagy, L., Dendoncker, N., Butler, A., Rounsevell, M., Gottfried, M., Grabherr, G., Pauli, H. (2009c) Land use change impacts on alpine biodiversity in Europe: past, present and future. *Atlas of Biodiversity Risks. From Europe to the Globe and from Stories to Maps* (eds Settele J et al.), in press. Pensoft Publishers, Sofia.

Nakhoutsrichvili, G. & Ozenda, P. (1998) Aspects géobotaniques de la haute montagne dans le Caucase essai de comparison avec les Alpes. *Écologie*, **29**, 139–144.

Nakhutsrishvili, G. (2003) High mountain vegetation of the Caucasus region. *Alpine Biodiversity in Europe* (eds L. Nagy, G. Grabherr, C. Körner & D.B.A. Thompson), pp. 93–103. Springer, Berlin, Heidelberg, New York.

Nara, K., Nakaya, H. & Hogetsu, T. (2003) Ectomycorrhizal sporocarp succession and production during early primary succession on Mount Fuji. *New Phytologist*, **158**, 193–206.

Navarro, G. & Maldonado, M. (2002) *Geografía Ecológica de Bolivia. Vegetación y Ambientes Acuaticos*. Fundación Simón I. Patiño, Cochabamba.

Navas, C.A. (2002) Herpetological diversity along Andean elevational gradients: links with physiological ecology and evolutionary physiology. *Comparative Biochemistry and Physiolgy—Part A: Molecular & Integrative Physiology*, **133**, 469–485.

Navas, C.A. (2006) Patterns of distribution of anurans in high Andean tropical elevations: Insights from integrating biogeography and evolutionary physiology. *Integrative and Comparative Biology*, **46**, 82–91.

Negi, G.C.S., Rikhari, H.C. & Singh, S.P. (1993) Plant regrowth following selective horse and sheep grazing and clipping in an Indian central Himalayan alpine meadow. *Arctic and Alpine Research*, **25**, 211–215.

Nemergut, D., Anderson, S., Cleveland, C., Martin, A., Miller, A., Seimon, A. & Schmidt, S. (2007) Microbial community succession in an unvegetated, recently deglaciated soil. *Microbial Ecology*, **53**, 110–122.

Nilsson, C. & Gotmark, F. (1992) Protected areas in Sweden—is natural variety adequately represented. *Conservation Biology*, **6**, 232–242.

Ninot, J.M., Batllori, E., Carillo, E., Ferré, A. & Gutiérrez, E. (2008) Vegetation structure from subalpine forest to alpine pastures in the Catalan Pyrenees. *Plant Ecology & Diversity*, **1**, 47–57.

Novis, P.M. (2002) Ecology of the snow alga *Chlainomonas kolii* (Chlamydomonadales, Chlorophyta) in New Zealand. *Phycologia*, **41**, 280–292.

Nullet, D. & Juvik, J.O. (1994) Generalized mountain evaporation profiles in the tropics and subtropics. *Singapore Journal of Tropical Geography*, **15**, 17–24.

Nunes, P.A.L.D. & van den Bergh, J.C.J.M. (2001) Economic valuation of biodiversity: sense or nonsense? *Ecological Economics*, **39**, 203–222.

Nüsser, M. (2000) Change and persistence: Contemporary landscape transformation in the Nanga Parbat Region, Northern Pakistan. *Mountain Research and Development*, **20**, 348–355.

Nylehn, J. & Totland, O. (1999) Effects of temperature and natural disturbance on growth, reproduction, and population density in the alpine annual hemiparasite Euphrasia frigida. *Arctic, Antarctic and Alpine Research*, **31**, 259–263.

Odland, A. & Birks, H.J.B. (1999) The altitudinal gradient of vascular plant richness in Aurland, western Norway. *Ecography*, **22**, 548–566.

Ohsawa, M. (1990) An interpretation of latitudinal patterns of forest limits in South and East Asian mountains. *Journal of Ecology*, **78**, 326–339.

Ohsawa, M. (1995) Latitudinal comparison of altitudinal changes in forest structure, leaf type, and species richness in humid monsoon Asia. *Vegetatio,* **121,** 3–10.

Ohtonen, R., Fritze, H., Pennanen, T., Jumpponen, A. & Trappe, J.M. (1999) Ecosystem properties and microbial community changes in primary succession on a glacier forefront. *Oecologia,* **119,** 239–246.

Okitsu, S. & Ito, K. (1984) Vegetation dynamics of the Siberian dwarf pine (*Pinus pumila* Regel) in the Taisetsu mountain range, Hokkaido, Japan. *Vegetatio,* **58,** 105–113.

Oliphant, A.J., Spronken-Smith, R.A., Sturman, A.P. & Owens, I.F. (2003) Spatial variability of surface radiation fluxes in mountainous terrain. *Jounal of Applied Meteorology,* **42,** 113–128.

Ollier, C.D. (1986) The origin of alpine landforms in Australasia. *Flora and Fauna of Alpine Australasia* (ed B.A. Barlow), pp. 3–26. CSIRO, Canberra.

Olofsson, J., Moen, J. & Oksanen, L. (1999) On the balance between positive and negative plant interactions in harsh environments. *Oikos,* **86,** 539–543.

Olsen, C.S. & Larsen, H.O. (2003) Alpine medicinal plant trade and Himalayan mountain livelihood strategies. *Geography Journal,* **169 PN 3,** 243–254.

Olsson, E.G.A., Austrheim, G. & Grenne, S.N. (2000) Landscape change patterns in mountains, land use and environmental diversity, Mid-Norway 1960–1993. *Landscape Ecology,* **15,** 155–170.

Omelon, C.R., Pollard, W.H. & Ferris, F.G. (2006) Environmental controls on microbial colonization of high Arctic cryptoendolithic habitats. *Polar Biology,* **30,** 19–29.

Onipchenko, V.G. (2005) *Alpine Ecosystems in the Northwest Caucasus.* Kluwer, Dordrecht.

Onipchenko, V.G., Semenova, G.V. & van der Maarel, E. (1998) Population strategies in severe environments: Alpine plants in the northwestern Caucasus. *Journal of Vegetation Science,* **9,** 27–40.

Ortiz, D. (2003) Ecuador. *Los Paramos del Mundo. Proyecto Atlas Mundial de los Paramos* (eds R. Hofstede, P. Segarra & P.M. Vasconez). Global Peatland Initiative/NC-IUCN/EcoCiencia, Quito.

Owens, P.N. & Slaymaker, O. (1997) Contemporary and post-glacial aeolian deposition in the Coast Mountains of Britsh Columbia, Canada. *Geografiska Annaler Series a-Physical Geography,* **79,** 267–276.

Ozenda, P. & Borel, J.L. (2003zx) The alpine vegetation of the Alps. *Alpine Biodiversity in Europe* (eds L. Nagy, G. Grabherr, C. Koerner & D.B.A. Thompson), pp. 53–64. Springer, Berlin, Heidelber, New York.

Ozenda, P. & Borel, J.L. (2006) La vegetation des Alpes occidentales. Un sommet de la biodiversite. *Braun-Blanquetia,* **41,** 1–45.

Ozenda, P. (1985) *La végétation de la Chaine Alpine dans l'ensemble montagnard Européen.* Masson, Paris.

Ozenda, P. (1994) *La Végétation du Continent Européen.* Delachaux et Niestlé, Lausanne.

Ozenda, P. (2002) *Perspectives pour une Geobiologie des Montagnes.* Presses Polytechniques et Universitaires Romandes, Lausanne.

Påhlsson, L. (1994) *Vegetationstyper i Norden.* Nordiska Ministerrådet, Copenhagen.

Pallaruelo, S. (1994) *Cuadernos de la Trashumancia N° 6: Pirineo Aragonés.* Icona, Madrid.

Palmer, S.C.F., Hester, A.J., Elston, D.A., Gordon, I.J. & Hartley, S.E. (2003) The perils of having tasty neighbors: grazing impacts of large herbivores at vegetation boundaries. *Ecology,* **84,** 2877–2890.

Pansu, M., Sarmiento, L., Metselaar Etselaar, K., Herve, D. & Bottner, P. (2007) Modelling the transformations and sequestration of soil organic matter in two contrating ecosystems of the Andes. *European Journal of Soil Science,* **58,** 775–785.

Parmuzin, Y.P. (1979) Rastitelnost [Vegetation]. *Tundrolesiye SSSR [Forest-tundra of the USSR]* (ed M.A. Platonov), pp. 83–91. Mysl Publishers, Moscow.

Paul, A.L., Schuerger, A.C., Popp, M.P., Richards, J.T., Manak, M.S. & Ferl, R.J. (2004) Hypobaric biology: Arabidopsis gene expression at low atmospheric pressure. *Plant Physiology,* **134,** 215–223.

Pauli, H., Gottfried, M. & Grabherr, G. (1999) Vascular plant distribution patterns at the low-temperature limits of plant life—the alpine-nival ecotone of Mount Schrankogel (Tyrol, Austria). *Phytocoenologia,* **29,** 297–325.

Pauli, H., Gottfried, M. & Grabherr, G. (2003) The Piz Linard (3411m), the Grisons, Switzerland—Europe's oldest mountain vegetation study site. *Alpine Biodiversity in Europe* (eds L. Nagy, G. Grabherr, C. Körner & D.B.A. Thompson), pp. 443–448. Springer, Berlin, Heidelberg, New York.

Pauli, H., Gottfried, M., Dirnböck, T., Dullinger, S. & Grabherr, G. (2003) Assessing the long-term dynamics of endemic plants at summit habitats. *Alpine Biodiversity in Europe* (eds L. Nagy, G. Grabherr, C. Körner & D.B.A. Thompson), pp. 195–207. Springer, Berlin, Heidelberg, New York.

Pauli, H., Gottfried, M., Hohenwallner, D., Reiter, K., Casale, R. & Grabherr, G. (2004) *The GLORIA Field Manual—Multi-Summit Approach.* European Commission DG Research, EUR 21213, Office for Official Publications of the European Communities, European Commission, Luxembourg.

Pauli, H., Gottfried, M., Nagy, L., Barancok, P., Bayfield, N., Borel, J.-L., Coldea, G., Erschbamer, B., Hohenwallner, D., Kazakis, G., Klettner, C., Michelsen, O., Moiseev, D., Molau, U., Molero Mesa, J., Nakhutsrishvili, G., Reiter, K., Rossi, G., Stanisci, A., Villar, L., Vittoz, P. & Grabherr, G. (2009). Aspect preferences of vascular plants on European alpine summits. *Plant Ecology & Diversity,* in review.

Pauli, H., Gottfried, M., Reiter, K., Klettner, C. & Grabherr, G. (2007) Signals of range expansions and contractions of vascular plants in the high Alps: observations (1994–2004) at the GLORIA master site Schrankogel, Tyrol, Austria. *Global Change Biology,* **13,** 147–156.

Payette, S., Eronen, M. & Jasinski, J.J.P. (2002) The circumboreal tundra-taiga interface: late Pleistocene and Holocene changes. *Ambio,* **Special Report Number XX,** 15–22.

Pedrotti, F. & Gafta, D. (2003) The high mountain flora and vegetation of the Apennines and the Italian Alps. *Alpine Biodiversity in Europe* (eds L. Nagy, G. Grabherr, C. Koerner & D.B.A. Thompson), pp. 73–84. Springer, Berlin, Heidelberg, New York.

Pendry, C.A., Dick, J., Pullan, M.R., Miller, A.G., Neale, S., Nees, S. & Watson, M.F. (2007) In search of a functional flora—towards a greater integration of ecology and taxonomy. *Plant Ecology,* **192,** 161–168.

Perez, Q.V. (1983) *Geografia de Chile. Tomo III. Biogeografia.* Instituto Geografico Militar de Chile, Santiago de Chile.

Pickering, C.M. & Buckley, R.C. (2003) Swarming to the summit—managing tourists at Mt Kosciuszko, Australia. *Mountain Research and Development,* **23,** 230–233.

Pielou, E.C. (1994) *A Naturalist's Guide to the Arctic.* Cambridge University Press, Cambridge.

Podwojewski, P., Poulenard, J., Zambrana, T. & Hofstede, R. (2002) Overgrazing effects on vegetation cover and properties of volcanic ash soil in the paramo of Llangahua and La Esperanza (Tungurahua, Ecuador). *Soil Use and Management,* **18,** 45–55.

Posch, M. (2002) Impacts of climate change on critical loads and their exceedances in Europe. *Environmental Science & Policy,* **5,** 307–317.

Poulenard, J., Podwojewski, P. & Herbillon, A.J. (2003) Characteristics of non-allophanic Andisols with hydric properties from the Ecuadorian páramos. *Geoderma,* **117,** 267–281.

Poulenard, J., Podwojewski, P., Janeau, J.L. & Collinet, J. (2001) Runoff and soil erosion under rainfall simulation of Andisols from the Ecuadorian Paramo: effect of tillage and burning. *Catena,* **45,** 185–207.

Pounds, J.A., Fogden, M.P.L. & Campbell, J.H. (1999) Biological response to climate change on a tropical mountain. *Nature,* **398,** 311–315.

Preston, C.D., Roy, D.B. & Hill, M.O. (1997) The phytogeography of Scotland. *Botanical Journal of Scotland* **49,** 191–204.

Preston, D., Fairbairn, J., Paniagua, N., Maas, G., Yevara, M. & Beck, S. (2003) Grazing and Environmental Change on the Tarija Altiplano, Bolivia. *Mountain Research and Development,* **23,** 141–148.

Price, J.P. (2004) Floristic biogeography of the Hawaiian Islands: influence of area, environment and palaeogeography. *Jounal of Biogeography,* **31,** 487–500.

Proctor, J., Edwards, I.D., Payton, R.W. & Nagy, L. (2007) Zonation of forest vegetation and soils of Mount Cameroon, West Africa. *Plant Ecology,* **192,** 251–269.

Pümpel, B. (1977) Bestandstruktur, Phytomassevorrat und Produktion verschiedener Pflanzengesellschaften im Glockenergebiet. *Veröffentlichungen des Österreichischen MaB-Hochgebirgsprogramms Hohe Tauern. Band 1. Alpine Grasheide Hohe Tauern Ergebnisse der Projektstudie 1976* (ed A. Cernusca), pp. 83–101. Universitätsverlag Wagner, Innsbruck.

Pyankov, V.I., Kondratchuk, A.V. & Shipley, W. (1999) Leaf structure and speciÆc leaf mass: the alpine desert plants of the Eastern Pamirs, Tadjikistan. *New Phytologist,* **143,** 131–142.

Qian, H., White, P.S., Klinka, K. & Chourmouzis, C. (1999) Phytogeographical and community similarities of alpine tundras of Changbaishan Summit, China, and Indian Peaks, USA. *Journal of Vegetation Science,* **10,** 869–882.

Quézel, P. (1957) *Peuplement Végétal des Hautes Montagnes de L'Afrique du Nord.* Editions Paul Lechevalier, Paris.

Quézel, P. (1965) La végétation du Sahara du Tchad á Mauritanie. Fisher Verlag, Stuttgart.

Quézel, P. (1967) La végétation des hauts sommets du Pinde et de l'Olympe de Thessalie. *Vegetatio,* **14,** 127–228.

Rada, F., Goldstein, G., Azocar, A. & Torres, F. (1987) Supercooling along an altitudinal gradient in *Espeletia schultzii,* a caulescent giant rosette species. *Journal of Experimental Botany,* **38,** 491–497.

Raffl, C., Mallaun, M., Mayer, R. & Erschbamer, B. (2006) Vegetation succession pattern and diversity changes in a glacier valley, Central Alps, Austria. *Arctic Antarctic and Alpine Research*, **38**, 421–428.

Ram, J., Singh, J.S. & Singh, S.P. (1989) Plant biomass, species diversity and net primary production in a central Himalayan high altitude grassland. *Jounal of Ecology*, **77**, 456–468.

Ramirez, O., Arana, M., Bazán, E., Ramirez, A. & Cano, A. (2007) Assemblages of bird and mammal communities in two major ecological units of the Andean highland plateau of southern Peru. *Ecologia Aplicada*, **6**, 139–148.

Ramsay, P.M. & Oxley, E.R.B. (2001) An assessment of aboveground net primary productivity in Andean grasslands of central Ecuador. *Mountain Research and Development*, **21**, 161–167.

Rangel-Ch, J.O. (2000) La region paramuna y franja aledana en Colombia. *Colombia. Diversidad Biotica III. La Region de Vida Paramuna* (ed J.O. Rangel-Ch), pp. 1–22. Colombia Universidad Nacional de Colombia, Bogota.

Raunkiaer, C. (1934) *The Life Forms of Plants*. Oxford University Press, Oxford.

Raven, P.H. (1973) Evolution of Subalpine and Alpine Plant Groups in New Zealand. *New Zealand Journal of Botany*, **11**, 177–200.

Rawat, G. & Adhikari, B. (2005) Floristics and distribution of plant communities across moisture and topographic gradients in Tso Kar Basin, Changthang Plateau, Eastern Ladakh. *Arctic, Antarctic, and Alpine Research*, **37**, 539–544.

Rawat, G.S. (1987) Floristic structure of snowline vegetation in central Himalaya, India. *Arctic and Alpine Research*, **19**, 195–201.

Read, D.J. & Haselwandter, K. (1981) Observations on the mycorrhizal status of some alpine plant communities. *New Phytologist*, **88**, 341–352.

Rebertus, A.J., Kitzberger, T., Veblen, T.T. & Roovers, L.M. (1997) Blowdown history and landscape patterns in the Andes of Tierra del Fuego, Argentina. *Ecology*, **78**, 678–692.

Rehder, H. (1994) Soil nutrient dynamics in East African alpine ecosystems. *Tropical Alpine Environments* (eds P.W. Rundel, A.P. Smith & F.C. Meinzer), pp. 223–228. Cambridge University Press, Cambridge.

Reichel, J.D. (1986) Habitat use by alpine mammals in the Pacific Northwest, U.S.A. *Arctic and Alpine Research*, **18**, 111–118.

Reimer, H. (1970) Vertically Differentiated Water Balance in Tropical High Mountains—with Special Reference to the Sierra Nevada de Santa Marta, Colombia. International Association of Scientific Hydrology. *Symposium on World Water Balance*, pp 262–273. Publication No 93.

Reisch, C., Poschlod, P. & Wingender, R. (2003) Genetic differentiation among populations of *Sesleria albicans* Kit. ex Schultes (Poaceae) from ecologically different habitats in central Europe. *Heredity*, **91**, 519–527.

Reisigl, H. & Keller, R. (1994) *Alpenpflanzen in Lebensraum*. Gustav Fisher Verlag, Stuttgart, Jena, New York.

Renvoize, S.A. (1998) *Gramineas de Bolivia*. Royal Botanic Gardens, Kew, London.

Revyakina, N. & Revyakin, V. (2004) Altai. *Gebirge der Erde. Landschaft, Klima, Pflanzenwelt* (eds C.A. Burga, F. Kloetzli & G. Grabherr), pp. 144–150. Ulmer, Stuttgart.

Reynolds, D.N. (1984) Alpine annual plants: phenology, photosynthesis, and growth of three Rocky Mountain species. *Ecology*, **65**, 759–766.

Reynolds, J.F. & Tenhunen, J.D. (1996) *Landscape Function and Disturbance in Arctic Tundra.* Springer, Berlin, Heidelberg, New York.

Riberon, A., Miaud, C., Grossenbacher, K. & Taberlet, P. (2001) Phylogeography of the Alpine salamander, *Salamandra atra* (Salamandridae) and the influence of the Pleistocene climatic oscillations on population divergence. *Molecular Ecolology,* **10,** 2555–2560.

Ricankova, V., Fric, Z., Chlachula, J., Stastna, P., Faltynkova, A. & Zemek, F. (2006) Habitat requirements of the long-tailed ground squirrel (*Spermophilus undulatus*) in the southern Altai. *Journal of Zoology,* **270,** 1–8.

Richter, M. (2001) *Vegetationszonen der Erde.* Klett-Perthes, Gotha, Stuttgart.

Rihm, B. & Kurz, D. (2001) Deposition and critical loads of nitrogen in Switzerland. *Water, Air and Soil Pollution,* **130,** 1223–1228.

Rikhari, H.C., Negi, G.C.S., Pant, G.B., Rana, B.S. & Singh, S.P. (1992) Phytomass and primary productivity in several communities of a Central Himalayan Alpine Meadow, India. *Arctic and Alpine Research,* **24,** 344–351.

Rikhari, H.C., Negi, G.C.S., Ram, J. & Singh, S.P. (1993) Human-induced secondary succession in an alpine meadow of central Himalaya, India. *Arctic and Alpine Research,* **25,** 8–14.

Rivas-Martinez, S. (1981) Les étages bioclimatiques de la végétation de la péninsule ibérique. *Anales del Jardin Botanico de Madrid,* **37,** 251–268.

Rivas-Martinez, S. (1995) Classificacion bioclimatica de la terra. *Folia Botánica Matritensis,* **16,** 1–29.

Rivas-Martinez, S. (1999) *Worldwide Bioclimatic Classification System.* http://www.ucm.es/info/cif/book/bioc/tabla.htm.

Rixen, C., Haeberli, W. & Stoeckli, V. (2004) Ground temperatures under ski pistes with artificial and natural snow. *Arctic, Antarctic and Alpine Research,* **30,** 419–427.

Rousseaux, M.C., Julkunen-Tiitto, R., Searles, P.S., Scopel, A.L., Aphalo, P.J. & Ballare, C.L. (2004) Solar UV-B radiation affects leaf quality and insect herbivory in the southern beech tree Nothofagus antarctica. *Oecologia,* **138,** 505–512.

Rudloff, W. (1981) *World Climates with Tables of Climatic Data and Practical Suggestions.* Wissentsschaftliche Verlagsgesellschaft, Stuttgart.

Ruiz, D., Moreno, H.A., Gutiérrez, M.E., Zapata, P.A. (2008) Changing climate and endangered high mountain ecosystems in Colombia. Science of the Total Environment **398**:122–132.

Ruiz, G. (1987) Respiratory and hematological adaptations to high altitude in *Telmatobius* frogs from the Chilean Andes. *Comparative Biochemistry and Physiology—Part A: Molecular & Integrative Physiology,* **76,** 109–113.

Ruiz, G., Rosemann, M. & Veloso, A. (1989) Altitudinal distribution and blood values in the toad, Bufo spinulosus. *Comparative Biochemistry and Physiology—Part A: Molecular & Integrative Physiology,* **94,** 643–646.

Rundel, P.W., Smith, A.P. & Meinzer, F.C. (1994) *Tropical Alpine Environments. Plant Form and Function.* Cambridge University Press, Cambridge.

Ruth, R.H. (1965) Sitka spruce (*Picea sitchensis* (Bong.) Carr.). *Silvics of Forest Trees of the United States. Agriculture Handbook 271* (ed H.A. Fowells), pp. 311–317. U.S. Department of Agriculture, Washington, DC.

Rutherford, G.K. (1968) Observations on a succession of soils on Mt Giluwe, eastern New Guinea. *Annals of the Association of American Geographers,* **58,** 304–312.

Sandvik, S.M. (2001) Somatic and demographic costs under different temperature regimes in the late-flowering alpine perennial herb Saxifraga stellaris (Saxifragaceae). *Oikos,* **93,** 303–311.

Saros, J.E., Interlandi, S.J., Wolfe, A.P. & Engstrom, D.R. (2003) Recent changes in the diatom community structure of lakes in the Beartooth Mountain Range, USA. *Arctic, Antarctic and Alpine Research,* **35,** 18–23.

Saunders, I.R. & Bailey, W.G. (1994) Radiation and energy budgets of alpine tundra environments of North America. *Progress in Physical Geography,* **18,** 517–538.

Saunders, I.R. & Bailey, W.G. (1996) The physical climatology of alpine tundra, Scout Mountain, British Columbia, Canada. *Mountain Research and Development,* **16,** 51–64.

Schadt, C.W., Martin, A.P., Lipson, D.A. & Schmidt, S.K. (2003) Seasonal dynamics of previously unknown fungal lineages in tundra soils. *Science,* **301,** 1359–1361.

Schenker, R. (1983) Effects of temperature-acclimation on cold-hardiness of alpine microarthropods. *Revue d'Ecologie et de Biologie du Sol,* **20,** 37–47.

Scherrer, P. (2003) *Monitoring vegetation change in the Kosciuszko alpine zone, Australia.* PhD, Griffith University, Gold Coast.

Schmidt, R., Kamenik, C., Kaiblinger, C. & Hetzel, M. (2004) Tracking Holocene environmental changes in an alpine lake sediment core: application of regional diatom calibration, geochemistry, and pollen. *Journal of Paleolimnology,* **32,** 177–196.

Schmidt, S.K., Reed, S.C., Nemergut, D.R., Grandy, A.S., Cleveland, C.C., Costello, E.K., Weintraub, M.N., Hill, A.W., Meyer, A.F., Martin, A.P. & Neff, J.C. (2008) The earliest stages of ecosystem succession in high-elevation (5000 meters above sea level), recently de-glaciated soils. Proceedings of the Royal Society, Series B 275, 2793–2802.

Schmidt, S.K., Sobeniak-Wiseman, L.C., Kageyama, S.A., Halloy, S., Schadt, C.W. (2008) Mycorrhizal and dark-septate fungi in plant roots above 4270 meters elevation in the Andes and Rocky Mountains. *Arctic, Antarctic and Alpine Research* **40**, 576–583.

Schmidt, S.K., Costello, E.K., Nemergut, D.R., Cleveland, C.C., Reed, S.C., Weintraub, M.N., Meyer, A.F. & Martin, A.M. (2007) Biogeochemical consequences of rapid microbial turnover and seasonal succession in soil. *Ecology,* **88,** 1379–1385.

Schönswetter, P., Lachmayer, M., Lettner, C., Prehsler, D., Rechnitzer, S., Reich, D.S., Sonnleitner, M., Wagner, I., Hülber, K., Schneeweiss, G.M., Travnicek, P. & Suda, J. (2007) Sympatric diploid and hexaploid cytotypes of *Senecio carnoliacus* (Asteraceae) in the Eastern Alps are separated along an altitude gradient. *Journal of Plant Research,* **120,** 721–725.

Schönwiese, C.D. & Rapp, J. (1997) *Climate Trend Atlas of Europe Based on Observations, 1891–1990.* Kluwer, Dordrecht.

Seagle, S.W. (2003) Can ungulates foraging in a multiple-use landscape alter forest nitrogen budgets? *Oikos,* **103,** 230–234.

Seastedt, T.R. & Vaccaro, L. (2001) Plant species richness, productivity, and nitrogen and phosphorus limitations across a snowpack gradient in Alpine Tundra, Colorado, USA. *Arctic, Antarctic and Alpine Research,* **33,** 100–106.

Seastedt, T.R., Walker, M.D. & Bryant, D.M. (2001) Controls on decomposition processes in alpine tundra. *Structure and Function of an Alpine Ecosystem,*

*Niwot Ridge, Colorado* (eds W.D. Bowman & T.R. Seastedt), pp. 222–236. Oxford University Press, Oxford.

Seimon, T.A., Seimon, A., Daszak, P., Halloy, S.R.P., Schloegel, L.M., Aguilar, C.A., Sowell, P., Hyatt, A.D., Konecky, B. & Simmons, J.E. (2007) Upward range extension of Andean anurans and chytridiomycosis to extreme elevations in response to tropical deglaciation. *Global Change Biology*, **13**, 288–299.

Selvaraj, K. & Chen, C.T.A. (2006) Moderate chemical weathering of subtropical Taiwan: Constraints from solid-phase geochemistry of sediments and sedimentary rocks. *Journal of Geology*, **114**, 101–116.

Seppa, H., Hannon, G.E. & Bradshaw, R.H.W. (2004) Holocene history of alpine vegetation and forestline on Pyhakero mountain, northern Finland. *Arctic Antarctic and Alpine Research*, **36**, 607–614.

Settele, J. & et al. (2009) *Atlas of Biodiversity Risks. From Europe to the Globe and from Stories to Maps. The Core Final Product of ALARM.* Pensoft, Sofia.

Shahgedanova, M., Mikhailov, N., Larin, S. & Bredikhin, A. (2002a) The mountains of southern Siberia. *The Physical Geography of Northern Eurasia* (ed M. Shahgedanova), pp. 314–349. Oxford University Press, Oxford.

Shahgedanova, M., Perov, V. & Mudrov, Y. (2002b) The mountains of northern Russia. *The Physical Geography of Northern Eurasia* (ed M. Shahgedanova), pp. 284–313. Oxford University Press, Oxford.

Sherrod, S.K. & Seastedt, T.R. (2001) Effects of the northern pocket gopher (Thomomys talpoides) on alpine soil characteristics, Niwot Ridge, CO. *Biogeochemistry*, **55**, 195–218.

Sieg, B., Drees, B. & Daniëls, J.A. (2006) Vegetation and altitudinal zonation in continental west Greenland. *Meddelelser om Gronland Bioscience*, **57**, 1–93.

Sievering, H. (2001) Atmospheric chemistry and deposition. *Structure and Function of an Alpine Ecosystem. Niwot Ridge, Colorado* (eds W.D. Bowman & T.R. Seastedt), pp. 32–44. Oxford University Press, Oxford.

Simmons, I.G. (2005) The mountains of Northern Europe: towards an environmental history for the last ten thousand years. *Mountains of Northern Europe: Conservation, Management, People and Nature* (eds D.B.A. Thompson, M.F. Price & C.A. Galbraith), pp. 141–150. TSO, Edinburgh.

Simpson, B.B. & Todzia, C.A. (1990) Patterns and processes in the development of the high Andean flora. *American Journal of Botany*, **77**, 1419–1432.

Singh, H. & Sundriyal, R. (2005) Composition, economic use, and nutrient contents of Alpine Vegetation in the Khangchendzonga Biosphere Reserve, Sikkim Himalaya, India. *Arctic, Antarctic, and Alpine Research*, **37**, 591–601.

Sipman, H.J.E. (1995) Preliminary review of the lichen biodiversity of the Colombian montane forests. *Biodiversity and Conservation of Neotropical Montane Forests* (eds S.P. Churchill, H. Balsev, E. Forero & J.L. Luteyn), pp. 313–320. New York Botanical Garden, New York.

Sklenár, P. & Jørgensen, P.M. (1999) Distribution patterns of páramo plants in Ecuador. *Journal of Biogeography*, **26**, 681–691.

Sklenár, P. (1999) Nodding capitula in superparamo Asteraceae: an adaptation to unpredictable environment. *Biotropica*, **31**, 394–402.

Skogland, T. (1980) Comparative feeding strategies of arctic and alpine Rangifer. *Journal of Animal Ecology*, **49**, 81–98.

Skotnicki, M.L., Selkirk, P.M., Broady, P., Adam, K.D. & Ninham, J.A. (2001) Dispersal of the moss *Campylopus pyriformis* on geothermal ground near the summits of Mount Erebus and Mount Melbourne, Victoria Land, Antarctica. *Antarctic Science,* **13,** 280–285.

Smith, A.P. & Young, T.P. (1994) Population biology of *Senecio keniodendron* (Asteraceae)—an Afroalpine giant rosette plant. *Tropical Alpine Environments. Plant Form and Function* (eds P.W. Rundel, A.P. Smith & F.C. Meinzer), pp. 273–293. Cambridge University Press, Cambridge.

Smith, C.A.S., Clark, M., Broll, G., Ping, C.L., Kimble, J.M. & Luo, G. (1999) Characterization of selected soils from the Lhasa region of Qinghai-Xizang Plateau, SW China. *Permafrost Periglacial Process,* **10,** 211–222.

Smith, J.M.B. & Cleef, A.M. (1988) Composition and origins of the worlds tropical alpine floras. *Journal of Biogeography,* **15,** 631–645.

Smith, J.M.B. & Klinger, L.F. (1985) Aboveground: Belowground phytomass ratios in venezuelan paramo vegetation and their significance. *Arctic and Alpine Research,* **17,** 189–198.

Smith, J.M.B. (1977) Man's impact upon some New Guinea mountain ecosystems. *Subsistence and Survival: Rural Ecology in the Pacific* (eds T.P. Bayliss-Smith & R.G.A. Feachem), pp. 185–213. Academic Press, London.

Smith, J.M.B. (1980) Ecology of the high mountains of New Guinea. *Alpine Flora of New Guinea. Volume 1: General Part* (ed P. van Royen), pp. 111–132. J. Cramer, Vaduz.

Smith, J.M.B. (1982) Origins of tropicalpine flora. *Biogeography and Ecology of New Guinea* (eds J.L. Gressitt, J. Illies & F.R.G. Schlitz), pp. 287–308. Junk, The Hague.

Smith, J.M.B. (1985) Vegetation patterns in response to environmental stress and disturbance in the Papua New Guinea Highlands. *Mountain Research and Development,* **5,** 329–338.

Smith, S.E. & Read, D.J. (1997) *Mycorrhizal Symbioses.* Academic Press, London.

Solfjeld, I. & Johnsen, O. (2006) The influence of root-zone temperature on growth of *Betula pendula* Roth. *Trees—Structure and Fuction,* **20,** 320–328.

Sømme, L. (1997) Adaptations to the alpine environment in insects and other terrestrial arthropods. *Ecosystems of the World. 3. Polar and Alpine Tundra* (ed F.E. Wielgolaski), pp. 11–25. Elsevier, Amsterdam.

Soriano, W.E. (1981) *Los Modos de Produccion en el Imperio de los Incas.* Segunda Edicion. Amararu Editores, Lima.

Spehn, E., Lieberman, M. & Körner, C. (2006) *Land Use Change and Mountain Biodiversity.* Taylor & Francis, CRC Press, Boca Raton, London

Spooner, I.S., Hills, L.V. & Osborn, G.D. (1997) Reconstruction of Holocene changes in Alpine vegetation and climate, Susie Lake, British Columbia, Canada. *Arctic and Alpine Research,* **29,** 156–163.

Spooner, I.S., Hills, L.V. & Osborn, G.D. (1997) Reconstruction of Holocene changes in Alpine vegetation and climate, Susie Lake, British Columbia, Canada. *Arctic and Alpine Research,* **29,** 156–163.

Stahr, A. & Hartmann, T. (1999) *Landschaftsformen und Lanschaftselemente im Hochgebirge.* Springer, Berlin, Heidelberg, New York.

Steinger, T., Körner, C. & Schmid, B. (1996) Long-term persistence in a changing climate: DNA analysis suggests very old ages of clones of alpine *Carex curvula. Oecologia,* **105,** 94–99.

Stern, M.J. & Guerrero, M. (1997) Sucesión primaria en el Volcán Cotopaxi y sugerencias para el manejo de hábitats frágiles dentro del Parque Nacional.

*Estudios Sobre Diversidad y Ecología de Plantas* (eds R. Valencia & H. Balslev), pp. 217–229. Pontificia Universidad Católica del Ecuador, Quito.

Stöcklin, J. & Bäumler, E. (1996) Seed rain, seedling establishment and clonal growth strategies on a glacier foreland. *Journal of Vegetation Science*, 7, 45–56.

Stohlgren, T.J., Chase, T.N., Pielke, R.A., Sr, Kittel, T.G.F. & Baron, J.S. (1998) Evidence that local land use practices influence regional climate, vegetation, and stream flow patterns in adjacent natural areas. *Global Change Biology*, 4, 495–504.

Strid, A., Andonoski, A. & Andonovski, V. (2003) The high mountain vegetation of the Balkan Peninsula. *Alpine Biodiversity in Europe* (eds L. Nagy, G. Grabherr, C. Koerner & D.B.A. Thompson), pp. 113–132. Springer, Heidelberg, Berlin, New York.

Sturša, J. (1998) Research and management of the Giant Mountains' arctic-alpine tundra (Czech Republic). *Ambio*, 27, 358–360.

Stutzer, A. (2000) Forestline and treeline on Saualpe: a comparison of old and new pictures. *Forstwissenschaftliche Centralblatt*, 119, 20–31.

Stuwe, M. & Nievergelt, B. (1991) Recovery of alpine ibex from near extinction—the result of effective protection, captive breeding, and reintroductions. *Applied Animal Behaviour Science*, 29, 379–387.

Sullivan, J.H., Teramura, A.H. & Ziska, L.H. (1992) Variation in uv-b sensitivity in plants from a 3,000-m elevational gradient in Hawaii. *American Journal of Botany*, 79, 737–743.

Suslov, S.P. (1961) *Physical Geography of Asiatic Russia.* W.H. Freeman and Co., San Francisco and London.

Svoboda, J. & Freedman, B. (1994) *Ecology of a Polar Oasis: Alexandra Fiord, Ellesmere Island, Canada.* Captus University Press, Toronto.

Svoboda, J. & Henry, G.H.R. (1987) Succession in marginal arctic environments. *Arctic and Alpine Research*, 19, 373–384.

Swan, L.W. (1981) The aeolian region of the Himalaya and the Tibetan Plateau. *Geological and Ecological Studies of Qinghai-Xizang Plateau* (ed L. Dong-sheng), pp. 1971–1976. Science Press, Beijing.

Swan, L.W. (1992) The aeolian biome. *BioScience*, 42, 262–270.

Takahashi, T. & Shoji, S. (2002) Distribution and classification of volcanic ash soils. *Global Environmental Research*, 6, 83–97

Tasser, E. & Tappeiner, U. (2002) Impact of land use changes on mountain vegetation. *Applied Vegetation Science*, 5, 173–184.

Tasser, E., Mader, M. & Tappeiner, U. (2003) Effects of land use in alpine grasslands on the probability of landslides. *Basic and Applied Ecology*, 4, 271–280.

Taylor, D.R., Aarssen, L.W. & Loehle, C. (1990) On the relationship between r/K selection and environmental carrying capacity: a new habitat templet for plant life history strategies. *Oikos*, 58, 239–250.

Taylor, D.W.M. (1977) Floristic relationships along the Cascade-Sierran axis. *American Midland Naturalist*, 97, 333–349.

Teich, M., Lardelli, C., Bebi, P., Gallati, D., Kytzia, S., Pohl, M., Putz, M. & Rixen, C. (2007) *Klimawandel und Wintertourismus: Okonomische und okologische Auswirkungen von technischer Beschneigung.* pp. 169. Eidgenossische Forschunganstalt fur Wald, Schnee und Landschaft WSL, Birmensdorf.

Tenow, O. (1996) Hazards to a mountain birch forest—Abisko in perspective. *Ecological Bulletins*, 45, 104–114.

Thaler, K. (1999) Nival invertebrate animals in the East Alps: a faunistic overview. *Cold-Adapted Organisms: Ecology, Physiology, Enzymlogy, and Molecular Biology* (eds R. Margesin & F. Schinner), pp. 165–179. Springer, Berlin, Heidelberg, New York.

Theodose, T.A. & Bowman, W.D. (1997) Nutrient availability, plant abundance, and species diversity in two alpine tundra communities. *Ecology*, **78**, 1861–1872.

Theurillat, J.-P., Schlüssel, A., Geissler, P., Guisan, A., Velluti, C. & Wiget, L. (2003) Vascular plant and bryophyte diversity along elevation gradients in the Alps. *Alpine Biodiversity in Europe* (eds L. Nagy, G. Grabherr, C. Körner & D.B.A. Thompson), pp. 185–193. Springer, Berlin, Heidelberg, New York.

Thomas, D.N., Fogg, G.E., Convey, P., Fritsen, C.H., Gili, J.-M., Laybourn-Parry, J., Reid, K., Walton, D.W.H. (2008) The Biology of Polar Regions. Oxford University Press, Oxford.

Thomas, W.D. & Duval, B. (1995) Sierra Nevada, California, U.S.A., snow algae: snow albedo changes, algal-bacterial relationship, and ultraviolet radiation effects. *Arctic and Alpine Research*, **27**, 389–399.

Thompson, D.B.A., Nagy, L., Johnson, S.M. & Robertson, P. (2005) The nature of mountains: an introduction to science, policy and management issues. *Mountains of Northern Europe. Conservation, Management, People and Nature* (eds D.B.A. Thompson & C.A. Galbraith), pp. 43–55. TSO Scotland, Edinburgh.

Thompson, D.B.A., Whitfield, D.P., Galbraith, C.A., Duncan, K., Smith, R.D., Murray, S. & Holt, S. (2003) Breeding bird assemblages and habitat use of alpine areas in Scotland. *Alpine Biodiversity in Europe* (eds L. Nagy, G. Grabherr, C. Körner & D.B.A. Thompson), pp. 327–338. Springer, Berlin, Heidelberg, New York.

Thomson, A. (2005) Transhumance and vegetation degradation in Iceland before 1900: an historical grazing model. *Mountains of Northern Europe: Conservation, Management, People and Nature* (eds D.B.A. Thompson, M.F. Price & C.A. Galbraith), pp. 179–184. TSO, Edinburgh.

Thornthwaite, C.W. (1933) The climates of the Earth. *Geographical Review*, **23**, 433–440.

Thuiller, W., Lavorel, S. & Araujo, M.B. (2005) Niche properties and geographical extent as predictors of species sensitivity to climate change. *Global Ecology and Biogeography*, **14**, 347–357.

Tisch, P.A. & Kelly, D. (1998) Can wind pollination provide a selective benefit to mast seeding in *Chionochloa macra* (Poaceae) at Mt Hutt, New Zealand? *New Zealand Journal of Botany*, **36**, 637–643.

Tokeshi, M. (1999) *Species Coexistence. Ecological and Evolutionary Perspectives.* Blackwell, Oxford.

Tol, G.J. & Cleef, A.M. (1994) Aboveground biomass structure of a *Chusquea tessellata* bamboo páramo, Chingaza National Park, Cordillera Oriental, Colombia. *Vegetatio*, **115**, 29–39.

Totland, O. & Alatalo, J.M. (2002) Effects of temperature and date of snowmelt on growth, reproduction, and flowering phenology in the arctic/alpine herb, *Ranunculus glacialis*. *Oecologia*, **133**, 168–175.

Totland, Ø. & Esaete, J. (2002) Effects of willow canopies on plant species performance in a low-alpine community. *Plant Ecology*, **161**, 157–166.

Totland, O. (1993) Pollination in alpine Norway—flowering phenology, insect visitors, and visitation rates in 2 plant-communities. *Canadian Journal of Botany—Revue Canadienne de Botanique*, **71**, 1072–1079.

Totland, O. (1997) Effects of flowering time and temperature on growth and reproduction in Leontodon autumnalis var. taraxaci a late- flowering alpine plant. *Arctic and Alpine Research,* **29,** 285–290.

Totland, O. (1999) Effects of temperature on performance and phenotypic selection on plant traits in alpine Ranunculus acris. *Oecologia,* **120,** 242–251.

Tovar, O. (1993) Las gramineas (Poaceae) del Peru. *Ruizia,* **13,** 1–480.

Tribsch, A. & Schönswetter, P. (2003) Patterns of endemism and comparative phylogeography confirm palaeoenvironmental evidence for Pleistocene refugia in the Eastern Alps. *Taxon,* **52,** 477–497.

Tscherko, D., Hammesfahr, U., Marx, M.C. & Kandeler, E. (2004) Shifts in rhizosphere microbial communities and enzyme activity of *Poa alpina* across an alpine chronosequence. *Soil Biology and Biochemistry,* **36,** 1685–1698.

Tscherko, D., Hammesfahr, U., Zeltner, G., Kandeler, E. & Bocker, R. (2005) Plant succession and rhizosphere microbial communities in a recently deglaciated alpine terrain. *Basic and Applied Ecology,* **6,** 367–383.

Türk, R. & Gärtner, G. (2003) Biological soil crusts of the subalpine, alpine and nival areas in the Alps. *Biological Soil Crusts* (eds J. Belnap & O.L.Lange), pp. 67–74. Springer, Heidelberg, Berlin, New York.

Väisänen, R. (1998) Current research and trends in mountain biodiversity in NW Europe. *Pirineos,* **151–152,** 131–156.

Vajda, A. & Venelainen, A. (2005) Feedback processes between climate, surface and vegetation at the northern climatological tree-line (Finnish Lappland). *Boreal Environment Research,* **10,** 299–314.

van der Heiden, M.G.A. & Sanders, I.R. (2003) Mycorrhizal ecology: synthesis and perspectives. *Mycorrhizal Ecology* (eds M.G.A. van der Heiden & I.R. Sanders), pp. 441–456. Springer, Berlin, Heilderberg, New York.

van der Wal, R. (2006) Do herbivores cause habitat degradation or vegetation state transition? Evidence from the tundra. *Oikos,* **114,** 177–186.

van der Wal, R., Bardgett, R.D., Harrison, K.A. & Stien, A. (2004) Vertebrate herbivores and ecosystem control: cascading effects of faeces on tundra ecosystems. *Ecography,* **27,** 242–252.

van Reenen, G. & Gradstein, S.R. (1983) Studies on Colombian cryptogams. 20. A transect analysis of the bryophyte vegetation along an altitudinal gradient on the Sierra Nevada de Santa Marta, Colombia. *Acta Botanica Nerlandica,* **32,** 163–175.

van Royen, P. (1980) *The Alpine Flora of New Guinea. Vol. 1. General Part.* J. Cramer, Vaduz.

van Royen, P. (1980) *The Alpine Flora of New Guinea. Vols. 1–4.* J. Cramer, Vaduz.

van Steenis, C.G.G.J. (1972) *The Mountain Flora of Java.* EJ Brill, Leiden.

Vane-Wright, R.I. (1996) Identifying priorities for the conservation of biodiversity: systematic biological criteria within a socio-political framework. *Biodiversity. A Biology of Numbers and Difference* (ed K.J. Gaston), pp. 309–344. Blackwell, Oxford.

Väre, H., Lampinen, R., Humphries, C. & Williams, P. (2003) Taxonomic diversity of vascular plants in the European alpine areas. *Alpine Biodiversity in Europe* (eds L. Nagy, G. Grabherr, C. Körner & D.B.A. Thompson). Springer Verlag, Berlin; Heidelberg, New York.

Veit, H. (2002) *Die Alpen—Geoökologie und Lanschaftsentwicklung.* Ulmer, Stuttgart.

Veit, H. (2004) Gebirgstypen. *Gebirge der Erde. Landschaft, Klima, Pflanzenwelt* (eds C.A. Burga, F. Klötzli & G. Grabherr), pp. 16–19. Ulmer, Stuttgart.

Vestal, J.R. (1993) Cryptoendolithic communities from hot and cold deserts: speculation on microbial colonization and succession. *Primary Succession on Land* (ed J.W. Miles, D.W.H.), pp. 5–16. Blackwell, Oxford.

Vetaas, O.R. & Grytnes, J.A. (2002) Distribution of vascular plant species richness and endemic richness along the Himalayan elevation gradient in Nepal. *Global Ecology and Biogeography,* **11,** 291–301.

Vetaas, O.R. (1994) Primary succession of plant assemblages on a glacier fore-land—Bodalsbreen, southern Norway. *Journal of Biogeography,* **21,** 297–308.

Viereck, L.A., Dyrness, C.T., Batten, A.R. & Wenzlick, K.J. (1992) *The Alaska Vegetation Classification.* United States Department of Agriculture Forest Service Pacific Northwest Research Station, Portland.

Villar, L., Sesé, J.A. & Ferrandez, J.V. (1997) *Atlas de la Flora del Pirineo Aragones I.-II.* Ediciones La Val de Onsera, Angues.

Vinebrooke, R.D. & Leavitt, P.R. (1999) Differential responses of littoral communities to ultraviolet radiation in an alpine lake. *Ecology,* **80,** 223–237.

Virtanen, R. (1998) Impact of grazing and neighbour removal on a heath plant community transplanted onto a snowbed site, NW Finnish Lapland. *Oikos,* **81,** 359–367.

Virtanen, R. (2000) Effects of grazing on above-ground biomass on a mountain snowbed, NW Finland. *Oikos,* **90,** 295–300.

Virtanen, R. (2003) The high mountain vegetation of the Scandes. *Alpine Biodiversity in Europe* (eds L. Nagy, G. Grabherr, C. Körner & D.B.A. Thompson), pp. 31–38. Springer, Berlin, Heidelberg, New York.

Virtanen, R., Dirnböck, T., Dullinger, S., Grabherr, G., Pauli, H., Staudinger, M. & Villar, L. (2003) Patterns in the plant species richness of European high mountain vegetation. *Alpine Biodiversity in Europe* (eds L. Nagy, G. Grabherr, C. Körner & D.B.A. Thompson), pp. 149–172. Springer, Berlin, Heidelberg, New York.

Virtanen, R., Eskelinen, A. & Gaare, E. (2003) Long-term changes in alpine plant communities in Norway and Finland. *Alpine Biodiversity in Europe* (eds L. Nagy, G. Grabherr, C. Körner & D.B.A. Thompson), pp. 411–422. Springer, Berlin, Heidelberg, New York.

Virtanen, R., Henttonen, H. & Laine, K. (1997) Lemming grazing and structure of a snowbed plant community—A long-term experiment at Kilpisjarvi, Finnish Lapland. *Oikos,* **79,** 155–166.

Volodicheva, N. (2002) The Caucasus. *The Physical Geography of Northern Eurasia* (ed M. Shahgedanova), pp. 350–378. Oxford University Press, Oxford.

Wade, L.K. & McVean, D.N. (1969) *Mt Wilhelm Studies 1: The Alpine and Subalpine Vegetation.* Australian National University, Canberra.

Walker, D.A. & Walker, M.D. (1996) Terrain and vegetation of the Imnavait Creek watershed. *Landscape Function and Disturbance in Arctic Tundra* (eds J.F. Reynolds & J.D. Tenhunen), pp. 73–108. Springer, Berlin, Heidelberg, New York.

Walker, D.A., Halfpenny, J.C., Walker, M.D. & Wessman, C.A. (1993) Long-term studies of snow-vegetation interactions. *BioScience,* **43,** 287–301.

Walker, L.R. (1993) Nitrogen fixers and species replacements in primary succession. *Primary Succession on Land* (eds J. Miles & D.W.H. Walton), pp. 249–272. Blackwell, Oxford.

Walker, M.D., Gould, W.A. & Chapin III, F.S. (2001a) Scenarios of biodiversity changes in arctic and alpine tundra. *Global Biodiversity in a Changing*

*Environment* (eds F.S. Chapin III, O.E. Sala & E. Huber-Sannwald), pp. 83–100. Springer, Heidelberg, Berlin, New York.

Walker, M.D., Ingersoll, R.C. & Webber, P.J. (1995) Effects of interannual climate variation on phenology and growth of two alpine forbs. *Ecology*, **76**, 1067–1083.

Walker, M.D., Walker, D.A., Theodose, T.A. & Webber, P.J. (2001b) The vegetation. Hierarchical species-environment relationships. *Structure and Function of an Alpine Ecosystem, Niwot Ridge, Colorado* (eds W.D. Bowman & T.R. Seastedt), pp. 99–127. Oxford University Press, Oxford.

Walker, M.D., Webber, P.J., Arnold, E.H. & Ebertmay, D. (1994) Effects of interannual climate variation on aboveground phytomass in alpine vegetation. *Ecology*, **75**, 393–408.

Walker, P.H. & Costin, A.B. (1971) Atmospheric dust accession in South-Eastern Australia. *Australian Journal of Soil Research*, **9**, 1–5.

Walter, H. (1973) *Vegetationszonen und Klima*. Ulmer, Stuttgart.

Walther, G.-R., Beissner, S. & Burga, C.A. (2005) Trends in the upward shift of alpine plants. *Journal of Vegetation Science*, **16**, 541–548.

Walton, D.W.H. (1993) The effects of cryptogams on mineral substrates. *Primary Succession on Land* (eds J. Miles & D.W.H. Walton), pp. 33–53. Blackwell, Oxford.

Wang, J.T. (1988) The steppes and deserts of the Xizang Plateau (Tibet). *Vegetatio*, **75**, 135–142.

Wardle, P., Ezcurra, C., Ramirez, C. & Wagstaff, S. (2001) Comparison of the flora and vegetation of the southern Andes and New Zealand. *New Zealand Journal of Botany*, **39**, 69–108.

Watt, A.S. & Jones, E.W. (1948) The ecology of the Cairngorms. I. The environment and the altitudinal zonation of the vegetation. *Journal of Ecology*, **35**, 105–129.

Webber, P.J. (1978) Spatial and temporal variation of the vegetation and its productivity. *Vegetation and Production Ecology of an Alaskan Arctic Tundra* (ed L.L. Tieszen), pp. 37–112. Springer, Berlin, Heidelberg, New York.

West, J.B. (2006) Human responses to extreme altitudes. *Integrative and Comparative Biology*, **46**, 25–34.

Whinam, J. & Chilcott, N.M. (2003) Impacts after four years of experimental trampling on alpine/sub-alpine environments in western Tasmania. *Journal of Environmental Management*, **67**, 339–351.

Whitmore, T.C. (1998) *An Introduction to Tropical Rain Forests*. Oxford University Press, Oxford.

Wielgolaski, F.E. (1997b) Fennoscandian tundra. *Polar and Alpine Tundra. Ecosystems of the World* (ed F.E. Wielgolaski), 27-85. Elsevier, Amsterdam.

Wielgolaski, F.E. (1997a) Introduction. *Polar and Alpine Tundra. Ecosystems of the World* (ed F.E. Wielgolaski). Elsevier, Amsterdam.

Wielgolaski, F.E. (1997c) Adapatation in plants. *Ecosystems of the World. 3. Polar and Alpine Tundra* (ed F.E. Wielgolaski), pp. 7–10. Elsevier Science, Amsterdam.

Wijk, S. (1986) Performance of *Salix herbacea* in an alpine snow-bed gradient. *Journal of Ecology*, **74**, 675–684.

Wilkinson, D.M. (2006) *Fundamental Processes in Ecology – an Earth Systems Approach*. Oxford University Press, Oxford.

Williams, M.W. & Caine, N. (2001) Hydrology and hydrochemistry. *Structure and Function of an Alpine Ecosystem, Niwot Ridge, Colorado* (eds W.D. Bowman & T.R. Seastedt), pp. 75–96. Oxford University Press, Oxford.

Williams, R.J. & Costin, A.B. (1994) Alpine and subalpine vegetation. *Australian Vegetation* (ed R.H. Groves), pp. 467–500. Cambridge University Press, Cambridge.

Willis, K.J. & McElwain, J.C. (2002) *The Evolution of Plants.* Oxford University Press, Oxford.

Wilson, C., Grace, J., Allen, S. & Slack, F. (1987) Temperature and stature: a study of temperatures in montane vegetation. *Functional Ecology,* **1,** 504–413.

Wilson, S.D. (1993) Competition and resource availability in heath and grassland in the Snowy Mountains of Australia. *Journal of Ecology,* **81,** 445–451.

Winkworth, R.C., Wagstaff, S.J., Glenny, D. & Lockhart, P.J. (2005) Evolution of the New Zealand mountain flora: origins, diversification and dispersal. *Organisms, Diversity and Evolution,* **5,** 237–247.

Wood, D. (1971) The adaptive significance of a wide altitudinal range for montane species. *Transactions of the Botanical Society of Edinburgh,* **41,** 119–124.

Wookey, P.A., Bol, R.A., Caseldine, C.J. & Harkness, D.D. (2002) Surface age, ecosystem development, and C isotope signatures of respired $CO_2$ in an alpine environment, North Iceland. *Arctic, Antarctic and Alpine Research,* **2002,** 76–87.

Woolgrove, C.E. & Woodin, S.J. (1996) Ecophysiology of a snow-bed bryophyte *Kiaeria starkei* during snowmelt and uptake of nitrate from meltwater. *Canadian Journal of Botany—Revue Canadien de Botanie,* **74,** 1095–1103.

Wu, R.G. & Tiessen, H. (2002) Effect of land use on soil degradation in alpine grassland soil, China. *Soil Science Society of America Journal,* **66,** 1648–1655.

Wynn-Williams, D.D. (1993) Microbial processes and initial stabilization of fellfield soil. *Primary Succession on Land* (eds J. Miles & D.W.H. Walton), pp. 17–32. Blackwell, Oxford.

Yapp, R.H. (1922) The concept of habitat. *Journal of Ecology,* **10,** 1–17.

Yoccoz, N.G., Nichols, J.D. & Boulinier, T. (2001) Monitoring of biological diversity in space and time. *Trends in Ecology and Evolution,* **16,** 446–453.

Yoshino, M.M. (1978) Altitudinal vegetation belts of Japan with special reference to climatic conditions. *Arctic and Alpine Research,* **10,** 449–456.

Young, K.R. & León, B. (2007) Tree-line changes along the Andes: implications of spatial patterns and dynamics. *Philosophical Transactions of the Royal Society of London Series B—Biological Sciences,* **362,** 263–272.

Young, K.R. & Lipton, J.K. (2006) Adaptive governance and climate change in the tropical highlands of western South America. *Climatic Change,* **78,** 63–102.

Young, T.P. & Evans, M.E. (1993) Alpine vertebrates of Mount Kenya, with particular notes on the rock hyrax. *Journal of the East African Natural History Society,* **82(202),** 54–79.

Young, T.P. & Peacock, M.M. (1992) Giant Senecios and alpine vegetation of Mount Kenya. *Journal of Ecology,* **80,** 141–148.

Young, T.P. & Smith, A.P. (1994) Alpine herbivory on Mount Kenya. *Tropical Alpine Environments. Plant Form and Function* (eds P.W. Rundel, A.P. Smith & F.C. Meinzer), pp. 319–335. Cambridge University Press, Cambridge.

Young, T.P. (1994) Population biology of Mount Kenya lobelias. *Tropical Alpine Environments. Plant Form and Function* (eds P.W. Rundel, A.P. Smith & F.C. Meinzer), pp. 251–272. Cambridge University Press, Cambridge.

Yue, T.X., Fan, Z.M. & Liu, J.Y. (2005) Changes of major terrestrial ecosystems in China since 1960. *Global and Planetary Change,* **48,** 287–302.

Zech, W. & Hintermaier-Erhard, G. (2002) *Böden der Welt: ein Bildatlas.* Spektrum, Heidelberg, Berlin.

Zellmer, I.D. (1995) UV-B-tolerance of alpine and arctic Daphnia. *Hydrobiologia,* **307,** 153–159.

Zhang, Y.M., Zhang, Z.B. & Liu, J.K. (2003) Burrowing rodents as ecosystem engineers: the ecology and management of plateau zokors *Myospalax fontanierii* in alpine meadow ecosystems on the Tibetan Plateau. *Mammal Review,* **33,** 284–294.

Zhao, X.Q. & Zhou, X.M. (1999) Ecological basis of Alpine meadow ecosystem management in Tibet: Haibei Alpine Meadow Ecosystem Research Station. *Ambio,* **28,** 642–647.

Ziska, L.H., Teramura, A.H. & Sullivan, J.H. (1992) Physiological sensitivity of plants along an elevational gradient to UV-B radiation. *American Journal of Botany,* **79,** 863–871.

Zlotin, R.I. (1978) Structure and productivity of high-altitude ecosystems in the Tien Shan, USSR. *Arctic and Alpine Research,* **10,** 425–427.

Zlotin, R.I. (1997) Geography and organization of high mountain ecosystems in the former USSR. *Ecosystems of the World. 3. Arctic and Alpine Tundra* (ed F. E. Wielgolaski). Elsevier, Amsterdam.

Zunckel, K. (2003) Managing and conserving Southern African grasslands with high endemism—The Maloti-Drakensberg Transfrontier Conservation Development Program. *Mountain Research and Development,* **23,** 113–118.

# Index

Note: Page numbers with "f" and "t" denote figures and tables, respectively.